Quantitative Modeling of Derivative Securities

From Theory to Practice

Marco Avellaneda

in collaboration with

Peter Laurence

CRC Press
Taylor & Francis Group
Boca Raton London New York

CRC Press is an imprint of the
Taylor & Francis Group, an **informa** business

A CHAPMAN & HALL BOOK

Chapman & Hall/CRC Press
Taylor & Francis Group
6000 Broken Sound Parkway NW, Suite 300
Boca Raton, FL 3487-2742

First issued in paperback 2020

© 2000 by Taylor & Francis Group, LLC
Chapman & Hall /CRC Press is an imprint of Taylor & Francis Group, an Informa business

No claim to original U.S. Government works

ISBN-13: 978-0-367-57914-2 (pbk)
ISBN-13: 978-1-58488-031-8 (hbk)

visit the Taylor & Francis Web site at
http://www.taylorandfrancis.com

and the CRC Press Web site at
http://www.crcpress.com

Library of Congress Cataloging-in-Publication Data

Avellaneda, Marco 1955-
    Quantitative modeling of derivative securities from theory to practice / Marco
Avellaneda in collaboration with Peter Laurence.
        p. cm.
    Includes bibliographical references and index.
        ISBN 1-58488-031-7 (alk. paper)
        1. Derivative securities.   2. Options (Finance).   3. Exotic options (Finance).   I. Laurence,
Peter. II. Title.
    P. II. Rovatti, Riccardo. III. Setti, Gianluca. IV. Series.
HG6024.A3 A93   1999
332.63′228—dc21   99-047242

Library of Congress Card Number 99-047242

Contents

To Cassandra and Magda

Introduction

This book originated in lecture notes for the courses *Mathematics of Finance I* and *II*, which I have taught at the Courant Institute since the fall of 1993. As the material evolved and the possibility of writing a book became more real, I joined forces with Peter Laurence, of the University of Rome, who provided the scholarship and technical expertise needed to develop these notes and shape them into a coherent text. *Quantitative Modeling of Derivative Securities* is the fruit of more than 2 years of close collaboration between us.

Our motivation for writing this book can be traced to the early 1990s when there was an increasing interest on the part of Wall Street firms in the so-called structured financial products or "second-generation derivatives." At that time, it had become commonplace for top-bracket investment banks to market new financial products with tailor-made payoffs.[1] The capability of designing these new financial derivatives, to bring them to the market and to manage their risk using financial engineering, is still seen as a competitive advantage. Another important aspect of quantitative analysis that developed strongly in the 1990s is the management of derivatives at the portfolio level using multifactor models. This "aggregate" approach to risk-management went far beyond the single-asset Black–Scholes model. For example, the 1990s saw U.S. dollar interest-rate derivatives markets reach their maturity. Quantitative models of the term-structure of interest rates—until then the realm of econometricians and Fed watchers—enabled Wall Street traders to warehouse and manage thousands of derivatives simultaneously. Asset pricing theory was a hammer that found its nail. This book is strongly influenced by the following two aspects of modeling derivatives: (1) pricing new financial products and measuring their market risk and (2) developing multifactor models that deal with several underlying securities—particularly in the realm of fixed-income derivatives.

This is a textbook on the theory behind modeling derivatives and their risk-management. The more theoretical portions of the book were drawn from several sources, among them Darrell Duffie's *Dynamic Asset Pricing Theory,* which provides a superb road map to the financial markets and asset-pricing, and the papers of Cox, Ingersoll, and Ross. Other sources included many readings in quantitative models, our own research on option volatility and risk-management and, last but not least, 2 years' experience in Wall Street: first at Banque Indosuez' foreign-exchange options department and later at Morgan Stanley Dean Witter's Derivative Products Group in the area of fixed-income derivatives.

Rather than attempting to write a "handbook" of derivatives with its mandatory list of mathematical formulas, we decided to focus on the valuation principles that are common to most derivative securities. This common thread is called *Arbitrage Pricing Theory.* We then focused on the analysis of the most widely traded structures and tried to link the theory with the practical aspects. This has the effect of showing the theory at work and, ultimately, of

---

[1] See, for instance, Mark Rubinstein, Exotic Options, working paper, Haas School of Business, U.C. Berkeley, 1991.

revealing its scope and limitations. We occasionally point out instances in which the standard theory does not explain how market risk occurs. For example, we believe in developing intuition about synthetic and dynamic hedging under real market conditions. The concepts of "pin risk," "Gamma risk," "volatility risk," and problems related to discontinuous payoffs and barriers are very important. Quite remarkably, however, the effects of market imperfections on derivatives pricing and risk-management *as perceived by traders* are seldom presented in financial economics texts.[2] We believe that they give important clues for understanding derivatives, as opposed to being "nuisances" that do not fit the theory. Financial modeling is very different from modeling in the natural sciences. Unlike physics, where we deal with reproducible experiments with well-defined initial conditions, the models and ideas presented in this book deal with phenomena for which we have only limited information and that are not necessarily reproducible.

The mathematical style of the book is informal. We avoided using a "theorem–proof" style or giving complicated definitions of things that otherwise seem clear. Some of the key theoretical results deserve to be theorems and are stated as such (regardless of whether their proof is difficult or not). By and large, we treat mathematics as a language. We have not attempted to make this book self-contained. We have also not attempted to trivialize the mathematical level, since this would defeat our purpose of doing good theory *and* good application. Fortunately for us and for the reader, *Quantitative Modeling of Derivative Securities* does not exist in a literary vacuum, either on the mathematical or the financial side of the equation. In the first part of the book, the main tools are linear algebra and elementary probability. In the second part, we introduce the main ideas of stochastic calculus and apply them to develop continuous-time finance. Here, the mathematical level is more demanding, although we stay away from measure theory and other topics that are only tangentially related to the subject. We feel that continuous-time finance is beautiful and that it can be an excellent tool to develop derivative pricing models. However, we also believe that an overly technical mathematical treatment has no place in a book on derivatives. A candid introduction to the main mathematical ideas and some solid bibliographical references should suffice.

The book is essentially divided into two parts. The first part (Chapters 1 through 8) can be seen as dealing mostly with discrete lattice models. The first chapter discusses the no-arbitrage theorem in the context of a one-period securities market with uncertainty. The following chapters deal with the multiperiod binomial model, the Black–Scholes formula, generalizations of the Black–Scholes model, and option price sensitivities. The Black–Scholes formula is derived as an approximation, or rather as a limit of the binomial Cox–Ross–Rubinstein model. We then discuss American options and lattice schemes for pricing general derivative securities. These results use only random walks and discrete models. The first part ends with an introduction to Brownian motion and Ito calculus followed with a long chapter on digital options and barrier options. This chapter uses essentially all the theory discussed until then as well as specific aspects of hedging barrier options and examples.

The second part (Chapters 9 through 15) deals with continuous-time finance, modeling the term-structure of interest rates and pricing fixed-income derivatives. Here, we find it useful to discuss powerful mathematical concepts, such as Ito calculus and Girsanov's theorem, which link the notions of "subjective" probability with the risk-neutral, or risk-adjusted, probability used for pricing. The connection between the computation of expected values under diffusion measures and the solution of partial differential equations is established. We then present the main financial instruments traded in fixed-income markets, and the role of the yield curve for pricing instruments such as swaps and bonds. The following chapters deal with the Heath–

---

[2]One major exception is the very informative book by Nassim Taleb, *Dynamic Hedging*, Wiley, New York, 1997.

Jarrow–Morton no-arbitrage conditions and present the main fixed income models in use today. The book ends with a discussion of the pricing of interest-rate options.

Writing this book was a learning experience. It is a pleasure to thank those that provided valuable suggestions and insights along the way. I particularly thank Nassim Taleb, Raphael Douady, Zhifeng (Frank) Zhang, Lewis Scott, Paul Wilmott, and Nicole El Karoui for sharing many interesting ideas on modeling derivatives. I also thank traders and practitioners Howard Savery (Republic National Bank), Peter Tselepas (Banque Indosuez/Credot Agricole), Sergio Kostek (Morgan Stanley), Pablo Calderon (Goldman Sachs), Jay Janer (Banco Fonte Cindam/BNP), Richard Pedde (Nomura Securitites), Carlos Korcarz (Banco Exprinter) Anna Raitcheva (Salomon Smith Barney), Thomas Artarit (CIBC), Philippe Burke (Morgan Stanley), Dino Buturovic (Bear Stearns), Adhil Reghai (Paribas), Marco Aurelio Teixeira (BM&F São Paulo), Gyorgy Varga (Financial Consultoria Economica), and Simon Altkorn Monti (MERVAL, Buenos Aires). I am grateful to my collaborators and former students Antonio Paras, Yingzi Zhu, Arnon Levy, Juan Carlos Porras, Dominick Samperi, Craig Friedman, Joshua Newman, Lucasz Kruk, Robert Buff, Nicolas Grandchamps, and to Joe Langsam and the fixed-income research team at Morgan Stanley. I also thank the attendants of the NYU Math Finance classes that worked through the material and honored me by attending the lectures. I am grateful to my colleagues at the Courant Institute of Mathematical Sciences for their support, to Tamar Arnon and Lisa Huntington, and especially to Dave McLaughlin. Finally, writing this book would not have been possible without the invaluable editorial assistance of Ms. Dawn Duffy.

Marco Avellaneda

New York

My participation in this book is an outgrowth of a longstanding scientific collaboration and friendship with principal Marco Avellaneda. Marco first introduced me to the subject of mathematical finance in the mid-nineties. As my knowledge of the field has grown, I discovered that my favorite areas of expertise in applied mathematics, such as free boundary problems for partial differential equations, had many fascinating applications to derivative pricing in the presence of an early exercise feature.

I then tried presenting parts of Marco's lecture notes in a doctoral course on finance at the University of Rome in 1996. The reaction I received confirmed the need for a good textbook on mathematical finance in which the powerful language of probability was used in a way that the underlying beautiful financial ideas would be clarified rather than obscured. It was about this time that Marco gave me the opportunity to work with him to improve upon the existing lecture notes and add new material, with the objective of creating a book that would be of value to both theoreticians and practitioners.

I would like to thank Nicole El Karoui, who over the last two years generously shared with me many fine insights into the subject as well as some unforgettable couscous. Lastly I thank my wife Magda and my mother Steffi for their unconditional support over the years.

Peter Laurence

Rome

Chapter 1

Arbitrage Pricing Theory: The One-Period Model

This chapter describes the basic principles of derivative security valuation. The ideas presented here can be applied to most valuation problems—from the simplest ones, involving straightforward compound interest calculations, to the most complicated, such as the valuation of exotic options. For simplicity, we will discuss a model for a securities market with finitely many final states and with a single trading period. In this model, the main definitions and results can be formulated with elementary mathematics. The key idea behind asset pricing in markets with uncertainty is the notion of *absence of arbitrage opportunities.*

Suppose that an investor takes a "position in the marketplace," by buying and selling securities, that has zero net cost and guarantees (i) no losses in the future and (ii) some chance of making a profit. In this hypothetical situation, the investor has a positive probability of realizing a profit without taking risk. This situation is known as an *arbitrage opportunity* or simply an *arbitrage*. Although arbitrage opportunities may arise sporadically in financial markets, they cannot last long. In fact, an arbitrage can be viewed as a relative mispricing between correlated assets. If this mispricing becomes known to sufficiently many investors, the prices will be affected as they move to take advantage of such opportunity. As a consequence, prices will change and the arbitrage will disappear. This principle can be stated as follows: *in an efficient market there are no permanent arbitrage opportunities.*

Example 1.1

 Suppose that the current (spot) price of an ounce gold is \$398 and that the 3-month forward price is \$390. Furthermore, suppose that the annualized 3-month interest rate for borrowing gold (known as the "convenience yield") is 10% and that the interest rate on 3-month deposits is 4% (annualized). This situation gives rise to an arbitrage opportunity. In fact, an arbitrager can borrow 1 ounce of gold, sell it at its current price of \$398 (go short 1 ounce), lend this money for 3 months, and enter into a 3-month forward contract to buy one ounce of gold at \$390. Since the cost of borrowing the ounce of gold is $\$398 \times 10\%/4 = 398 \times 2.5\% = \$9.95$ and the interest on the 3-month deposit amounts to $\$398 \times (.01) = \$3.98$, the total financing cost for this operation is \$5.97. He will therefore have $398 - 5.97 = 392.03$ dollars in his bank account after 3 months. By purchasing the ounce of gold in 3 months at the forward price of \$390 and returning it, he will make a profit of \$2.03. (This argument neglects transaction costs and assumes that interests are paid after the lending period.) ⬜

1.1   The Arrow–Debreu Model

We will consider the simplest possible model of a securities market with uncertainty, the *Arrow–Debreu model.* We assume that there are N securities, s_1, s_2, ..., s_N, that can be held long or short by any investor. Initially, an investor takes a position in the market by acquiring a portfolio of securities. He holds this position during a trading period (a specified period of time), which gives him the right to claim/owe the dividends paid by the securities (capital gains/losses). He liquidates the position after the trading period and incurs a net profit/loss from the change in the market value of his holdings (market gains/losses). The model is completely specified in terms of the *price vector* for the N securities

$$\mathbf{p} = (p_1, p_2 \cdots p_N) \tag{1.1}$$

and the *cash-flows matrix*

$$\mathbf{D} = \left(D_{ij}\right), \ 1 \le i \le N, 1 \le j \le M \tag{1.2}$$

where M is the number of all possible "states" of the market at the end of the trading period. The jth row of the matrix $\mathbf{D}$ represents the M possible cash-flows associated with holding one unit of the jth security, including dividend payments and market profit/losses.[1] We assume that the matrix D is known to all investors but that the final state of the market, represented by the M different columns, is not known in advance.

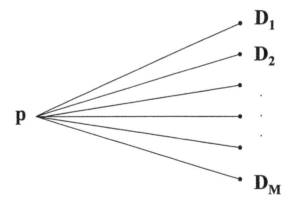

FIGURE 1.1
 Schematic representation of the Arrow-Debreu model for a securities market with uncertainty. In this "one-period model" there are N securities available for trading at prices $p = (p_1, \ldots, p_N)$. The investor takes a position in these securities. After that, a "final state" of the market is revealed. The values of the securities in each state are given by columns of the dividend matrix $D_{.j}$, $j = 1, \ldots M$.

A *portfolio* of securities is represented by a vector

$$\theta = (\theta_1, \theta_2 \cdots \theta_N) \ . \tag{1.3}$$

---

[1] Prices and cash-flows are expressed in the same unit of account. For simplicity, we take this unit to be a dollar.

Here, θ_i represents the number of units of the ith security held in the portfolio. If θ_i is positive, the investor is *long* the security and hence has acquired the right to receive the corresponding cash-flow $\theta_i D_{ij}$ at the end of the period. If θ_i is negative, the investor is *short* the security and thus will have a liability at the end of the trading period. (Short positions are taken by borrowing securities and selling them at the market price.) It is assumed that all investors can take short and long positions in arbitrary amounts of securities. Transaction costs, commissions and tax implications associated with trading are neglected. For simplicity, we assume that the amounts θ_i held long or short are not necessarily integers, but instead arbitrary real numbers.

The price of a portfolio θ is

$$\theta \cdot \mathbf{p} = \sum_1^N \theta_i \, p_i \,,$$

and the cash-flow for this portfolio in the jth "state" of the market will be

$$\theta \cdot \mathbf{D}_{.j} = \sum_1^N \theta_i \, D_{ij} \,. \tag{1.4}$$

We can express mathematically the concept of arbitrage opportunity within this simple model.

DEFINITION 1.1   *An arbitrage portfolio is a portfolio θ such that either*
(i)

$$\theta \cdot \mathbf{p} = 0 \,,$$

$$\theta \cdot \mathbf{D}_{.j} \geq 0 \quad \textit{for all} \quad 1 \leq j \leq M \,,$$

and
$$\theta \cdot \mathbf{D}_{.j} > 0 \quad \textit{for some} \quad 1 \leq j \leq M$$

or
$$\theta \cdot \mathbf{p} < 0 \,,$$

and
$$\theta \cdot \mathbf{D}_{.j} \geq 0 \quad \textit{for all} \quad 1 \leq j \leq M \,.$$

In plain words, an arbitrage portfolio is a position in the market that either (i) has zero initial cost, has no "down side" regardless of the market outcome, and offers a possibility of realizing a profit, or (ii) realizes an immediate profit for the investor and has no down side. We remark here that the distinction between the two cases is not really important: it is a consequence of the general form of the model, in which the nature of the "securities" is not specified. If it is possible to lend money (buy bonds), then the second case reduces to the first, because the investor can lend out the initial profit and then realize a positive cash-flow at the end of the trading period.

THEOREM 1.1
If there exists a vector of positive numbers

$$\pi = (\pi_1 \,, \pi_2 \,, \ldots \pi_M)$$

such that

$$\mathbf{p} = \mathbf{D} \cdot \pi \,, \tag{1.5a}$$

i.e., if

$$p_i = \sum_{1}^{M} D_{ij}\pi_j \quad for\ all\ 1 \le i \le N\,, \tag{1.5b}$$

there exist no arbitrage portfolios.[2] *Conversely, if there are no arbitrage portfolios, there exists a vector π with positive entries satisfying (1.5a) and (1.5b).*

PROOF [3] The first statement is easy to verify. If (1.5a) and (1.5b) holds, then for any portfolio θ,

$$\theta \cdot \mathbf{p} = \theta \cdot (\mathbf{D} \cdot \pi)$$

$$= (\theta \cdot \mathbf{D}) \cdot \pi \,. \tag{1.6}$$

Suppose that θ is an arbitrage portfolio. By definition, its initial value is nonpositive and its cash-flows are nonnegative for all final states. Furthermore, either (i) at least one cash flow is positive, or (ii) the initial cost is negative. Clearly, Eq. (1.6) tells us that neither case can occur. In fact, since the π_j's are all positive, the initial cost will be positive if at least one of the cash-flows is positive and it will be zero if all the cash-flows are zero.

We pass to the proof of the converse statement: no-arbitrage implies the existence of a vector with positive entries satisfying (1.5a) and (1.5b). Let $\mathbf{R}^{M+1}$ denote the vector space of $M + 1$-tuples $\mathbf{x} = (x_0 \ldots, x_M)$ and let $\mathbf{R}_+^{M+1}$ represent the *closed convex cone*

$$\mathbf{R}_+^{M+1} \equiv \left\{ \mathbf{x} : x_j \ge 0\,, \text{ for all } 0 \le j \le M \right\}\,.$$

Recall that a subset C of $\mathbf{R}^q$ is called a cone if, given $X \in C$ and given an arbitrary nonnegative real number λ, $\lambda X \in C$, i.e., all rays through the point X are contained in C.

Let L be the linear subspace of $\mathbf{R}^{M+1}$ defined by

$$L \equiv \left\{ (-\theta \cdot \mathbf{p}\,, \theta \cdot \mathbf{D}_{.1}\,, \ldots, \theta \cdot \mathbf{D}_{.M})\,, \theta \in \mathbf{R}^N \right\}\,.$$

As is easily checked from the definition, nonexistence of arbitrage portfolios implies that the subspace L and the cone $\mathbf{R}_+^{M+1}$ intersect only at the origin, $(0, \ldots, 0)$.

From Convex Analysis, (cf. Rockafellar, Princeton University Press, 1990), it is known that there must exist a *separating hyper-plane*, i.e., a linear subspace H of $\mathbf{R}_+^{M+1}$ of dimension M,

---

[2]Equation (1.5) states that the price vector is a linear compilation of the columns of the cash-flow matrix with positive coefficients.

[3]From D. Duffie, *Dynamical Asset Pricing Theory*, Princeton University Press, Princeton, NJ, 1992.

which separates $\mathbf{R}^{M+1} \setminus \{0\}$ and L.[4] The general equation for a hyper-plane in $\mathbf{R}^{M+1}$ is

$$H \equiv \left\{ \mathbf{x} : \sum_0^M \lambda_j x_j = 0 \right\},$$

where $\lambda = (\lambda_0 \ldots \lambda_M)$ is a vector in $\mathbf{R}^{M+1}$. In terms of the normal λ to the hyper-plane, the concept of separation can be expressed as follows:

$$\lambda \cdot X > \lambda \cdot Z \quad \text{for all } X \in \mathbf{R}_+^{M+1} \text{ and for all } Z \neq 0 \in L, \qquad (*)$$

Clearly, since L is a linear space the latter condition can hold only if

$$\lambda \cdot Z = 0 \quad \text{for all } Z \in L,$$

which means that the subspace L is contained in H. On the other hand the condition $\lambda \cdot X > 0$ for all $X \in \mathbf{R}_+^{M+1}$, $X \neq 0$ is equivalent to having $\lambda_j > 0$ for all j. But, since L is contained in H, we conclude that for all $\theta = (\theta_1, \ldots, \theta_N)$ in $\mathbf{R}^N$

$$-\lambda_0 \theta \cdot \mathbf{p} + \sum_1^M \lambda_j \theta \cdot \mathbf{D}_{.j} = 0.$$

This implies that

$$-\lambda_0 \mathbf{p} + \sum_1^M \lambda_j \mathbf{D}_{.j} = \mathbf{0},$$

or

$$\mathbf{p} = \sum_1^M \frac{\lambda_j}{\lambda_0} \mathbf{D}_{.j}$$

$$= \sum_1^M \pi_j \mathbf{D}_{.j},$$

with $\pi_j = \frac{\lambda_j}{\lambda_0}$, and where the positivity of π_j for all j follows immediately from the positivity of λ_j. This is precisely what we wanted to show.[5] ∎

Theorem 1.1 implies that prices and cash-flows must satisfy certain relations in a no-arbitrage economy. The positive coefficients π_j, $1 \leq j \leq M$ are usually called *state-prices*. To give a

---

[4]A more standard form of the separation theorem that is valid for arbitrary closed disjoint convex sets does not guarantee the strict separation necessary for the proof we give here. The strict separation, (*), follows by assuming that one of the two convex sets (i.e., $\mathbf{R}_+^{M+1}$) is, in addition, a cone that does not contain a nontrivial linear subspace. See Duffie, 1992.

[5]A more intuitive geometric interpretation of this theorem is given in Section 1.2.

financial interpretation to this theorem, we define the *risk-neutral probabilities* or *risk-adjusted probabilities* (the terminology will become clear later).

$$\hat{\pi}_j = \frac{\pi_j}{\left(\sum_1^M \pi_k\right)} \qquad 1 \le j \le M .$$

These coefficients are all positive and have sum 1 so, mathematically, they can be viewed as probabilities. Also, set

$$1 + R = 1/\left(\sum_1^M \pi_j\right) .$$

Now, suppose that there exists an investment opportunity which guarantees a payoff of \$1 at the end of the period—a bond or money-market deposit. In terms of the model, the payoff can be represented as the vector $(1, 1, \ldots, 1)$ in $\mathbf{R}^M$. According to (1.5a) and (1.5b), the value of this riskless bond must be

$$p_{bond} = \sum_1^M \pi_j = 1/(1 + R) ,$$

so we have

$$R = \text{interest rate prevailing over the period} .$$

We can rewrite relation (1.5a) as

$$p_i = \frac{1}{1 + R} \sum_1^M D_{ij} \hat{\pi}_j \qquad (1.7)$$

or

$$p_i = \frac{1}{1 + R} \mathbf{E}\{\mathbf{D}_i\} ,$$

where $\mathbf{E}$ is the expectation-value operator associated with the probabilities $\hat{\pi}_j$, $1 \le j \le M$.[6] We have established the following corollary of Theorem 1.1:

THEOREM 1.2
Assume that the market admits no arbitrage portfolios and that there exists riskless lending/borrowing at rate R%. Then, there exists a probability measure defined on the set of possible market outcomes, $\{1, 2, \ldots, M\}$, such that the value of any security is equal to the expected value of its future cash flows discounted at the riskless lending rate.

This is an important general principle of the Arrow–Debreu model. It has several remarkable implications. First of all, notice that in our original model, we did not make any assumptions about the frequency at which each of the M "states" occurred. These frequencies could, in principle, be determined statistically, by observing the market over many time periods. One could then write

$$\text{Prob.}\{\text{state } j \text{ occurs}\} = f_j$$

---
[6]Note: within the generality of the Arrow–Debreu theory, we could have $R \le 0$.

for $1 \leq j \leq M$. This raises the following question: what is the relation between the risk-neutral probabilities $\hat{\pi}_j$ of the no-arbitrage theorem and the probabilities that arise by observing the frequency of the different states? Interestingly enough, the two probabilities can be quite different. The market value of a given security will *not* be equal, in general, to its discounted expected cash-flows under the *frequential* probabilities. This has to do with investors' perception of the *risk* of holding different securities given the present information. Thus, the market may attach *economic values* to future states that are not proportional to their observed frequency in the past. If we had

$$p_i = \frac{1}{1+R} \sum_1^M D_{ij}\, f_j \,, \tag{1.8}$$

the importance attached by the investors to the cash flows in the different future states would be proportional to their frequency, i.e., the different states are equally "important" after adjusting for frequency. On the other hand, writing the pricing equation (1.7) in the form

$$p_i = \frac{1}{1+R} \sum_1^M D_{ij} \left(\frac{\hat{\pi}_j}{f_j} \right) f_j \,,$$

we see that the prices of securities are *weighted* statistical averages of future cash-flows discounted at the riskless rate. The "weights" $\frac{\hat{\pi}_j}{f_j}$ reflect investor's preferences toward the different states; they are usually called *state-price deflators*.

One consequence of Eq. (1.7) is that

$$\sum_{j=1}^M \hat{\pi}_j \left(\frac{D_{i,j}}{p_i} - 1 \right) = R \,,$$

that is, *under the risk-neutral probabilities, the expected return of any traded security is equal to the riskless interest rate.*

To make this interpretation of state-prices more specific, suppose that an additional set of M state-contingent "elementary securities" $s_{N+1}, s_{N+2}, \ldots, s_{N+M}$ is introduced in the market. For each j, the security s_j has cash-flow \$1 in state j and \$0 otherwise. The matrix D is enlarged to a matrix of dimensions $(N + M) \times M$. Notice that a portfolio containing D_{i1} units of s_{N+1}, D_{i2} units of s_{N+2}, etc., has cash-flows $(D_{i1}, D_{i2}, \ldots, D_{iM})$ according to the M possible final states. Thus, it provides the *same return* as the ith "standard" security s_i. If there are no arbitrage opportunities, then the value of such portfolio should be equal to the value of the security. This can be seen as follows: if the price of the portfolio is less than p_i, then an investor can short the portfolio and purchase the security, making an immediate profit. After the trading period, the cash-flows from the security exactly compensate the short position in the portfolio, and hence the investor will be able to make a profit without taking risk. A similar arbitrage can be constructed if the portfolio is traded at a lower price than the security. Therefore, risk-neutral probabilities are consistent with the statement

$$\hat{\pi}_j = \text{price of } s_{N+j} \quad j = 1, \ldots, M \,,$$

in the sense that we must have

$$p_i = \sum_1^M D_{ij} \, \hat{\pi}_j$$

for all i, which is precisely Eq. (1.5b). We conclude that *state-prices can be interpreted as a set of market prices for "state-contingent claims" that pay* \$1 *in state j and zero otherwise, for* $1 \leq j \leq M$, supporting the notion that state-prices correspond to the prices of wealth in the different states. Notice then that the risk-neutral probabilities are those that "make the investor risk-neutral," given the prices of wealth of the different states.

Example 1.2

Let us assume a hypothetical presidential election with candidates from two political parties, A and B. Historically, there have been equal numbers of presidents from either party. On the other hand, a well-known and reputable book-maker (English, of course) is giving the following odds for each candidate: candidate from party A: 3–5; candidate from Party B: 1–4 (i.e., a gambler wins \$4 for every dollar he risks if Party B wins). Given the historical information, the *frequential* probabilities are, of course, $f_A = f_B = .5$. However, from the point of view of the bookie, the "odds-probabilities" are

$$\pi_A = \text{const.} \times \frac{5}{3}, \quad \pi_B = \text{const.} \times 4,$$

which gives, after computing the unknown constant,

$$\pi_A = .2942, \quad \pi_B = .7058.$$

The latter can be interpreted as "risk-neutral" or "state-price" probabilities. Suppose, for instance, that an individual who is unaware of the bookmaker's odds offers you equal odds on candidate of Party B winning (you bet \$1 and receive \$2 if B wins and \$0 if A wins). You can immediately create an "arbitrage" by betting \$2 × .2942 = \$.5882 on candidate A with the book-maker. In this case, you have guaranteed payoff \$2 regardless of who wins the election at a cost of \$1.5882, or a net profit of \$.4117. ▢

In conclusion, the f_j's are probabilities in a statistical sense, whereas the risk-neutral probabilities π_j's are "mathematical probabilities" used to calculate the market values of all securities, including "state-contingent" claims (i.e., derivatives), from their cash-flows. These two notions of probability should not be confused. The correct market values of securities are determined from the risk-neutral probabilities.

1.2 Security-Space Diagram: A Geometric Interpretation of Theorem 1.1

The Arrow–Debreu model with N traded securities and M final states can be visualized geometrically. This interpretation is slightly different than the one presented in the proof of Theorem 1.1: let $\mathbf{R}^N$ represent N-dimensional Euclidean space. Since there are M final states, we can represent the final outcomes as M vectors $\mathbf{D}._1, \ldots, \mathbf{D}._M$, where the entries of each vector represent the final cash-flows of each traded security. (See Figure 1.2).

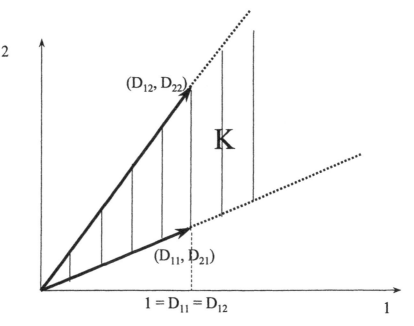

FIGURE 1.2
The Arrow-Debreu model viewed geometrically. The "ambient space" has dimension
N **(the number of securities). The convex set** K **is the span of the** M **dividend (column)**
vectors which give the cash-flows of the securities in each of the M **final states.**

In addition, we can draw the price vector $\mathbf{p}$ as an element of this space. The absence of arbitrage corresponds to the price vector $\mathbf{p}$ lying in the interior of the convex cone $\mathbf{K}$ generated by the M cash-flow vectors, as shown in Figure 1.3.

In fact, if $\mathbf{p}$ would lie in the exterior or on the boundary of $\mathbf{K}$, the separating hyper-plane theorem ensures that there exists linear subspace described by the equation

$$\theta \in \mathbf{R}^N : \quad \theta \cdot \mathbf{n} = 0 ,$$

where $\mathbf{n}$ is an N-vector, which *separates* $\mathbf{p}$ from the interior of the convex cone generated by the $\mathbf{D}._j$'s. The separation property would imply (possibly after a change of the sign of $\mathbf{n}$) the inequalities

$$\mathbf{p} \cdot \mathbf{n} \leq 0, \quad \mathbf{D}._j \cdot \mathbf{n} \geq 0$$

with at least one inner product positive for some j—i.e., an arbitrage.

As in the discussion preceding Theorem 1.2, the existence of a bond or money-market deposit can be used to pass from a convex cone to a bounded convex set in dimension $N - 1$. In fact, assume without loss of generality that the bond corresponds to the security with $i = 1$, so that all the $\mathbf{D}$-vectors have the first entry equal to one ($D_{1j} = 1$). Denoting a generic vector in $\mathbf{R}^N$ by

$$\theta = (\theta_1, \hat{\theta}) , \quad \hat{\theta} = (\theta_2, \ldots \theta_N) ,$$

we can consider the intersection of the convex cone $\mathbf{K}$ with the hyper-plane in $\mathbf{R}^N$ of vectors with first coordinate equal to p_1, the price of the bond. This intersection is the convex set $\hat{\mathbf{K}}$

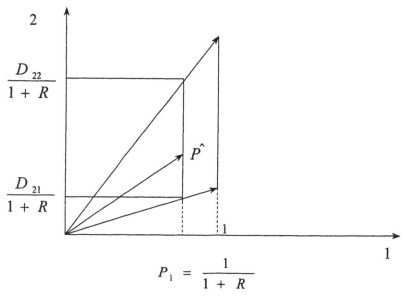

FIGURE 1.3

"No arbitrage" means that the price vector must lie in the convex cone K generated by the M cash-flow vectors corresponding the different final states. In the case $M = 2$, when one security is a riskless bond, the no-arbitrage condition means that the price of the stock, P_2, satisfies $\frac{D_{21}}{(1+R)} < P_2 < \frac{D_{22}}{(1+R)}$.

generated by the vectors

$$p_1 \widehat{\mathbf{D}}_{.1}, \; p_1 \widehat{\mathbf{D}}_{.2}, \; \dots p_1 \widehat{\mathbf{D}}_{.M}.$$

The no-arbitrage condition is then equivalent to stating that $\hat{\mathbf{p}}$ lies in the interior of the convex set generated by these vectors,[7] i.e.,

$$\hat{\mathbf{p}} = \sum_{j=0}^{M} \hat{\pi}_j \, p_1 \, \widehat{\mathbf{D}}_{.j}$$

$$= p_1 \sum_{j=0}^{M} \hat{\pi}_j \, \widehat{\mathbf{D}}_{.j}$$

where the weights $\hat{\pi}_j$ are positive and sum to 1. (In the special case of Figure 1.1(b) the convex set $\widehat{\mathbf{K}}$ is a line segment.) Since, by definition, we have $p_1 = \frac{1}{1+R}$, the last formula is equivalent to Eq. (1.7).

---

[7]Some readers will recognize this result as a well-known result in game theory known as *Farkas' lemma.*

1.3 Replication

The interpretation of state-prices as the values of elementary state-contingent claims is the basis for the valuation of derivative securities in a no-arbitrage market.

Given a security s, and a set of securities $s_1, s_2, \ldots, s_K$, we say that the portfolio $(\theta_1, \theta_2, \ldots, \theta_K)$, (representing holdings in each of the K securities) *replicates* s if the security and the portfolio have identical cash-flows. Under no-arbitrage conditions, the value of the security and of the replicating portfolio must be the same. Otherwise, an arbitrage could be realized either by shorting the portfolio and buying the security or, alternatively, by shorting the security and buying the portfolio. If the value of the portfolio is less than the value of the security, the first strategy is an arbitrage. If the portfolio is worth more than the security, the second strategy is an arbitrage. This argument gives rise to a simple but important valuation principle.

PROPOSITION 1.1

In a no-arbitrage market, if a security admits a replicating portfolio of traded securities, its value is equal to the value of the replicating portfolio.

Here are some elementary applications of this principle.

Forward Prices. Suppose that a security has spot price P and that the yield for riskless lending over the trading period is R. Consider a *forward contract*, which consists of an agreement to purchase the security at the end of the trading period at price K. Assume that the security pays no dividends over the trading period. The no-arbitrage price of the forward contract is

$$Q = P - K/(1 + R).$$

To see this, consider a portfolio consisting of being long one unit of the security and short $K/(1 + R)$ worth of riskless bonds. After the trading period, the holder of this portfolio will own the security and will owe $\$K$. Therefore, if he receives K at the end of the final period he will be able to meet his cash obligation and deliver the security; he will have no profit/loss. The portfolio is equivalent to having long position in the forward contract. In practice, forward contracts are designed so that they have zero initial cost ($Q = 0$). The *forward price*, which is price for delivery of the security *after* the trading period, is

$$F \equiv (1 + R) P,$$

because $K = F$ makes the initial cost zero.

Put–Call Parity. This important example of replication involves options (see Figure 1.4). Suppose that a security with price S is traded as well as a call option and a put option with exercise price (strike price) K. Recall, from 1.1, that a *call option* is a contract that gives the holder the right, but not the obligation, to purchase (one unit of) the security at price K, at a stipulated date (expiration date). A *put option* gives the holder the right to sell the underlying security at price K at the maturity date. In this example, we assume that the call and the put have same strike prices and expiration dates. Denote the market prices of the call and the put by C and P, respectively. Let R be the interest paid for riskless lending over the duration of the option (compounded simply). Suppose that an investor has the following position: long one call, short one unit of the underlying security and long $K/(1+R)$ worth of riskless bonds. Let us examine the cash-flows arising from this position at the maturity date of the call. If, on the one hand, the price of the security at the expiration date, S_T, is greater than K, the investor can exercise the call, purchasing the security at the strike price K and then returning the security held short. This leaves him with zero profit/loss. On the other hand, if the price of the underlying security is less than K, he will not exercise the call. At the option's expiration date his new position is short one unit of underlying security and he has K in cash, for a total value of $K - S_T$. If, instead, the investor would have held a put struck at K initially, his position after the trading period would have been neutral if $S_T \geq K$ (the put goes unexercised) and worth $K - S_T$ if $S_T < K$, since he can exercise the put. Therefore, the portfolio "long call, short underlying asset, long $K/(1+R)$ in bonds" replicates the put. We conclude that

$$P = C - S + K/(1+R).$$

This is the *put–call parity* relation. It shows how to construct a "synthetic put" via a portfolio. Since there are four variables in this equation, we can produce similar portfolios to replicate other assets as well. Accordingly, the position "long underlying security, long put, short $K/(1+R)$ in bonds" replicates a call. The position "long one call, short one put, long $K/(1+R)$ in bonds" replicates the cash-flows of the underlying security. Finally, the position "long one unit of security, short one call, long one put" replicates a riskless bond paying $$K$ at the expiration date of the option.

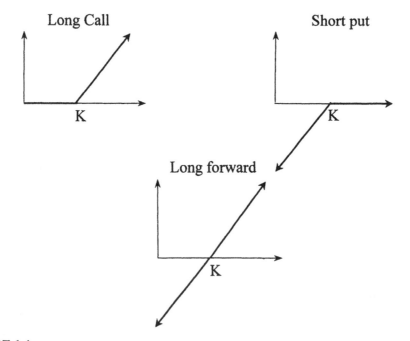

FIGURE 1.4
Interpretation of put–call parity using the option payoff functions. If we add the graphs corresponding to a long call position and short put position, we obtain a straight line with slope 1 passing through K. Thus a long call/short put position is equivalent to holding a forward contract to buy the stock at K dollars.

1.4   The Binomial Model

We present the simplest case of the Arrow–Debreu model (see Figure 1.5). Consider an ideal situation in which there are only two states ($M = 2$) and two securities: a bond with yield $R\%$ and a security (s) with price P. We assume that the cash-flows of this security in states 1 and 2 are $P\,U$ and $P\,D$, respectively, where U and D are given numbers and $D < U$. If there are no arbitrage opportunities, we must have

$$P = \frac{1}{1+R}\left(\hat{\pi}_1\,P\,U + \hat{\pi}_2\,P\,D\right)$$

where $\hat{\pi}_1$ and $\hat{\pi}_2$ are positive and $\hat{\pi}_1 + \hat{\pi}_2 = 1$. Thus, the risk-neutral probabilities satisfy

$$\begin{cases} \hat{\pi}_1 + \hat{\pi}_2 = 1 \\[2mm] \hat{\pi}_1\,U + \hat{\pi}_2\,D = 1 + R. \end{cases}$$

It is easy to see that this system will have positive solutions if and only if

$$D < 1 + R < U. \tag{1.9}$$

This is intuitively clear: if $1 + R \geq U$, the return on riskless lending is greater than or equal to the return for investing in the risky security, regardless of the final state. An arbitrage could then be achieved by shorting the security and lending out the proceeds at the riskless rate. Similarly, if $1 + R \leq D$, the investor can make a riskless profit by borrowing at the riskless rate and purchasing the security. Suppose that (1.9) holds. The solution of the linear system is

$$\hat{\pi}_1 = \frac{1 + R - D}{U - D} \quad , \quad \hat{\pi}_2 = \frac{U - 1 - R}{U - D}. \tag{1.10}$$

Thus, the risk-neutral probabilities are entirely determined from the parameters of the binomial model. The actual probabilities of occurrence of each state are irrelevant in the pricing process (as long as neither one is zero—we must assume that both states can occur).

This has interesting consequences. Suppose now that we augment the number of traded securities (the number of final states is still $M = 2$). Then, the price of any security is completely determined by future cash-flows because the risk-neutral probabilities are still given by (1.10). A "state-contingent" security which has cash-flows D_1 in state 1 and D_2 in state 2 must have value V, where

$$V = \frac{1}{1+R}\left(\hat{\pi}_1\,D_1 + \hat{\pi}_2\,D_2\right). \tag{1.11}$$

This idea can be applied to price any security contingent on the value of s. For instance, a "call option" on s with exercise price K (with $P\,D < K < P\,U$) has cash-flows

$$D_1 = P\,U - K, \quad D_2 = 0$$

since the holder will gain the difference between the market value and the exercise price if this difference is positive. We conclude that the arbitrage-free value of this call option on s is

$$V_{call} = \frac{1}{1+R} \cdot \frac{1+R-D}{U-D} \cdot (PU - K).$$

We can also illustrate the idea of replicating portfolios in the binomial model. Consider a portfolio (θ_1, θ_2) representing an investor's holdings in the security s (with price P) and the $1 discount bond (with price $\frac{1}{1+R}$), respectively. This portfolio yields exactly the same returns as a security with cash-flows D_1 and D_2 provided that

$$\begin{cases} \theta_1 \, PU + \theta_2 = D_1 \\ \theta_1 \, PD + \theta_2 = D_2. \end{cases}$$

Solving for θ_1 and θ_2, we find that

$$\theta_1 = \frac{D_1 - D_2}{PU - PD} \quad \text{and} \quad \theta_2 = \frac{UD_2 - DD_1}{U-D}. \tag{1.12}$$

Since holding the portfolio (θ_1, θ_2) gives the investor the same returns as holding the security with cash-flows D_1 and D_2, the two positions should have equal value. We conclude that $V = \theta_1 P + \theta_2/(1 + R)$. Substituting the values of the θs from (1.12), we recover the price (1.11). Notice that this calculation, based on replicating the payoff of the contingent claim, did not require knowing the risk-neutral probabilities. This is because the binomial model admits a *unique* set of state-prices/risk-neutral probabilities.

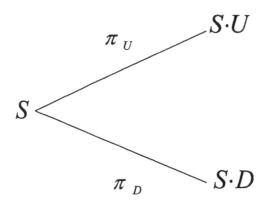

FIGURE 1.5
The one-period binomial model ($N = 2$, $M = 2$).

1.5   Complete and Incomplete Markets

In the binomial model, *any* vector of future cash flows (D_1, D_2) (the index labels the final state) can be replicated in terms of a portfolio of the basic security and a riskless bond. This property can be generalized to the setting of the N-securities/M-states model.

DEFINITION 1.2 *A securities market with M states is said to be complete if, for any cash-flow vector ($D_1, D_2, \ldots, D_M$), there exists a portfolio of traded securities ($\theta_1, \theta_2, \ldots, \theta_N$) that has cash-flow D_j in state j, for all $1 \leq j \leq M$.*

Market completeness is therefore equivalent to having a cash-flows matrix $\mathbf{D} = (D_{ij})$ with the property that the system of linear equations

$$\theta \cdot \mathbf{D} = E,$$

or

$$\sum_1^N \theta_i D_{ij} = E_j, \quad 1 \leq j \leq M,$$

has a solution $\theta \in \mathbf{R}^N$ for any $E \in \mathbf{R}^M$.

From Linear Algebra, we know that this property is satisfied if and only if

$$rank\ \mathbf{D} = M,$$

which is equivalent to saying that the column vectors of the matrix D span the entire space $\mathbf{R}^M$. Market completeness is a very strong assumption, which greatly simplifies the valuation of derivative securities. Derivative securities can be represented by general cash-flow vectors ($D_1, D_2, \ldots, D_M$), as opposed to the N "standard securities" that have cash-flow vectors ($D_{i1}, D_{i2}, \ldots, D_{iM}$), for $1 \leq i \leq N$. Since any derivative security is equivalent to a portfolio of standard traded assets, its price is fully determined from $\mathbf{E}$ in the absence of arbitrage. More formally, we have

PROPOSITION 1.2

Suppose that the market is complete and that there are no arbitrage opportunities. Then there is a unique set of state-prices $(\pi_1, \pi_2, \ldots, \pi_M)$ satisfying (1.5a) and (1.5b), and hence a unique set of risk-neutral probabilities $(\hat{\pi}_1, \hat{\pi}_2, \ldots, \hat{\pi}_M)$. Conversely, if there is a unique set of state-prices, then the market is complete.

PROOF The no arbitrage hypothesis implies the existence of a vector π such that $\mathbf{p} = \mathbf{D}\pi$. If the market is complete the rank of $\mathbf{D}^t$ (the superscript "t" indicates the transposed matrix) is M. Therefore, the kernel of $\mathbf{D}$ is trivial and the equation $\mathbf{p} = \mathbf{D}\pi$ has a unique solution for π. This fact together with the interpretation that we gave below Theorem 1.2 of the state-price vector π as the price of the state-dependent contingent claims implies that the price of a state contingent claim that pays \$1 in the jth state of the world and zero otherwise is completely determined for all j. Therefore, there can be at most one set of state-prices. If they exist, state-prices are unique.

To prove the converse statement, suppose that there exists a unique vector of state-prices $\pi = (\pi_1, \pi_2, \ldots, \pi_M)$ (with strictly positive entries) such that (1.5a) and (1.5b) are satisfied. We will argue that the market is complete by contradiction. In fact, if the market is not complete, then $rank\ \mathbf{D} < M$. From Linear Algebra, we know that the matrix $\mathbf{D}$ must have a nonempty *right-null space*, i.e., there exists $\lambda = (\lambda_1, \lambda_2, \ldots, \lambda_M)$ such that

$$\mathbf{D} \cdot \lambda = \mathbf{0}, \tag{1.13}$$

or, equivalently,

$$\sum_1^M D_{ij}\lambda_j = 0, \quad \text{for all } 1 \le i \le N.$$

Using the no-arbitrage relation (1.5a) and (1.13), we conclude that

$$\mathbf{D} \cdot (\pi + \rho\lambda) = \mathbf{p},$$

for all real numbers ρ. Since the entries of π are strictly positive, we can choose ρ sufficiently small so that $\pi_j + \rho\lambda_j$ is positive for all j. Therefore, we have constructed a new state-price vector, contradicting our hypothesis. We conclude that, in a no-arbitrage market, uniqueness of state-prices implies that the market is complete. ∎

The above argument can also be used to characterize the set of state-price vectors corresponding to an incomplete market with given price vector $\mathbf{p}$ and cash-flow matrix $\mathbf{D}$. Let $k = rank\ \mathbf{D}$. In an incomplete market, we have $M - k > 0$. Notice that k is the dimension of the right-null space of $\mathbf{D}$. Given two state-price vectors $\pi^{(1)}$ and $\pi^{(2)}$ and ρ such that $0 \le \rho \le 1$, the convex combination $\rho\pi^{(1)} + (1 - \rho)\pi^{(2)}$ is also a state-price vector. Hence, the set of all state-price vectors is convex. Moreover, in the proof of the Proposition, we showed that each state-price vector is contained in a $(M - k)$-dimensional neighborhood inside this cone. Consequently, *the set of admissible state-price vectors is an open cone in an $(M - k)$-dimensional affine subspace of* $\mathbf{R}^M$.

What is the financial meaning of a complete/incomplete market? In a complete market there is a unique set of state-prices. However, in real markets it is usually impossible to completely identify the set of final states or quantify the investors' preferences toward different states. Market completeness is a convenient idealization of the behavior of securities markets. Incomplete markets—with many possible price structures satisfying the no-arbitrage condition—are the rule rather than the exception.

1.6 The One-Period Trinomial Model

We describe a simple example of an incomplete market (Figure 1.6). Assume that there are two securities—a riskless bond with yield R paying \$1 at the end of the trading period and a security s. There are three states, which correspond to different cash-flows for s. We assume that the price of s is P and that the cash-flows of s are PU in state 1, PM in state 2, and PD in state 3, with

$$D < M < U.$$

Clearly, this market is incomplete, because the dimension of the cash-flows matrix is 3×2 (if there are more states than traded securities, the market is always incomplete). We can investigate the conditions for the existence of state-prices. Since we have assumed that there is riskless lending, we have

$$\frac{1}{1 + R} = \pi_1 + \pi_2 + \pi_3, \tag{1.14}$$

for any admissible set of state-prices. Since there is no arbitrage, we must have

$$P = PU\pi_1 + PM\pi_2 + PD\pi_3,$$

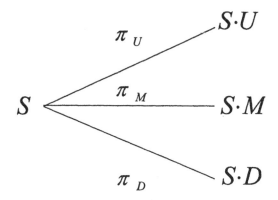

FIGURE 1.6
One-period trinomial model. If the only traded securities are a stock and riskless money, we have $N = 2$, $M = 3$. The market is then incomplete, because there are many possible risk-neutral probability measures.

or

$$1 = U \pi_1 + M \pi_2 + D \pi_3 . \tag{1.15}$$

Admissible state-price vectors π must satisfy equations (1.14) and (1.15) and have positive entries. These equations show that there is no arbitrage if and only if

$$D < 1 + R < U , \tag{1.16}$$

the interpretation of which was given in the discussion of the binomial model. Assuming that condition (1.16) holds, the set of state prices can be visualized as a line segment corresponding to the intersection of the planes described by (1.14) and (1.15) and the positive quadrant in $\mathbf{R}^3$. The two extreme points of this segment are

$$\pi_1 = \frac{(1+R) - D}{(1+R)(U-D)} , \quad \pi_2 = 0 , \quad \pi_3 = \frac{U - (1+R)}{(1+R)(U-D)} \tag{1.17}$$

and

$$\pi_1 = 0 , \quad \pi_2 = \frac{M - (1+R)}{(1+R)(M-D)} , \quad \pi_3 = \frac{(1+R) - D}{(1+R)(M-D)} , \tag{1.18a}$$

if $M \geq (1+R)$, or

$$\pi_1 = \frac{(1+r) - M}{(1+R)(U-M)} , \quad \pi_2 = \frac{U - (1+R)}{(1+R)(U-M)} , \quad \pi_3 = 0 , \tag{1.18b}$$

if $M < (1+R)$.

Because the prices of traded assets are linear functions of the state prices, according to Eq. (1.5b), this calculation can be used to derive *bounds* on the values of contingent claims based on the available information. In fact, we know that a linear function defined on a closed convex set attains its maximum and minimum values at the extreme points of this set. (Actually, since the set of state-prices is an *open* convex set, the maxima and minima are attained at "degenerate" state-price vectors that have entries equal to zero and hence are not, strictly speaking, state-prices.) To illustrate this, consider the case of a call option on the basic security s with strike price K. To fix ideas, assume that $PM < K < PU$. Then, the cash-flows for this option are $PU - K$ in state 1 and 0 in states 2 and 3. Its no-arbitrage value is $C = \pi_1 (PU - K)$. Therefore, using (1.17), (1.18a) and (1.18b), we find that

$$C^+ = \frac{(1+R) - D}{(1+R)(U - D)} (PU - K)$$

is an upper bound for the price of the option. If $M \geq (1 + R)$ the lower bound on the price is $C^- = 0$, and if $M < (1 + R)$, the lower bound is

$$C^- = \frac{(1+R) - M}{(1+R)(U - M)} (PU - K).$$

(Of course, the upper and the lower bounds coincide when $M = D$, and we recover the result of the binomial model.)

This example shows an important application of state prices as a tool for contingent claim valuation in incomplete markets. State prices are not unique but can nevertheless be used to obtain partial information about fair prices. This result can be interpreted financially in terms of *risk-aversion* of different agents. A market participant that does not want to incur *any* risk will bid (be willing to buy) the security at the price corresponding to the lower bound C^- and will offer (be willing to sell) the security at the price corresponding to the upper bound C^+. Since there is incomplete information about Arrow–Debreu prices, transactions made between the bounds imply a risk for the buyer as well as for the seller.

1.7   Exercises

1. Consider a hypothetical country where the government has declared a "currency band" policy, in which the exchange rate between the domestic currency, denoted by XYZ, and the U.S. dollar is guaranteed to fluctuate in a prescribed band, namely

$$USD\ 0.95 \leq XYZ \leq USD\ 1.05 ,$$

for at least 1 year. Suppose also that the government has issued 1-year notes denominated in the domestic currency that pay a simply compounded annualized rate of 30%. Assuming that the corresponding interest rate for U.S. deposits is 6%, show that this market is not arbitrage-free in the "pure" sense. Describe the situation in terms of the Arrow–Debreu model. Propose some realistic scenarios that could make this pure arbitrage disappear in practice.

2. (i) Show that the set of all probability measures on a finite state-space of M elements can be represented as a convex subset $\mathbf{P}$ of the Euclidean space $\mathbf{R}^M$. Given a security s defined by its price and its cash-flows, verify that the set of measures which are risk-neutral for this security corresponds to the intersection of the set $\mathbf{P}$ with a hyper-plane in $\mathbf{R}^M$. Similarly, show that the set of admissible risk-neutral measures for a securities market with N securities corresponds to the intersection of $\mathbf{P}$ with N hyper-planes.

(ii) Apply this analysis to the trinomial model of Section 1.6—assuming that $S = \$100$, $U = 1.10$, $M = 1.00$, $D = .80$, $R = .05$ and that a call option with strike $105 is trading at a premium of $C = \$3.80$. Show that if, instead, $C = \$1.00$, there is an arbitrage opportunity.

3. On the week of September 7, 1996, Ladbroke, a London bet-maker, gave the following odds regarding the upcoming U.S. presidential election: Clinton 1–6, Dole 7–2, Perot 1–50. (For instance, Ladbroke pays £1 for every £6 bet on Clinton if he wins.) Calculate the corresponding risk-neutral probabilities for the victory of each candidate assuming that one of them will necessarily win.

References and Further Reading

[1] Arrow, K. (1952), *Le Role des valeurs boursieres pour la répartition des risques*, Econometrie, Colloque International C.N.R.S.

[2] Arrow, K. (1970), *Essays in the Theory of Risk Bearing*, North-Holland, London.

[3] Arrow, K. and Debreu, G. (1954), Existence of an Equilibrium for a Competitive Economy, *Econometrica* 22, pp. 265–290.

[4] Debreu, G. (1959), *Theory of Value: An Axiomatic Analysis of Economic Equilibrium*, Yale University Press, New Haven, CT.

[5] Drèze, J. (Winter 1970-1), Market Allocation under Uncertainty, *European Economic Review*, pp. 133–165.

[6] Duffie, D. (1992), *Dynamical Asset Pricing Theory*, Princeton University Press, Princeton, NJ.

[7] Ross, S. (1978), A Simple Approach to the Valuation of Risky Streams, *Journal of Business* 51, pp. 453–475.

[8] Varian H.R. (1992), *Microeconomic Analysis*, W.W. Norton & Co., New York, 3rd ed.

Chapter 2

The Binomial Option Pricing Model

The simplest model for pricing derivative securities is the binomial model. It generalizes the one-period "up–down" model of Chapter 1 to a multiperiod setting, assuming that the price of the underlying asset follows a random walk.

In the binomial model, there are N trading periods and $N + 1$ trading dates, $t_0, t_1, \ldots, t_N$. It is possible to invest in a risky security (for instance stock) and to lend or borrow money at a riskless rate R. The price of the risky security at date t_n is denoted by S_n, $n = 0, 1, \ldots, N$. This price varies according to the rule

$$S_{n+1} = S_n\, H_{n+1} \ , \ 0 \leq n \leq N - 1 \,, \tag{2.1}$$

where H_{n+1} is a random variable such that

$$H_{n+1} = \begin{cases} U & \text{with probability } p \\ \\ D & \text{with probability } q \end{cases} \tag{2.2}$$

with $p + q = 1$.

The situation can be visualized in terms of a *binomial tree,* shown in Figure 2.1. Each node of the tree is labeled by a pair of integers $(n, j), n = 0, \ldots, N$, and $j = 0, \ldots, n$. We use the convention that node (n, j) leads to nodes $(n + 1, j)$ and $(n + 1, j + 1)$ at the next trading date with the "up" move corresponding to $(n + 1, j + 1)$ and the "down" move to $(n + 1, j)$. The price of the underlying asset at the node (n, j) is therefore as shown in Figure 2.1.

Assume first that the risky asset is a stock that pays no dividends for $0 \leq n \leq N$. Let us determine a probability measure on the set of paths $(S_0, S_1, \ldots, S_n, S_N)$ which makes the model arbitrage-free.[1] Since the asset price divided by the returns on riskless investment must be a martingale under the pricing measure, we should have

$$S_n = \frac{1}{1 + R}\, E\{\, S_{n+1} \mid S_n\, \}$$

or,

$$1 = \frac{1}{1 + R}\, (\, P_U\, U + P_D\, D\,) \,, \tag{2.3}$$

---

[1] The terms "probability measure," "probability distribution," "probability assignment," etc. , are equivalent for our purposes.

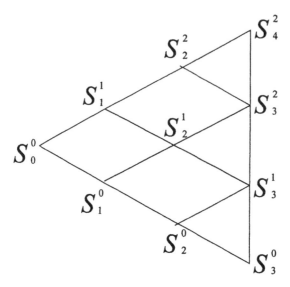

FIGURE 2.1
Multi-period binomial tree. The stock price goes up or down by a factor $H = U$ or D at each time.

where P_D and P_U represent risk-neutral conditional probabilities for up or down moves given the spot price S_n. Using Eq. (2.3) and the constraint $P_U + P_D = 1$, we can determine the probabilities P_U and P_D in terms of the parameters of the model. They are given by

$$P_U = \frac{1 + R - D}{U - D} \; , \quad P_D = \frac{U - (1 + R)}{U - D} . \tag{2.4}$$

In particular, P_U and P_D are independent of the spot price. Thus, we have

PROPOSITION 2.1
Given the parameters U, D, and R, there is a unique probability measure that makes the binomial model arbitrage-free. This probability is the one for which the successive price shocks H_n, $n = 1, \ldots, N$ are independent and identically distributed with

$$\text{Prob.} \{ H_n = U \} = \frac{1 + R - D}{U - D}$$

and

$$\text{Prob.} \{ H_n = D \} = \frac{U - (1 + R)}{U - D} .$$

2.1 Recursion Relation for Pricing Contingent Claims

We give a general method for calculating the value of a "European-style contingent claim," i.e., a derivative security that gives the holder the right to a payoff contingent on the value of the stock at some fixed date in the future. This date is taken to be $t_n = t_N$ and the payoff is represented by a known function of the stock price $F(S_N^j)$.[2]

We denote the value of the contingent claim at the trading date n conditional on $S_n = S_n^j$, i.e., its value at the node (n, j), by

$$V_n^j = V_n(S_n^j).$$

Since this derivative security has no intermediate cash-flows or coupons, its value must satisfy

$$V_n^j = \frac{1}{1+R} E\{ V_{n+1}(S_{n+1}) \mid S_n = S_n^j \}$$

or

$$\begin{cases} V_n^j = \frac{1}{1+R} \left[P_U V_{n+1}^{j+1} + P_D V_{n+1}^j \right] \\ \\ V_N^j = F(S_N^j). \end{cases} \tag{2.5}$$

This recursion relation determines the arbitrage-free value of the contingent claim inductively. From the values at maturity, one can derive the values at the date t_{N-1} and proceed backward in time until the present date. This procedure is sometimes called "rolling back the tree." Despite its apparent simplicity, relation (2.5) is very important. It will resurface under different forms in the course of this chapter.

The value of the derivative security can be obtained in closed form. For this, we view the arbitrage-free price as the discounted expected cash-flows of the security, so that

$$V_n^j = \frac{1}{(1+R)^{N-n}} E\left\{ F(S_N) \mid S_n = S_n^j \right\}$$

$$= \frac{1}{(1+R)^{N-n}} \sum_{k=0}^{N} \text{Prob.}\left\{ S_N = S_N^k \mid S_n = S_n^j \right\} \cdot F(S_N^k). \tag{2.6}$$

Let us calculate the conditional probabilities appearing in the latter equation. For this, we must "count" the number of paths going from S_n^j to S_N^k. Notice that if a path starts at S_n^j at time n it can end up at most at position S_N^{j+N-n}. This will happen only if $H_i = U$ for $i = n+1, \ldots, N-1$. Similarly, the lowest possible value is S_N^j, which corresponds to $H_i = D$ for all i. For any k such that $j < k < N-n+j$, the path will end up at S_N^k if and only if $H_i = U$ for $k-j$ periods out of $N-n+j$ and $H_i = D$ for the rest of the periods. It is important to notice that only the *number* of "up-shocks" matters and not the order in which they occur. Therefore, since the random variables H_i are independent, we have

$$\text{Prob.}\left\{ S_N = S_N^k \mid S_n = S_n^j \right\} = \binom{N-n}{k-j} \cdot P_U^{k-j} P_D^{N-n-k+j}, \tag{2.7}$$

where

$$\binom{N-n}{k-j} = \frac{(N-n)!}{(k-j)! \, (N-n-k+j)!}$$

---

[2] We will sometimes use the expression "date n" instead of "date t_n."

is the number of combinations of $k - j$ elements in a set of $N - n$ elements. Thus, in the binomial model, the Arrow–Debreu (or risk-neutral) conditional probability distribution for S_N given S_n is the *multinomial distribution* (Figure 2.2), well known from elementary Probability Theory.

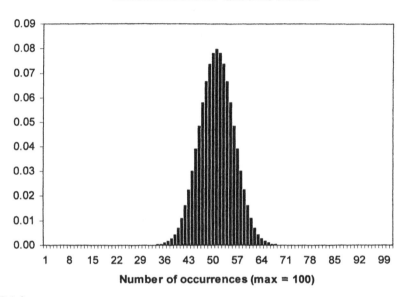

Multinomial distribution

FIGURE 2.2
Multinomial probability distribution generated by a binomial tree with $N = 100$.

The value (2.6) can be rewritten using Eq. (2.7) as

$$V_n^j = \frac{1}{(1 + R)^{N-n}} \sum_{k=j}^{N-n+j} \binom{N-n}{k-j} \cdot P_U^{k-j} \, P_D^{N-n-k+j} \cdot F(S_N^k). \qquad (2.8a)$$

In particular, if $n = j = 0$, we obtain the value in explicit form:

$$V_0^0 = \frac{1}{(1 + R)^N} \sum_{k=0}^{N} \binom{N}{k} \cdot P_U^k \, P_D^{N-k} \cdot F(S_N^k). \qquad (2.8b)$$

2.2 Delta-Hedging and the Replicating Portfolio

The pricing formula can be derived in a different way, using the idea of replicating portfolios.

Suppose that at the nth trading date, the stock price is S_n^j and that an agent holds (long or short) a portfolio of Δ_n^j shares and B_n^j dollars in an interest-bearing account, with total value

$$V_n^j = \Delta_n^j \, S_n^j + B_n^j.$$

The value of the agent's holdings after the next trading period would be

$$
\begin{cases}
\Delta_n^j S_{n+1}^{j+1} + B_n^j (1 + R) \equiv V_{n+1}^{j+1} & \text{in the "up" state} \\[2mm]
\Delta_n^j S_{n+1}^j + B_n^j (1 + R) \equiv V_{n+1}^j & \text{in the "down" state}
\end{cases}
$$

A relation between $\left(\Delta_n^j \, B_n^j \right)$ and the subsequent portfolio values is obtained by solving this system of equations. The result is

$$
\Delta_n^j = \frac{V_{n+1}^{j+1} - V_{n+1}^j}{S_{n+1}^{j+1} - S_{n+1}^j} \quad \text{shares} \tag{2.9a}
$$

and

$$
B_n^j = -\frac{1}{1 + R} \frac{S_{n+1}^j V_{n+1}^{j+1} - S_{n+1}^{j+1} V_{n+1}^j}{S_{n+1}^{j+1} - S_{n+1}^j} \tag{2.9b}
$$

Using algebra we see that

$$
V_n^j = \frac{1}{1 + R} \left[\frac{S_n^j(1 + R) - S_{n+1}^j}{S_{n+1}^{j+1} - S_{n+1}^j} \cdot V_{n+1}^{j+1} + \frac{S_{n+1}^{j+1} - S_n^j(1 + R)}{S_{n+1}^{j+1} - S_{n+1}^j} \cdot V_{n+1}^j \right]
$$

$$
= \frac{1}{1 + R} \left[P_U \, V_{n+1}^{j+1} + P_D \, V_{n+1}^j \right] \tag{2.10}
$$

where P_U and P_D are given in (2.4). Notice the similarity between this last equation and the recursion relation of the previous section.

Assume now that V_n^j is a function defined on the nodes of the binomial tree which satisfies the recursion relation (2.5) with final condition $V_N^j = F(S_N^j)$. Consider the following **trading strategy:**

(i) at time n take the position $\left(\Delta_n^j, \, B_n^j \right)$.

(ii) at subsequent times, maintain a number of shares in the portfolio equal to

$$
\Delta_k^j = \frac{V_{k+1}^{j+1} - V_{k+1}^j}{S_{k+1}^{j+1} - S_{k+1}^j} \, , \quad n \le k \le N \, . \tag{2.11}
$$

Equation (2.10) shows that the value of the agent's portfolio (stocks plus money-market account) will be equal to V_n^j, regardless of price movements, and that the share adjustments are such that the money-market account will be equal to $B_k^j = V_k^j - \Delta_k^j S_k^j$ for all $k \ge n$. The situation is described by Figure 2.3.

We conclude that

PROPOSITION 2.2

Suppose an agent takes an initial position in shares and bonds given by (2.9a) and (2.9b) and subsequently holds Δ_k^j shares in his portfolio if $S(t_k) = S_k^j$, for all $t_k \nmid t_N$. Then, the final value of the agent's portfolio is $F(S_N)$ at time t_N. Thus, if $S(t_0) = S_0^j$, the fair value at time t_0 of a contingent claim that pays $F(S_N)$ at time t_N is V_0^j.

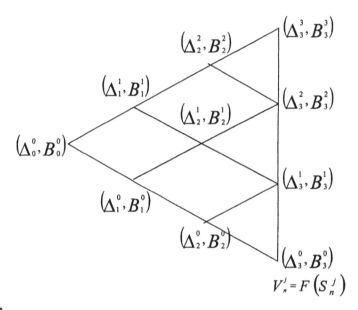

FIGURE 2.3
The replicating strategy is such that $B_n^j = V_n^j - \Delta_n^j S_n^j$ at each node of the tree. The final portfolio value is $F(S_N^j)$. (Here $N = 3$.)

2.3   Pricing European Puts and Calls

Let us calculate the fair price a European-style call option maturing after N periods using the binomial model. We shall use the payoff function

$$F(S_N) = \text{Max}(S_N - K, 0),$$

where K is the strike price. According to (2.8a) and (2.8b), the arbitrage-free value of the call is (with $S_0^0 = S$)

$$C(S, K; N) = \frac{1}{(1+R)^N} \sum_{k=0}^{N} \binom{N}{k} \cdot P_U^k \, P_D^{N-k} \cdot \text{Max}(S_N^k - K, 0)$$

$$= \frac{1}{(1+R)^N} \sum_{\substack{k \\ S_N^k > K}}^{N} \binom{N}{k} \cdot P_U^k P_D^{N-k} \cdot \left(S_N^k - K \right) \tag{2.12}$$

$$= S \cdot \left\{ \sum_{k > k_0}^{N} \binom{N}{k} \cdot Q_U^k Q_D^{N-k} \right\} - \frac{K}{(1+R)^N} \cdot \left\{ \sum_{k > k_0}^{N} \binom{N}{k} \cdot P_U^k P_D^{N-k} \right\},$$

where

$$Q_U = \frac{U}{1+R} P_U, \qquad Q_D = \frac{D}{1+R} P_D$$

and where k_0 is the smallest integer that is greater than or equal to

$$\ln \left(\frac{K}{S D^N} \right) / \ln \left(\frac{U}{D} \right). \tag{2.13}$$

The numerical evaluation of the call price is possible using tabular values for the multinomial distribution [notice that $Q_U + Q_D = 1$ from (2.3)], but, in practice, solving the recursion (2.5) numerically is easier.

Portfolio Delta

The equivalent portfolio (Δ, B) for the call option can be derived in closed form. One way to do this is to observe, using (2.11), that the *equity portion of the portfolio*, $E_k^j = \Delta_k^j S_k^j$, is a deterministic linear function of V_{n+1} and hence satisfies a recursion relation of type (2.5). Namely, we have

$$\begin{cases} E_n^j = \frac{1}{1+R} \cdot \left[P_U E_{n+1}^{j+1} + P_D E_{n+1}^j \right] & \text{if } n \leq N-1 \\[2mm] E_N^j = S_N^j \text{ if } S_N^j \geq K \text{ and } 0 \text{ if } S_N^j < K. \end{cases} \tag{2.14}$$

Therefore, we have

$$\Delta = \Delta_0^0 = E_0^0/S = \frac{1}{S(1+R)^N} \sum_{k > k_0}^{N} \binom{N}{k} \cdot S_N^k P_U^k P_D^{N-k}$$

$$= \sum_{k > k_0}^{N} \binom{N}{k} Q_U^k Q_D^{N-k}. \tag{2.15}$$

Money-Market Account

Similarly, the *cash component* of the equivalent portfolio satisfies the recursion relation (2.5) with final values $B_N^j = -K$ if $S_N^j \geq K$ and $B_N^j = 0$ if $S_N^j < K$. We conclude that

$$B = B_0^0 = -\frac{K}{(1+R)^N} \cdot \left\{ \sum_{k > k_0}^{N} \binom{N}{k} \cdot P_U^k P_D^{N-k} \right\} \tag{2.16}$$

It should be clear form these calculations that
- the equivalent portfolio for a call consists of a long equity position and a short cash position.
- the roles of stock and cash are symmetric: a call on the stock is equivalent to a "put on cash" (exchange stock for cash when the latter is undervalued with respect to the stock).[3]
- Δ approaches unity for $S \gg K$ ($k_0 \approx 0$).
- Δ approaches zero for $S \ll K$ ($k_0 \approx N$).

Puts

The value for a European put option predicted by the model is

$$P(S, K; N) =$$

$$\frac{K}{(1+R)^N} \cdot \left\{ \sum_{k < k_0}^{N} \binom{N}{k} \cdot P_U^k P_D^{N-k} \right\} - S \left\{ \sum_{k < k_0}^{N} \binom{N}{k} \cdot Q_U^k Q_D^{N-k} \right\}, \qquad (2.17)$$

This formula follows from direct calculation or from put–call parity. Notice also that (2.17) can be obtained from (2.12) by exchanging S by $K/(1+R)^N$ and the probabilities $P_\bullet$ and $Q_\bullet$, consistently with the remark made above. (Thus, we have yet another way of deriving the value of a put from that of a call.)

The equivalent portfolio of a put is

$$\Delta = - \sum_{k < k_0}^{N} \binom{N}{k} \cdot Q_U^k Q_D^{N-k}$$

$$B = \frac{K}{(1+R)^N} \cdot \left\{ \sum_{k < k_0}^{N} \binom{N}{k} \cdot P_U^k P_D^{N-k} \right\}.$$

It is important to notice that (independently of the model used)

$$\Delta \, (\text{put}) - \Delta \, (\text{call}) = -1 \, .$$

To proceed further with the analysis, we need to study how the parameters of the model are adjusted, or calibrated.

2.4 Relation Between the Parameters of the Tree and the Stock Price Fluctuations

The parameters N and R depend on the choice of time-interval between portfolio adjustments and the interest rate. Let T represent the duration of the contract in years and let r denote the

---

[3]This remark is important, for instance, in the case of currency options.

continuously compounded annualized interest rate. We assume equal periods between trading dates. The duration of each period is then

$$dt = \frac{T}{N}$$

and the interest rate for one period is

$$R = e^{r\,dt} - 1 \approx r\,dt \;.$$

The parameters U and D model the variability of the price of the underlying asset. For instance, if the asset were risk-less we would have, trivially,

$$U = D = 1 + R = e^{r\,dt} \;.$$

In order to reflect in the model the variability of the underlying asset, we will relate the parameters U and D to the statistics of the price.

By definition, the continuously compounded annualized **growth rate** of the stock is

$$Y = \frac{1}{T} \ln\left(\frac{S_N}{S_0}\right) \;. \tag{2.18}$$

(Thus, $Y = r$ for a riskless bond.) From Proposition 2.1, we have

$$\ln\left(\frac{S_N}{S_0}\right) = \sum_{n=1}^{N} \ln\left(\frac{S_n}{S_{n-1}}\right)$$

$$= \sum_{n=1}^{N} \ln(H_n) \;.$$

The expected growth rate is therefore

$$\mu = \mathrm{E}\{Y\}$$

$$= \sum_{1}^{N} E\left\{\frac{1}{T}\ln\left(\frac{S_i}{S_{i-1}}\right)\right\}$$

$$= \frac{1}{T}\sum_{1}^{N}(\ln U\, P_U + \ln D\, P_D)$$

$$= \frac{1}{dt}[(\ln U)\, P_U + (\ln D)\, P_D] \tag{2.19}$$

The variance of the growth rate is[4]

$$\text{Var } Y = \text{Var} \left\{ \sum_{n=1}^{M} \ln(H_n) \right\}$$

$$= \frac{1}{T^2} N \text{ Var } \ln(H_1) \quad \text{(by independence)}$$

$$= \frac{1}{T \, dt} \left\{ \left[(\ln U)^2 P_U + (\ln D)^2 P_D \right] - \left[(\ln U) P_U + (\ln D) P_D \right]^2 \right\}$$

$$= \frac{1}{T \, dt} \left[\ln \left(\frac{U}{D} \right) \right]^2 P_U \, P_D \tag{2.20}$$

In particular, the variance of the annual growth rate, obtained by setting $T = 1$ in the last equation, is

$$\sigma^2 \equiv \frac{1}{dt} \left[\ln \left(\frac{U}{D} \right) \right]^2 P_U \, P_D . \tag{2.21}$$

DEFINITION 2.1   *The standard deviation of the annualized growth rate is known as the* **volatility** *(of the underlying asset).*

We can also reason using **returns**. For an investment of T years, the return on investment is

$$\frac{S_N - S_0}{S_0} .$$

Therefore, the annualized expected return is

$$\frac{1}{T} \mathbf{E} \left\{ \frac{S_N - S_0}{S_0} \right\} = \frac{1}{T} \left(\mathbf{E} \left\{ \prod_{n=1}^{N} H_n \right\} - 1 \right)$$

$$= \frac{1}{T} \left(\left(\mathbf{E} \{ (H_1) \}^N \right) - 1 \right)$$

$$= \frac{1}{T} \left(e^{r T} - 1 \right)$$

where we used the fact that $\mathbf{E}(H_1) = e^{r \, dt}$. Thus, we have

PROPOSITION 2.3
 Under the risk-neutral probability measure induced by the binomial tree model, the expected return of the stock is identical to that of a riskless bond.[5]

---

[4] $\text{Var} X = E((X - E(X))^2) = E(X^2) - E(X)^2.$
[5] We have encountered this in the one-period model. Recall that we are measuring here the expected returns in

The variance of the expected returns or of the growth rate (they are essentially the same quantity for $T \gg 1$) are the primary quantities of interest in the binomial model.

Calibration of the Volatility Parameter

Let us determine the parameters U and D so that they are consistent with a given σ. For this, it is convenient to set

$$\begin{cases} U = U' e^{r\,dt} \\ D = D' e^{r\,dt} . \end{cases} \tag{2.22}$$

The no-arbitrage condition (2.3) then becomes

$$U' P_U + D' P_D = 1 . \tag{2.23}$$

Define a nonnegative parameter ρ by letting

$$\frac{U}{D} = \frac{U'}{D'} = e^{2\rho\,\sqrt{dt}} ,$$

Using (2.23) and the fact that P_U and P_D are probabilities, we obtain

$$\begin{cases} P_U + P_D = 1 \\ P_U\,P_D = \frac{\sigma^2}{4\,\rho^2} . \end{cases}$$

This system of equations has positive solutions if and only if $\rho \geq \sigma$. They are given by

$$P_U = \frac{1}{2} \left[1 \pm \sqrt{1 - \frac{\sigma^2}{\rho^2}} \right] \tag{2.24a}$$

and

$$P_D = \frac{1}{2} \left[1 \mp \sqrt{1 - \frac{\sigma^2}{\rho^2}} \right] . \tag{2.24b}$$

To find the corresponding values of U' and D' we use the formulas

$$P_U = \frac{1 - D'}{U' - D'} \, , \quad P_D = \frac{U' - 1}{U' - D'} \, ,$$

which follow from (2.22) and (2.4). Accordingly, we have

$$U' = \frac{e^{\rho\,\sqrt{dt}}}{P_D\,e^{-\rho\,\sqrt{dt}} + P_U\,e^{\rho\,\sqrt{dt}}} \tag{2.25a}$$

---

the "risk-neutral world," not in the "real world." In the risk-neutral world, the actual expected returns of the asset are irrelevant.

and

$$D' = \frac{e^{-\rho \sqrt{dt}}}{P_D \, e^{-\rho \sqrt{dt}} + P_U \, e^{\rho \sqrt{dt}}} \cdot \tag{2.25b}$$

Thus, the parameters U and D are given by

$$U = \frac{e^{\rho \sqrt{dt} + r\,dt}}{P_D \, e^{-\rho \sqrt{dt}} + P_U \, e^{\rho \sqrt{dt}}} \tag{2.26a}$$

and

$$D = \frac{e^{-\rho \sqrt{dt} + r\,dt}}{P_D \, e^{-\rho \sqrt{dt}} + P_U \, e^{\rho \sqrt{dt}}} \cdot \tag{2.26b}$$

Given dt and σ, we obtain several risk-neutral measures consistent with these parameters, which depend on the parameter ρ. The effect of changing ρ is to generate trees with different slopes for the branches.

Expected Growth Rate

We now discuss the expected growth rate under the risk-neutral measure, μ. From Eqs. (2.19), (2.24a), (2.24b), (2.26a), and (2.26b) we obtain

$$\mu = r + \frac{\rho}{\sqrt{dt}} \cdot (P_U - P_D) - \frac{1}{dt} \ln\left[P_D \, e^{-\rho \sqrt{dt}} + P_U \, e^{\rho \sqrt{dt}} \right]. \tag{2.27}$$

μ depends therefore on the choice of ρ. However, the effect of ρ diminishes as the tree is refined, i.e., in the asymptotic limit $dt \to 0$. In fact, using the identity

$$P_U - P_D = \pm \sqrt{1 - \frac{\sigma^2}{\rho^2}} \ ,$$

we obtain

$$\mu = r - \frac{1}{2}\sigma^2 + O\left((P_U - P_D)\rho^3 \, (dt)^{1/2} \right). \tag{2.28}$$

Thus, in the limit $dt \ll T$, we have

$$\mu \approx r - \frac{1}{2}\sigma^2 . \tag{2.29}$$

This is an important fact. In the limit $dt \ll 1$, *the (risk-neutral) expected growth rate is a function of the interest rate and the volatility.* The risk-neutral expected growth rate is independent of the "subjective" expected return.

Implementation of Binomial Trees

The most common choices for U and D to implement the binomial model are the following:

1. Symmetric probabilities. Take $\rho = \sigma$. Then

$$P_U = P_D = \frac{1}{2} \tag{2.30a}$$

and

$$U = \frac{e^{\sigma \sqrt{dt} + r\,dt}}{\cosh(\sigma \sqrt{dt})} \approx e^{\sigma \sqrt{dt} + \left(r - \frac{1}{2}\sigma^2\right)dt}$$

$$D = \frac{e^{-\sigma \sqrt{dt} + r\,dt}}{\cosh(\sigma \sqrt{dt})} \approx e^{-\sigma \sqrt{dt} + \left(r - \frac{1}{2}\sigma^2\right)dt}. \tag{2.30b}$$

It will be shown below that the approximation of $\cosh(\sigma \sqrt{dt})$ made in these two formulas has a negligible effect on the value of derivative securities if $dt \ll T$.

Example 2.1
Suppose volatility is 15% and that the interest rate is 6%. We wish to construct a binomial tree with 10 periods to value a 6-month instrument. Then

$$dt = 0.5/10 = 0.05.$$

Using method **1** ($\rho = \sigma$), we have

$$\begin{cases} U = e^{\sigma \sqrt{dt} + (r - 0.5\sigma^2)\,dt} = e^{0.15 \times \sqrt{0.05} + 0.06 \times 0.05} = 1.0372 \\[2mm] D = e^{-0.15 \times \sqrt{0.05} + 0.06 \times 0.05} = 0.9700. \end{cases}$$

☐

---

2. Another construction. Given a parameter ν we can choose

$$\begin{cases} U = e^{\sigma \sqrt{dt} + \nu\,dt} \\[2mm] D = e^{-\sigma \sqrt{dt} + \nu\,dt}. \end{cases} \tag{2.31a}$$

The corresponding risk-neutral probabilities are given by

$$P_U = \frac{1}{2}\left(1 + \frac{r - \nu - \frac{1}{2}\sigma^2}{\sigma}\sqrt{dt}\right)$$

$$P_D = \frac{1}{2} \left(1 - \frac{r - v - \frac{1}{2}\sigma^2}{\sigma} \sqrt{dt} \right). \tag{2.31b}$$

The probabilities were obtained by expanding formula (2.4) in powers of $dt^{1/2}$ and keeping terms up to order $\sqrt{dt}$.

A possible explanation for the choice (2.31a) resides in the interpretation of v as a "subjective" expected return (assuming that subjective probabilities $p_U = p_D = 0.5$). Observe that the expectation of S_1 under the *risk-neutral* probability is

$$\mathbf{E}(S_1) = S_0 (1 + r\,dt) + \text{lower-order terms}.$$

This agrees with the arbitrage-free condition to order dt of approximation.

2.5   The Limit for $dt \to 0$: Log-Normal Approximation

The limiting behavior of the model as $dt \to 0$ is of particular interest.

Recall that, under the risk-neutral measure, the logarithm of the price of the underlying asset is a sum of independent and identically distributed random variables:

$$Y = \frac{1}{T} \sum_{j=1}^{N} \ln H_j.$$

Since the parameters of the model have been adjusted so that, to leading order for $dt \ll T$, the variance and the mean of Y are given by

$$\mathbf{E}\{Y\} = r - \frac{1}{2}\sigma^2 \quad \text{and} \quad \operatorname{Var} Y = \frac{\sigma^2}{T},$$

we conclude from the Central Limit Theorem that the asymptotic probability distribution of the random variable Y as $dt \to 0$ is a Gaussian with mean $r - \sigma^2/2$ and variance σ^2/T. In particular, the price of the underlying asset at time t is *log-normally distributed*, i.e.,

$$S_t = S_0\, e^{\sigma \sqrt{t}\, Z + (r - 1/2\sigma^2)t}, \tag{2.32}$$

where Z is $N(0, 1)$ (the standard normal distribution). See Figures 2.4 and 2.5.
Consider now a European-style derivative security with a payoff function $F(S)$ maturing in T years. Its value under the arbitrage-free probability measure is given by

$$V = e^{-rT}\, \mathbf{E}\{ F(S_N) \}.$$

Therefore, by the Central Limit Theorem, under mild regularity assumptions on the function F (linear growth of $F(S)$ is sufficient) we have

$$\lim_{dt \to 0} V = e^{-rT}\, \frac{1}{\sqrt{2\pi}} \int_{-\infty}^{+\infty} F\left(S\, e^{z\sigma \sqrt{T}\,(r - \sigma^2/2)\,T} \right) e^{-\frac{z^2}{2}}\, dz, \tag{2.33}$$

Multinomial Distribution

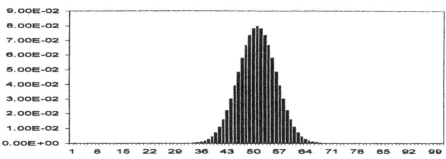

Gaussian Distribution

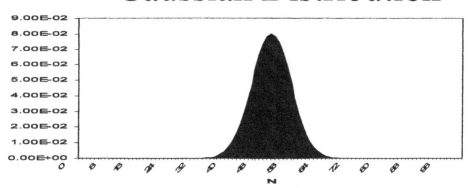

FIGURE 2.4
The multinomial distribution with $N = 100$, $p(0) = p(1) = 1/2$, and the Gaussian distribution with mean $\mu = N/2 = 50$ and variance $\sigma^2 = N/4$. Both graphs give rise to very similar probabilities.

where S is the spot price and we used the explicit form of the normal probability distribution. The differences in the fair prices that result from the normal approximation for the binomial model can be estimated numerically. They are generally small, of order $\sigma \sqrt{dt}$ for $dt \ll T$.[6]

2.6   The Black–Scholes Formula

We apply the latter result to option pricing. Assuming a volatility σ and an interest rate r, the log-normal approximation to the binomial pricing model gives

$$C(S, K; T)$$

$$= e^{-rT} \frac{1}{\sqrt{2\pi}} \int_{-\infty}^{+\infty} \mathrm{Max}\left(e^{\sigma \sqrt{T} z + (r - \sigma^2/2) T} - K, 0 \right) e^{-\frac{z^2}{2}} \, dz$$

---

[6]This error estimate will be made precise later.

Superimposed Distributions

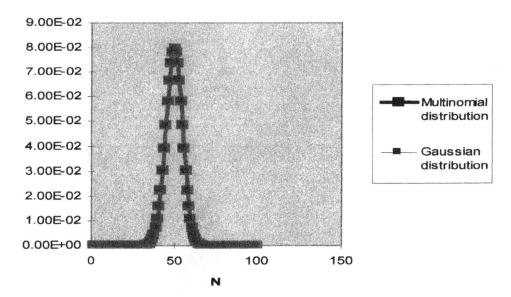

FIGURE 2.5
Superposition of the multinomial distribution and the Gaussian probabilities of the intervals $(j - 1/N,\ j + 1/N)$, **for** $j = 1, 2, \ldots, N - 1$. ($N = 100$).

$$= e^{-rT}\,\frac{1}{\sqrt{2\pi}}\int_{-d_2}^{+\infty} S\,e^{\,\sigma\,\sqrt{T}\,(r-\sigma^2/2)\,T}\;e^{-\frac{z^2}{2}}\,dz$$

$$-K\,e^{-rT}\,\frac{1}{\sqrt{2\pi}}\int_{-d_2}^{+\infty} e^{-\frac{z^2}{2}}\,dz\,, \tag{2.34}$$

where the lower limit of integration $-d_2$ satisfies the relation

$$-d_2 \;=\; \frac{1}{\sigma\sqrt{T}}\,\ln\!\left(\frac{S\,e^{rT}}{K}\right) \;-\; \frac{1}{2}\,\sigma\,\sqrt{T}$$

(which is obtained by solving the equation $e^{\sigma\sqrt{T}z+(r-\sigma^2/2)T} = K$ for d_2). In the first integral appearing in (2.34) we change variables

$$u = z - \sigma\sqrt{T}$$

and obtain the celebrated **Black–Scholes formula:**

$$C(S, K; T) \;=\; S\,N(d_1) \;-\; K\,e^{-rT}\,N(d_2)\,, \tag{2.35}$$

where

$$N(Z) = \frac{1}{\sqrt{2\pi}} \int_{-\infty}^{Z} e^{-\frac{z^2}{2}} \, dz \tag{2.36a}$$

is the *cumulative normal distribution* or *error function*,[7]

$$d_1 = \frac{1}{\sigma \sqrt{T}} \ln\left(\frac{S e^{rT}}{K}\right) + \frac{1}{2} \sigma \sqrt{T}, \tag{2.36b}$$

and

$$d_2 = \frac{1}{\sigma \sqrt{T}} \ln\left(\frac{S e^{rT}}{K}\right) - \frac{1}{2} \sigma \sqrt{T}. \tag{2.36c}$$

Puts can be valued in the same way. Using put–call parity and the relation

$$N(-Z) = 1 - N(Z),$$

we find that

$$P(S, K, T) = K e^{-rT} N(-d_2) - S N(-d_1).$$

The reader should note the strong parallel between the Black–Scholes formula and the binomial option pricing formula in (2.12). In fact, the Black–Scholes formula could have been derived by passing to the limit in (2.12) using the so-called *de Moivre–Laplace theorem* (Central Limit Theorem for standard random walk) to estimate the sums by integrals.

Example 2.2
A volatility of 15% corresponds to a daily standard deviation of $\sigma_{daily} = 0.15/\sqrt{365} = 0.008 = 0.8\%$. Sometimes people take a 250-day year (removing weekends). The estimate for the daily standard deviation is then $\sigma_{daily} = 0.15/\sqrt{250} \approx 1\%$. It is useful to memorize this particular volatility value to get a "feeling" for the relation between daily movements and annualized volatility. ▯

---

[7]Values of the error function are available in standard scientific calculators and software packages such as MATLAB.

Example 2.3

Suppose that the interest rate is 6% and assume a volatility of 15%. The value of a 3-month European call option on a stock that pays no dividends, with strike at 90% of the share price, is calculated as follows:

$$d_1 = \frac{1}{0.15 \times \sqrt{0.25}} \cdot \ln\left(e^{0.25 \times 0.06}/0.9\right) + 0.5 \times 0.15 \times \sqrt{0.25} = 1.6415$$

$$d_2 = \frac{1}{0.15 \times \sqrt{0.25}} \cdot \ln\left(e^{0.25 \times 0.06}/0.9\right) - 0.5 \times 0.15 \times \sqrt{0.25} = 1.5673$$

$$N(1.6415) = 0.9497 , \quad N(1.5673) = 0.9414 , \quad e^{-0.25 \times 0.06} = 0.9851$$

Call value $= 0.9497 - 0.9 \times 0.9851 \times 0.9414 = 0.1150 \approx 11.5\%$ of the share value. □

The key variable that enters the Black–Scholes formula is the volatility parameter. Needless to say, the value of an option can vary significantly according to the volatility. Figure 2.6 shows the Black–Scholes value of a call as a function of the spot price using different volatilities. The next chapter analyzes in greater detail the Black–Scholes formula and its implications.

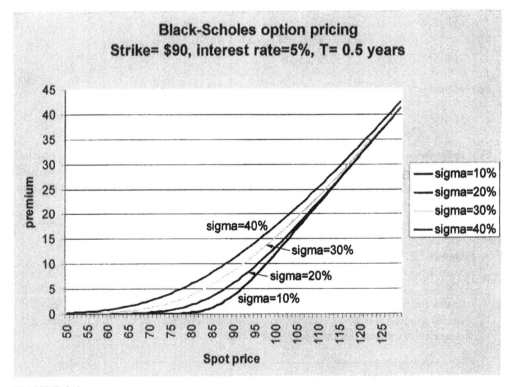

FIGURE 2.6
Value of a 6-month call option for different volatility values. The strike price is $90, the interest rate is 5% and the volatilities are 10%, 20%, 30% and 40%. The value of the call increases with the volatility.

In particular, we will focus on the effect of volatility on the value of an option and on the composition of the equivalent portfolio.

References and Further Reading

[1] Black, F., and Scholes, M. (1973), The pricing of options and corporate liabilities, *Journal of Political Economy,* **81**, pp. 637–659.

[2] Cox, J. and Rubinstein, M. (1985), *Option Theory,* Prentice-Hall, Englewood Cliffs, NJ.

[3] Cox, J., Ross, S., and Rubinstein, M. (1979), Option pricing: a simplified approach, *Journal of Financial Economics,* **7**, pp. 229–264.

[4] Duffie, D. and Protter, P. (1988), From discrete to continuous time finance: Weak convergence of the financial gain process, *Mathematical Finance,* **1**, pp. 1–16.

[5] Merton, R.C. (1973), Theory of rational option pricing, *Bell Journal of Economics and Management Science,* **4**, pp. 141–183.

Chapter 3

Analysis of the Black–Scholes Formula

The appeal of the Black–Scholes formula lies in the simple relation expressed between the price of the underlying asset and the price of options written on it. Another important feature of Black–Scholes is that it gives a *dynamical* relation between the underlying instrument, the option value, and the *hedge-ratio in the cash market* (Δ) that can be used to offset the risk of the option. The formula is the basis for pricing and hedging a variety of financial derivatives.[1]

Recall that the Black–Scholes formulas for pricing calls and puts are

$$C(S, K; T) = S N(d_1) - K e^{-rT} N(d_2) \tag{3.1}$$

and

$$P(S, K; T) = K e^{-rT} N(-d_2) - S N(-d_1). \tag{3.2}$$

Here S is spot price, K is the strike price, T is the time-to-maturity, r is the interest rate, and σ is the volatility. The "percentiles" d_i are given by

$$
\begin{cases}
d_1 = \frac{1}{\sigma \sqrt{T}} \ln \left(\frac{S e^{rT}}{K} \right) + \frac{1}{2} \sigma \sqrt{T} \\[2mm]
d_2 = \frac{1}{\sigma \sqrt{T}} \ln \left(\frac{S e^{rT}}{K} \right) - \frac{1}{2} \sigma \sqrt{T}.
\end{cases}
\tag{3.3}
$$

More generally, the no-arbitrage value of a European-style derivative security maturing in T years with stock-contingent value $F(S_T)$ at expiration is

$$V(S, T) = e^{-rT} \frac{1}{\sqrt{2\pi}} \int_{-\infty}^{+\infty} F \left(S e^{z \sigma \sqrt{T} + (r - \sigma^2/2) T} \right) e^{-\frac{z^2}{2}} dz$$

$$= e^{-rT} \mathbf{E}^{(l)} \{ F(S_T) \} \tag{3.4}$$

where $\mathbf{E}^{(l)}$ is the expectation value operator for the log-normal distribution with variance $\sigma^2 T$ and drift $(r - \frac{1}{2}\sigma^2) T$. The latter formula can be used to price option portfolios or other European-style contingent claims.

Figure 3.1 shows the Black–Scholes value of call option for different maturities (1, 3, and 9 months), with 15% volatility and 6% interest rate. Notice that the value of the call approaches

---

[1]It is remarkable that equity options have existed for 100 years but the Black–Scholes formula was discovered only in 1973. This is recognized as a major breakthrough in modern finance.

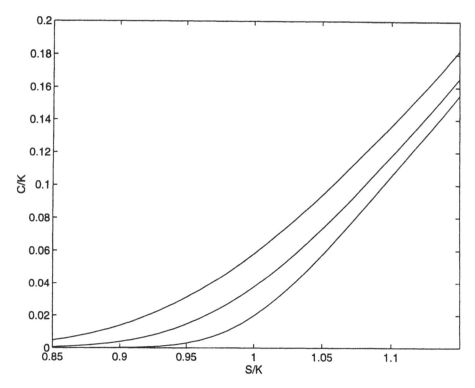

FIGURE 3.1
Values of a call option with different maturities (1, 3, and 9 months). We assumed
$\sigma = 15\%$ **and** $r = 6\%$.

zero for $S \ll K$ (the call is deep **out-of-the-money**) and is asymptotic to $S - K e^{-rT}$ —
the value of a forward contract at price K — for $S \gg K$ (the call is **deep-in-the-money**).
Thus, a deep-in-the money call is essentially equivalent to a forward contract, whereas a
deep-out-of-the-money call is equivalent to a bet on the rise of the stock price.

The *premium* $C(S, K; T)$ is often "decomposed" into three parts:

• the "intrinsic value" $Max(S - K, 0)$

• the "interest rate premium", $K\left(1 - e^{-rT}\right)$ if $S > K e^{-rT}$

• the "risk premium" $C(S, K; T) - Max\left(S - Ke^{-rT}, 0\right)$.

Clearly, the risk premium is sensitive to the value of the volatility used to price the call.

3.1 Delta

From the analysis of the binomial model, we know that the Delta of the equivalent portfolio
(or, more simply, the Delta of the derivative security) satisfies

$$\Delta_n^j = \frac{V_{n+1}^{j+1} - V_{n+1}^j}{S_{n+1}^{j+1} - S_{n+1}^j}, \tag{3.5}$$

which may be viewed as a "discrete derivative" of the value function $V_n^\bullet$ with respect to the spot price. Hence, we expect the Delta of a derivative security to be equal to $\partial V/\partial S$ in the log-normal approximation.

PROPOSITION 3.1
Consider a European-style derivative security with payoff $F(S_T)$. Then, under the assumptions of the binomial model and its log-normal approximation, we have

$$\lim_{dt \to 0} \Delta_0^0 \equiv \Delta(S, T) = \frac{\partial V(S, T)}{\partial S},$$ (3.6)

where $V(S, T)$ is the value of the derivative security given in (3.4).

PROOF Assume, for simplicity, that F is a smooth function and let $\mathbf{E}^b$ denote the expectation operator with respect to the *binomial* risk-neutral probability measure. Then, from (3.5) with $n = j = 0$, we have, if $dt \ll 1$ that

$$\Delta_0^0 = \frac{e^{-rT}}{S\,(U - D)}\,\mathbf{E}^b\,\{F(S_N)\,|S_1 = SU\,\} - \frac{e^{-rT}}{S\,(U - D)}\,\mathbf{E}^b\,\{F(S_N)\,|S_1 = SD\,\}$$

$$= \frac{e^{-rT}}{S\,(U - D)}\,\mathbf{E}^b\,\{F(S_{N-1}\,U)\,|S_0 = S\,\} - \frac{e^{-rT}}{S\,(U - D)}\,\mathbf{E}^b\,\{F(S_{N-1}\,D)\,|S_0 = S\,\}$$

$$\approx \frac{e^{-rT}}{S(U - D)}\mathbf{E}^b\{F'(S_{N-1})S_{N-1}\,(U - D)\} + \quad \text{lower order terms}$$

$$= \frac{e^{-rT}}{S}\mathbf{E}^b\{F'(S_{N-1})\,S_{N-1}\}$$ (3.7)

The second equality follows from the "homogeneity" of the tree. To obtain the last equality, we applied the Intermediate Value theorem to the function F.

Passing to the limit as $dt \to 0$, we have

$$\Delta(S, T) = e^{-rT}\,\mathbf{E}^{(l)}\left\{F'(S_T)\left(\frac{S_T}{S}\right)\right\}$$

$$= e^{-rT}\,\mathbf{E}^{(l)}\left\{\frac{\partial}{\partial S}F(S_T)\right\}$$

$$= e^{-rT}\,\frac{\partial}{\partial S}\,\mathbf{E}^{(l)}\{F(S_T)\}$$

$$= \frac{\partial}{\partial S}\,V(S, T).$$

(Here, we used the fact that $S_T = S\,e^{\sigma\,Z\,\sqrt{T} + (r - 1/2\sigma^2)\,T}$ and thus $\partial S_T/\partial S = S_T/S$.) The proof is complete. $\blacksquare$

This is an important result. For given values of interest rate and volatility, the discounted expectation with respect to the risk-neutral log-normal measure gives the theoretical value of the derivative security. The *sensitivity* of this theoretical value with respect to the spot price, i.e., $\partial V/\partial S$, gives the *hedge-ratio,* or number of units of the underlying security that, combined with a short position in the derivative, will offset immediate market risk. Thus,

> Long Δ units of the underlying asset, short 1 derivative $\Longleftrightarrow$ market-neutral .   (3.8)

Similarly,

> Short Δ units of the underlying security, long 1 derivative $\Longleftrightarrow$ market-neutral .   (3.9)

These rules follow from Proposition 2.2 of the previous chapter, where we showed that the above portfolios were equivalent, at any given date, to rolling over a position in riskless bonds for one period.

> **REMARK 3.1**   The actual number of cash instruments needed to hedge the market exposure of an option depends on the **notional amount** specified in the contract. For instance, stock option contracts traded on CBOT and AMEX are written for a notional amount of **100 shares**. Prices are quoted on a per-share basis. Therefore, the number of shares required to achieve a "market-neutral" position is $\pm\Delta \times 100$.   ∎

Another example worth considering is the case of *exchange-rate options.* Exchange-rate options for dollars against yen or deutschemark (DEM) are generally quoted in *percentage of dollar notional.* On the other hand, exchange rates and option strikes are commonly quoted using the value of a dollar in local currency: e.g., 1 dollar = 1.4134 DEM, dollar put with strike 1.40 DEM, etc. In the valuation of option premia and Deltas, there is a choice of which currency should be used as the "local" currency and which currency is the "underlying asset." For instance, if the foreign currency (say DEM) is chosen as the underlying asset (this is natural for a U.S.-based investor), then the Delta represents the amount of DEM spot that must be carried in the hedge per DEM notional. The amount of DEM *per dollar notional* is obtained by multiplying by the exchange rate (marks/dollar), e.g., by 1.4134. Some aspects of currency options will be considered in more detail in Chapter 8.

Option Deltas

To derive an expression for the Delta of a call, we differentiate (3.1) with respect to S. The result is

$$\Delta_{call}(S, K; T) = \frac{\partial}{\partial S} C(S, K; T)$$

$$= N(d_1) + S N'(d_1) \frac{\partial d_1}{\partial S} - K e^{-rT} N'(d_2) \frac{\partial d_2}{\partial S}$$

$$= N(d_1) .   \tag{3.10}$$

The last equality is obtained after some algebra[2] using the formulas

$$N'(d) = \frac{1}{\sqrt{2\pi}} e^{-\frac{d^2}{2}}$$

and

$$\frac{\partial d_i}{\partial S} = \frac{1}{S\sigma\sqrt{T}}.$$

The Delta for a put is obtained immediately from put–call parity. In fact, since

$$P(S, K; T) = C(S, K; T) - S + K e^{-rT},$$

we have

$$\Delta_{put} = \Delta_{call} - 1$$

$$= N(d_1) - 1$$

$$= N(-d_1). \tag{3.11}$$

In Figure 3.2 we present the graph of $\Delta_{call}(S, K; T)$ for the same parameter values as in Figure 3.1: ((—) = 6 months, (*)= 3 months, (- -)= 1 month). Notice that Δ is an increasing function of S (for both puts and calls).

Example 3.1

Calculate the Delta of a 6-month call option on an asset that pays no dividends, assuming a volatility of 16%, an interest rate of 6% and a strike price equal to 90% of spot.

Solution: Using the formula for d_1 given in (3.3), we find that $d_1 = 1.1348$. Hence $\Delta = N(1.1348) \approx 0.8719$.

3.2 Practical Delta Hedging

The holder of a portfolio of options with different maturities and strikes can dynamically hedge his exposure to price movements using Δ. To fix ideas, suppose that the portfolio consists of M different types of options (put/call, strike, maturity) with n_j options of type j, $1 \leq j \leq M$. At some initial time ($t = 0$), we denote the M maturities by $T_1, T_2, \ldots, T_M$, and the strikes by $K_1, K_2, \ldots, K_M$. The number of options of each type is denoted by n_j, $j = 1, \ldots, M$.

---

[2]This calculation is recommended to those readers who are not familiar with the Black–Scholes formula and want to practice some elementary manipulations.

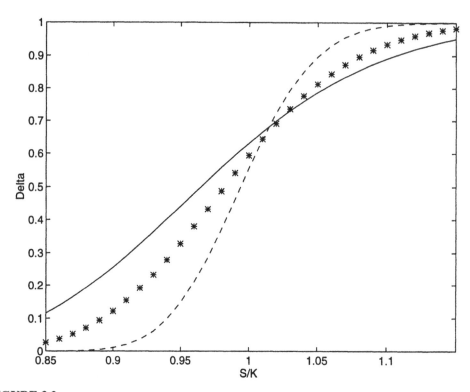

FIGURE 3.2
Delta as a function of the price of the underlying stock for the options of Figure 3.1. (—
= 9 months, * = 3 months, - - - = 1 month.)

According to Eq. (3.9), the holder of this portfolio can hedge dynamically the risk due to price changes by holding

$$\Delta_t = -\sum_{j=1}^{M} n_j \Delta(S_t, K_j, T_j - t)$$

units of the underlying asset for times $t = dt, \ldots, t_n$ where $n = 0, 1, 2, \ldots, \max\{T_j\}/dt$. (Here, dt represents a small interval of time which, in practice, must be determined by the hedger.) Assuming that the model assumptions are correct and, in particular, that **volatility has been correctly estimated,** the above strategy would have approximately riskless returns over time.

Example 3.2
Assume that $S = 100$, $\sigma = 0.16$, $r = 0.07$. Calculate the portfolio Delta for
+5   100 calls expiring in 60 days
−3   90 puts expiring in 90 days
−4   85 puts expiring in 120 days   ⬚

Solution: Individual calculation of the Deltas for each option gives:

60 day call with strike 100   Delta = 0.5826
90 day put with strike 90   Delta = −0.0571
120 day put with strike 85   Delta = −0.0194
<u>Combined Delta:</u> $\Delta = 5 \times 0.5826 + 3 \times 0.0571 + 4 \times 0.0194 = 3.1619$.

Example 3.3
  Calculate the Delta of the above portfolio if $\sigma = 30\%$. ⬚

Solution:
60 day call with strike 100   Delta = 0.5617
90 day put with strike 90   Delta = −0.1857
120 day put with strike 85   Delta = −0.1228
<u>Combined Delta:</u> $\Delta = 5 \times 0.5617 + 3 \times 0.1857 + 4 \times 0.1228 = 3.8568$.

Several factors may contribute to making Δ-hedging risky in practice. Namely,

- The log-normal assumption may not be valid (model risk).

- The hedge may not be done frequently enough to prevent losses due to price movements (hedge-slippage risk).

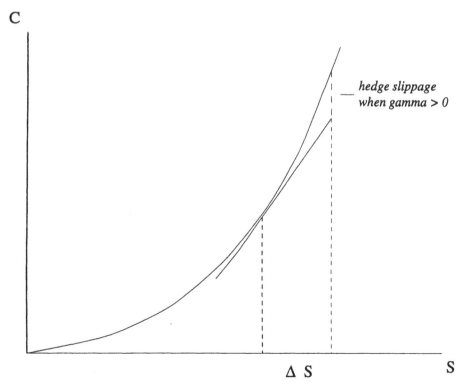

FIGURE 3.3
Illustration of hedge-slippage.

The first type of risk is complex to control—its discussion is outside the scope of the basic Black–Scholes model. Model risk is the most fundamental risk in option risk-management and will be studied in more detail later, as we relax some of the assumptions of the Black–Scholes model.[3] One important dimension of model risk is the *mis-specification of the volatility parameter* σ.

Hedge-slippage risk (Figure 3.3), on the other hand, stems from the fact that if the portfolio is not rebalanced frequently enough, the strategy is no longer riskless. Changes in the value of the hedge portfolio are then governed by higher-order derivatives of the value function with respect to the spot price. This question is of crucial importance in applications. In fact, even if we believe that the underlying price follows approximately a binomial process under the risk-neutral probability, rehedging at every time-step is virtually impossible. Traders are obliged to rebalance their portfolios after "macroscopic," rather than "microscopic" time-intervals.

3.3 Gamma: The Convexity Factor

DEFINITION 3.1   *Let $V(S, T)$ represent the value of a derivative security in the log-normal approximation. The Gamma (Γ) of the derivative security is the sensitivity of Δ with respect to S, i.e.,*

$$\Gamma(S, T) \;=\; \frac{\partial^2 V(S, T)}{\partial S^2} \,. \tag{3.12}$$

The concept of Gamma is important when the position cannot be adjusted *exactly* after each "microscopic" time period dt. One way of analyzing the problem is to assume that rehedging is not done at the smallest micro-shocks, but instead at some intermediate time-scale δt, such that $dt \ll \delta t \ll T$. The accuracy of the hedge then depends on the rate of change of Δ as S changes, which is precisely Gamma.

The primary example of a **positive-Gamma position** is a long option position. In fact, according to the Black–Scholes formula, we have

$$\Gamma_{call}(S, K; T) \;=\; \Gamma_{put}(S, K; T) \;=\; \frac{1}{S\sqrt{2\pi \sigma^2 T}} \, e^{-\frac{d_1^2}{2}} \,, \tag{3.13}$$

a formula that follows immediately by differentiating $\Delta(S, K, T) = N(d_1)$ with respect to S. The graph of Gamma as a function of the spot price is given in Figure 3.4. The parameter values are as in Figures 3.1 and 3.2.

To better understand the influence of convexity in hedging, it is useful to visualize the profit/loss that arises from hedging in terms of the graph of $V(S, T)$. Given that the spot price is S, an agent who is short the derivative security and takes a position in the equivalent portfolio will have a profit/loss in the cash market that varies on a straight line **tangent to the graph of the value function**. On the other hand, the value of the derivative security will vary along the graph of V. The effect of risk-neutral valuation is to achieve a position in the cash market after

---

[3]It is important for us to recognize, even at this early stage, the nonparametric nature of financial markets. The random-walk model is consistent with the no-arbitrage hypothesis but because markets are fundamentally incomplete, it is not sufficient to encompass changes in volatility expectations due to the arrival of new information.

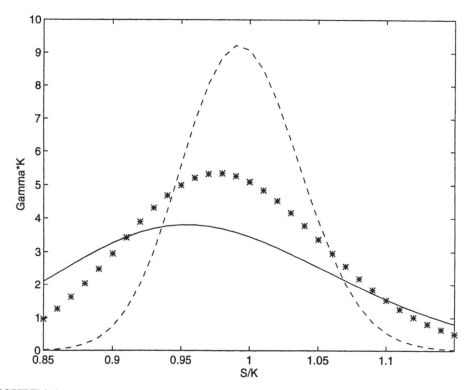

FIGURE 3.4
Gamma as a function of underlying asset price. (Same options as for Figures 3.1 and 3.2.)

time δt, which *on average* is equal to the value of the derivative at that new time/price level. However, unless the curvature of $V(S, T)$ is identically zero (in which case the contingent claim is equivalent to a static portfolio of cash instruments), there are instances in which

$$V(S_{t+\delta t}, T - (t + \delta t)) > S_{t+\delta t} \Delta_t + B_t e^{r \delta t} \tag{3.14a}$$

and other instances in which

$$V(S_{t+\delta t}, T - (t + \delta t)) < S_{t+\delta t} \Delta_t + B_t e^{r \delta t} \tag{3.14b}$$

according to the magnitude of the price change of the period δt.

It is graphically clear, see Figure 3.5, that inequality (3.14a) will hold for *large* price movements and inequality (3.14b) will hold for *small* price movements when V is convex (and thus $\Gamma > 0$).[4]

We conclude that

PROPOSITION 3.2

(i) The holder of an option who is short Δ units of the underlying asset will achieve a positive

---

[4]A Taylor expansion of the portfolio value shows that if it is Delta-neutral at time t, then at time $t + \delta t$ its value will be $\Delta \Pi = \Theta \delta t + \frac{1}{2} (\Delta S)^2$.

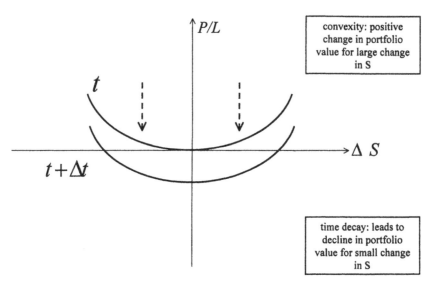

FIGURE 3.5
Convexity and time-decay are the main features of an option's value. The above diagram represents the value of a delta-hedged long option position at times t and $t + \Delta t$. Convexity is in favor of the option's holder: if the stock price moves by a sufficiently large amount, the profit/loss (P/L) of the position is positive. If the stock moves only by a small amount, the position has negative value at time $t + \Delta t$.

cash-flow if subsequently the price movement is sufficiently large and a negative cash-flow if the price movement is sufficiently small.

(ii) The writer of an option who is long Δ units of the underlying security will achieve a positive cash-flow if subsequently the price movement is sufficiently small and negative cash-flow if the price movement is sufficiently large.

The use of Gamma becomes particularly relevant when managing option portfolios. Agents can, for instance, "buy" or "sell" Gamma in order to achieve a position that is consistent with their views for the near future.[5] Buying or selling Gamma should be understood, of course, as buying or selling options.

Example 3.4
Calculate the Gamma of the portfolio of Examples 3.2 and 3.3.   ☐

Solution: Using formula (3.14a) and (3.14b), we find that if $\sigma = 0.16$,

60 day call with strike 100	Gamma ≈ 0.05
90 day put with strike 90	Gamma ≈ 0.02
120 day put with strike 85	Gamma $\approx$ negligible

---

[5]Needless to say, the value of Gamma depends crucially on the volatility parameter—more on this later.

Combined Gamma of the portfolio: $\Gamma = 5 \times 0.05 - 3 \times 0.02 = 0.19$

If $\sigma = 0.30$, then

60 day call with strike 100	Gamma ≈ 0.04
90 day put with strike 90	Gamma $\approx$ negligible
120 day put with strike 85	Gamma ≈ 0.01

Combined Gamma of the portfolio: $\Gamma = 5 \times 0.04 - 4 \times 0.01 = 0.16$

3.4 Theta: The Time-Decay Factor

The expression "an option is a wasting asset" is part of the options trading lore and is commonly used in option trading manuals. Consider, for instance, the case of call options. Since the interest-rate premium and the risk-premium are nonnegative, the option is worth more than its intrinsic value at any time before expiration.[6]

To evaluate the time-decay of the option, we can differentiate the option price with respect to T.

DEFINITION 3.2   *The Theta (Θ) of a European-style contingent claim with value function $V(S, T)$ is defined as*

$$\Theta(S, T) = -\frac{\partial V(S\ T)}{\partial T}. \tag{3.15}$$

The negative sign in this definition is due to the fact that T represents the time to maturity. Thus, Θ represents the variation in the fair value of a contingent claim for a small increment of time (decrease in time-to-expiration).

PROPOSITION 3.3

For any European-style contingent claim, we have

$$\Theta(S, T) = -\frac{1}{2}\sigma^2 S^2 \Gamma(S, T) - r S \Delta(S, T) + r V(S, T). \tag{3.16}$$

PROOF   Consider the basic formula

$$V(S, T) = e^{-rT} \frac{1}{\sqrt{2\pi}} \int_{-\infty}^{+\infty} F\left(S\, e^{z\sigma\sqrt{T} + (r - \sigma^2/2)\, T}\right) e^{-\frac{z^2}{2}}\, dz.$$

Differentiating with respect to T, we obtain

$$\frac{\partial V(S\ T)}{\partial T} \tag{3.17}$$

---

[6]We are assuming here that the underlying asset pays no dividends. Otherwise, this statement is not generally true.

$$-r\,V(S,T) + e^{-rT}\int_{-\infty}^{+\infty}\frac{\partial}{\partial T}F\left(S\,e^{z\,\sigma\,\sqrt{T}\,+\,(r\,-\,\sigma^2/2)\,T}\right)e^{-\frac{z^2}{2}}\frac{dz}{\sqrt{2\pi}}\,.$$

We can calculate the T-derivative of the function inside the integral sign. The result is

$$F'\left(S\,e^{z\,\sigma\,\sqrt{T}\,+\,(r\,-\,\sigma^2/2)\,T}\right)$$

$$\cdot S\,e^{z\,\sigma\,\sqrt{T}\,+\,(r\,-\,\sigma^2/2)\,T}\cdot\left(\frac{z\,\sigma}{2\,\sqrt{T}} + r - \sigma^2/2\right)$$

$$= r\,S\,\frac{\partial}{\partial S}F\left(S\,e^{z\,\sigma\,\sqrt{T}\,+\,(r\,-\,\sigma^2/2)\,T}\right)$$

$$+S\,\frac{\partial}{\partial S}F\left(S\,e^{z\,\sigma\,\sqrt{T}\,+\,(r\,-\,\sigma^2/2)\,T}\right)\cdot\left(\frac{z\,\sigma}{2\,\sqrt{T}} - \sigma^2/2\right). \qquad (3.18)$$

Let us substitute this expression into (3.17). This gives

$$\frac{\partial V(S\,T)}{\partial T} = -r\,V(S,T) + r\,S\,\Delta(S,T)$$

$$+e^{-rT}\int_{-\infty}^{+\infty}S\,\frac{\partial}{\partial S}F\left(S\,e^{z\,\sigma\,\sqrt{T}\,+\,(r\,-\,\sigma^2/2)\,T}\right)$$

$$\cdot\left(\frac{z\,\sigma}{2\,\sqrt{T}} - \sigma^2/2\right)e^{-\frac{z^2}{2}}\frac{dz}{\sqrt{2\pi}}\,. \qquad (3.19)$$

The desired result is now obtained by integration by parts, using the fact that

$$\frac{d}{dz}e^{-\frac{z^2}{2}} = -z\,e^{-\frac{z^2}{2}}\,.$$

We can rewrite the last integral in (3.19) as

$$e^{-rT}\int_{-\infty}^{+\infty}S\,\frac{\partial}{\partial z}\frac{\partial}{\partial S}F\left(S\,e^{z\,\sigma\,\sqrt{T}\,+\,(r\,-\,\sigma^2/2)\,T}\right)\cdot\frac{\sigma}{2\,\sqrt{T}}e^{-\frac{z^2}{2}}\frac{dz}{\sqrt{2\pi}}$$

$$-e^{-rT}\int_{-\infty}^{+\infty}S\,\frac{\partial}{\partial S}F\left(S\,e^{z\,\sigma\,\sqrt{T}\,+\,(r\,-\,\sigma^2/2)\,T}\right)\cdot\frac{\sigma^2}{2}e^{-\frac{z^2}{2}}\frac{dz}{\sqrt{2\pi}}$$

$$= e^{-rT}\int_{-\infty}^{+\infty}S\,\frac{\partial}{\partial S}\frac{\partial}{\partial z}F\left(S\,e^{z\,\sigma\,\sqrt{T}\,+\,(r\,-\,\sigma^2/2)\,T}\right)\cdot\frac{\sigma}{2\,\sqrt{T}}e^{-\frac{z^2}{2}}\frac{dz}{\sqrt{2\pi}}$$

$$-e^{-rT}\int_{-\infty}^{+\infty}S\,\frac{\partial}{\partial S}F\left(S\,e^{z\,\sigma\,\sqrt{T}\,+\,(r\,-\,\sigma^2/2)\,T}\right)\cdot\frac{\sigma^2}{2}e^{-\frac{z^2}{2}}\frac{dz}{\sqrt{2\pi}}$$

$$= e^{-rT} \int_{-\infty}^{+\infty} S \frac{\partial}{\partial S} S \frac{\partial}{\partial S} F\left(S e^{z\sigma\sqrt{T} + (r - \sigma^2/2) T}\right) \cdot \frac{\sigma^2}{2} e^{-\frac{z^2}{2}} \frac{dz}{\sqrt{2\pi}}$$

$$- e^{-rT} \int_{-\infty}^{+\infty} S \frac{\partial}{\partial S} F\left(S e^{z\sigma\sqrt{T} + (r - \sigma^2/2) T}\right) \cdot \frac{\sigma^2}{2} e^{-\frac{z^2}{2}} \frac{dz}{\sqrt{2\pi}}$$

$$= e^{-rT} \int_{-\infty}^{+\infty} S^2 \frac{\partial^2}{\partial S^2} F\left(S e^{z\sigma\sqrt{T} + (r - \sigma^2/2) T}\right) \cdot \frac{\sigma^2}{2} e^{-\frac{z^2}{2}} \frac{dz}{\sqrt{2\pi}}$$

$$= \frac{\sigma^2}{2} S^2 \Gamma(S, T) . \tag{3.20}$$

Combining (3.20) with (3.19), we conclude that Eq. (3.16) holds.   ∎

As an application, consider the case of an agent hedging a position in a derivative security. We can assume, without loss of generality that, initially, both $\Delta(S, T)$ and $V(S, T)$ are zero, since this can be achieved by assuming a position in shares and riskless bonds. In this case, Eq. (3.16) reduces to

$$\Theta(S, T) = -\frac{1}{2} \sigma^2 S^2 \Gamma(S, T) . \tag{3.21}$$

In other words, *if a portfolio of derivative securities and cash instruments has zero initial cost and is Delta-neutral, the time-decay factor* Θ *and the convexity factor* Γ *multiplied by* $\frac{\sigma^2}{2} S^2$ *are exactly opposite to each other.*

This is just a reformulation of Proposition 3.2: the *owner of Gamma* (who is net long options) is subject to time-decay value if the spot does not move but benefits from price movements. Conversely, the *seller of Gamma* (who is net short options) is subject to hedge-slippage risk if the spot price moves but gains if spot does not move.

Proposition 3.3 can be viewed as a derivation of the equation satisfied by the value of a contingent claim under the log-normal approximation. In fact, from (3.16), we have

PROPOSITION 3.4

Consider a European derivative security, contingent on the value of a security that pays no dividends, with final value $F(S_T)$, *where* T *is the time to maturity. In the log-normal approximation, its value* $V(S, T)$ *satisfies the* **Black–Scholes partial differential equation**

$$\frac{\partial V(S, T)}{\partial T} = \frac{\sigma^2}{2} S^2 \frac{\partial^2 V(S, T)}{\partial S^2} + r S \frac{\partial V(S, T)}{\partial S} - r V(S, T) \tag{3.22}$$

with

$$V(S, 0) = F(S) .$$

This equation will be derived again later in the context of Continuous-Time Finance. In the next section we show that the recursion relation for contingent-claim pricing on the binomial lattice,

$$V_n^j = \frac{1}{1 + R} \left[P_U V_{n+1}^{j+1} + P_D V_{n+1}^j \right] , \tag{3.23}$$

can be seen as a finite-difference approximation to the Black–Scholes equation, which is the "discrete analogue" of (3.22).

3.5   The Binomial Model as a Finite-Difference Scheme for the Black–Scholes Equation

We consider a solution of the Black–Scholes equation $V(S, t)$ with final condition $F(S)$ at $t = T$. At the same time, we consider a binomial tree with parameters

$$U = e^{\sigma \sqrt{dt}}, \quad D = e^{-\sigma \sqrt{dt}},$$

(using the notation of Chapter 2.) Define the auxiliary function

$$\tilde{V}_n^j = V\left(S_n^j, (N - n)\, dt\right),$$

which is obtained by evaluating $V(S, T - t)$ at the nodes of the tree.

Recall that the risk-neutral probabilities are (to leading order)

$$P_U = \frac{1}{2}\left(1 - \frac{\sigma \sqrt{dt}}{2}\right) + \frac{r \sqrt{dt}}{2\sigma}$$

$$P_D = \frac{1}{2}\left(1 + \frac{\sigma \sqrt{dt}}{2}\right) - \frac{r \sqrt{dt}}{2\sigma}.$$

Using Taylor's expansion, we have

$$\tilde{V}_{n+1}^{j+1} = \tilde{V}_n^j + (S\, V_S)_n^j \cdot (U - 1) + \frac{1}{2}(S^2\, V_{SS})_n^j\, (U - 1)^2 + (V_t)_n^j\, dt$$

$$= \tilde{V}_n^j + (S\, V_S)_n^j\, \sigma \sqrt{dt} + \frac{1}{2}\left(S^2\, V_{SS})_n^j + S\, V_S)_n^j\right) \sigma^2\, dt$$

$$+ (V_t)_n^j\, dt + O(dt^{3/2})$$

with a similar expansion for $\tilde{V}_{n+1}^j$.

Then, we conclude from this that

$$P_U\, \tilde{V}_{n+1}^{j+1} + P_D\, \tilde{V}_{n+1}^j$$

$$= \tilde{V}_n^j + \left(\frac{1}{2}(S^2\, V_{SS})_n^j + \frac{1}{2}(S\, V_S)_n^j\right) \sigma^2\, dt + (V_t)_n^j\, dt$$

$$- \frac{1}{2}(S\, V_S)_n^j\, \sigma^2\, dt + (S\, V_S)_n^j\, r\, dt + O(dt^{3/2})$$

$$= \tilde{V}_n^j + \frac{1}{2}\sigma^2 (S^2\, V_{SS})_n^j\, dt + r\, (S\, V_S)_n^j\, dt + (V_t)_n^j\, dt + O(dt^{3/2}).$$

Thus, since $1/(1 + R) = e^{-r\,dt} = 1 - r\,dt + O(dt^2)$, we have

$$\tilde{V}_n^j - \frac{1}{1 + R}\left[P_U\,\tilde{V}_{n+1}^{j+1} + P_D\,\tilde{V}_{n+1}^j \right] = \tilde{V}_n^j + \tilde{V}_n^j\,r\,dt - \tilde{V}_n^j$$

$$- \left(\frac{1}{2}\sigma^2\,(S^2\,V_{SS})_n^j\,dt + r\,(S\,V_S)_n^j\,dt + (V_t)_n^j\,dt \right) + O(dt^{3/2})$$

$$= \left(r\tilde{V}_n^j - \frac{1}{2}\sigma^2\,(S^2\,V_{SS})_n^j - r\,(S\,V_S)_n^j - (V_t)_n^j \right) dt + O(dt^{3/2})$$

$$= \left(r\tilde{V}_n^j - \frac{1}{2}\sigma^2\,(S^2\,V_{SS})_n^j - r\,(S\,V_S)_n^j + (V_T)_n^j \right) dt + O(dt^{3/2})$$

$$= O(dt^{3/2}),$$

because $V(S, T)$ satisfies the Black–Scholes PDE.[7] Therefore, $V(S, T-t)$ (with T held fixed and t varying from 0 to T) gives rise to a function $\tilde{V}_n^j$, defined on the binomial tree, which can be regarded as an "approximate solution" of the basic backward-induction Eq. (3.23). The point is that at each time step, the error is of order at most $dt^{3/2}$. We conclude that the total discrepancy between V_n^j and $\tilde{V}_n^j$ is of order $\sum dt^{3/2} = O(dt^{1/2})$. This result is important, since it shows that the binomial model can be viewed as a numerical scheme for solving the Black–Scholes equation.

References and Further Reading

[1] Black, F. and Scholes, M. (1973), The Pricing of Options and Corporate Liabilities, *Journal of Political Economy*, 81, pp. 637–659.

[2] Figlewski, S., Silber, W., and Subrahmanyam, M. (1990), *Financial Options: From Theory to Practice*, Business One Irwin, Homewood, IL.

[3] Hull, J. (1998), *Introduction to Futures and Options Markets*, Prentice-Hall, Upper Saddle River, NJ.

[4] Jarrow, R. (1983), *Option Pricing*, R.D. Irwin, Homewood, IL.

---

[7] Notice that this argument uses the fact that V_t is the derivative with respect to *calendar time* and $V_T = -V_t$ is the derivative with respect to the time to maturity. Also, the error estimate of order $dt^{3/2}$ is derived assuming that higher-order derivatives of $V(S, t)$ are bounded, which is true if $F(S)$ is smooth. The extension to $F(S) = \max(S - K, 0)$ requires a slightly more involved argument near $t = T$.

Chapter 4

Refinements of the Binomial Model

The binomial model discussed in Chapter 2 used two input parameters: the interest rate and the volatility. Until now, we assumed implicitly that these parameters were constant. In this chapter, we remove these assumptions, allowing for variations of these parameters with time.

A *term-structure of interest rates* is introduced to model a (more realistic) economy in which rates can vary with the duration of loans. We will also study time-inhomogeneity of the volatility process by introducing a *term-structure of volatilities*. Time-dependent volatilities are useful to incorporate into the pricing model the market's expectations about risk across time. Information about the temporal behavior of volatility is contained in the prices of liquid option instruments with different maturities written on a given underlying asset.

We will also discuss refinements of the binomial model that will permit us to extend the theory to derivative contracts of practical interest. These include derivatives contingent on underlying assets that pay dividends, options on futures and "structured" derivative instruments providing a stream of uncertain cash-flows across time.

4.1 Term-Structure of Interest Rates

We incorporate into the model different interest rates for different trading periods. Usually, interest rates are not constant in time. For example, the following table gives market values for interbank dollar deposits on August 23, 1995:[1]

Maturity	Bid	Offer
1 month	5.8700	6.0000
2 months	5.7800	5.9000
3 months	5.7800	5.9000
6 months	5.8100	5.9500
9 months	5.8100	5.9300

Implied **forward interest rates** can be obtained from such a "strip" of deposit rates, or from the markets in Eurodollar futures or Treasury bill futures.[2]

In the case of Eurodollar futures, the December 1997 Eurodollar 90-day futures contract gives the expected London Inter-Bank Offered Rate (LIBOR) for the period of January 1998

---

[1]Unless otherwise specified, interest rates are quoted in "bond-equivalent," or continuously compounded form. Bid = interest rate paid on deposits, offer = interest rate charged for loans.

[2]A forward interest rate (from date t_1 to date t_2) is the interest rate that a borrower/lender can obtain in the market for a loan for the period t_1 to t_2.

through March 1998. The March 1998 contract gives the expected 90-day LIBOR from April 1998 through June 1998, and so forth. These values can then be input in the model at different time periods.[3]

Example 4.1

The table given above can be used to obtain a 1-month interest rate, a 1-month forward interest rate for a loan starting in 1 month, a 1-month forward interest rate for a loan starting in 2 months, a 3-month interest rate for a loan starting in 3 months, and a 3-month interest rate for a loan starting in 6 months. For instance, to compute the 3-month interest rate for *lending* 6 months from now, $r_{6,9}^l$, we note that

• borrowing $1 for 9 months
• lending $e^{-r_{6,9}^l 0.25}$ dollars from month 6 to month 9
• lending $e^{-r_{0,6}^l 0.50} e^{-r_{6,9}^l 0.25}$ dollars from month 0 to month 6

results in a cash-flow of zero dollars 9 months from today. For borrowing for 9 months we take the 9 months offer rate, 5.93, and for lending for 6 months we take the bid rate, 5.81. Therefore, the effective rate for lending from 6 to 9 months satisfies

$$5.93 \times 0.75 = 5.81 \times 0.50 + r_{6,9}^l \times 0.25,$$

which gives an *offer* rate of 6.17%. To calculate the effective rate for borrowing over the same period, we observe that

• lending $1 for 9 months
• borrowing $e^{-r_{6,9}^b 0.25}$ from month 6 to month 9
• borrowing $e^{-r_{0,6}^b 0.50} e^{-r_{6,9}^b 0.25}$

results in a cash-flow of zero dollars 9 months from today. Hence,

$$5.81 \times 0.75 = 5.95 \times 0.50 + r_{6,9}^b \times 0.25,$$

which gives an effective *bid* rate of 5.53%.

In the simple models considered hereafter, differences between bid and offer prices will not be taken into account. Instead, we will consider the "riskless rate" to be the average between bid and offer rates. If we follow this rule, the equation for the effective 6-to-9 rate is

$$5.87 \times 0.75 = 5.88 \times 0.50 + r_{6,9} \times 0.25,$$

which gives an effective rate of 5.85%. Notice that this is just the average between the bid and offer rates obtained above. ☐

The notion of a unique "riskless" rate is an idealization, since it does not distinguish between borrowing and lending. In general, when we refer to a (single) "riskless rate," we understand

---

[3]Modeling the future interest rates as the forward rates implied by a strip of interest rate futures or deposit rates is not entirely correct for pricing and hedging derivative securities. The reason is that forward rates are just *expected* future rates, whereas the future short-term rates are not predictable in advance. However, this "forward rate approximation" is extremely useful to obtain a first-order approximation to varying interest-rate environments.

Term Rates

1M	5.935
2M	5.840
3M	5.840
6M	5.880
9M	5.870

Forward Rates

0-1M	5.935
1-2M	5.745
2-3M	5.840
3-6M	5.920
6-9M	5.850

FIGURE 4.1a
Interest rates: deriving forward rates from term rates.

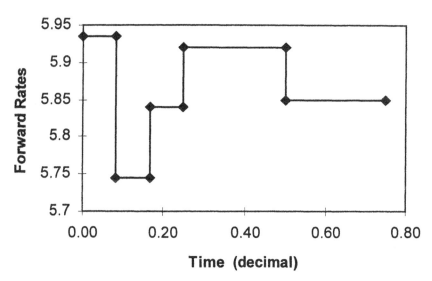

FIGURE 4.1b
Graphical representation of forward rates as a function of time.

this to mean, for instance, the average between the bid and the offered rates for that period. In other contexts it could mean either the bid or the offer.

Generalizing from the examples, we can state the following principle. Given a sequence of (continuously compounded) **term deposits**

$$\bar{r}_{t_1}, \bar{r}_{t_2}, \ldots \bar{r}_{t_N}$$

the corresponding forward rates are obtained by the formula

$$r_{t_i t_j} = \frac{t_j \bar{r}_{t_j} - t_i \bar{r}_{t_i}}{t_j - t_i} \ .$$

Suppose that we have, as in Chapter 2, Section 2.1, a binomial model with N trading periods. We consider a sequence of forward interest rates for the N periods quoted on a continuously compounded (bond-equivalent) basis:

$$r_0 \ , \ r_1 \ , \ r_2 \ , \ \ldots \ , \ r_{N-1} \ . \tag{4.1}$$

(This sequence could have been obtained from the procedure outlined above or otherwise.) The gross interest rates R_n, $0 \leq n \leq N - 1$ for the different periods are[4]

$$R_n = e^{r_n \, dt} - 1 \approx r_n \, dt \ , \tag{4.2}$$

where $dt = T/N$ represents the duration of each period. We wish to incorporate the term-structure of interest rates (4.1) into the binomial pricing model assuming, for simplicity, that the *local volatility*—the standard deviation of returns—remains constant through time. This can be done by defining, for each n, up–down parameters U_n and D_n as follows:

$$U_n = e^{r_n \, dt} \cdot U'$$

$$D_n = e^{r_n \, dt} \cdot D' \ , \tag{4.3}$$

where U' and D' will be specified later in terms of the volatility. An arbitrage-free measure on the space of price paths is defined by setting

$$S_{n+1} = H_{n+1} S_n \quad , \quad \text{for } 0 \leq n \leq N - 1 \ ,$$

where the random variables H_n are independent and satisfy

$$\text{Prob.}\{ H_{n+1} = U_n \} = P_U \ , \quad \text{Prob.}\{ H_{n+1} = D_n \} = P_D \ .$$

Here, P_U and P_D are the probabilities

$$P_U = \frac{1 + R_n - D_n}{U_n - D_n} = \frac{1 - D'}{U' - D'}$$

---

[4]We make the convention that r_n is the interest rate that applies to the $(n + 1)^{st}$ period.

and

$$P_D = \frac{U_n - 1 - R_n}{U_n - D_n} = \frac{U' - 1}{U' - D'} . \tag{4.4}$$

The reader will recognize here the usual arbitrage-free probabilities for the one-period model. Note that P_U and P_D are independent of n. In particular, the **single-period** or **local** volatility is also independent of n and is given by

$$\sigma_{loc}^2 = \ln\left(\frac{U'}{D'}\right) P_U \, P_D .$$

The annual volatility is therefore

$$\sigma^2 = \frac{1}{dt} \ln\left(\frac{U'}{D'}\right) P_U \, P_D .$$

The analysis of the model is very similar to the case treated in Chapter 2. In particular, the probabilities P_U and P_D are given by

$$P_U = \frac{1}{2}\left[1 \pm \sqrt{1 - \frac{\sigma^2}{\rho^2}} \right]$$

and

$$P_D = \frac{1}{2}\left[1 \mp \sqrt{1 - \frac{\sigma^2}{\rho^2}} \right],$$

where $\rho = \frac{1}{2\sqrt{dt}} \ln (U'/D')$. The parameters U' and D' are given by

$$U' = \frac{e^{\rho \sqrt{dt}}}{P_D \, e^{-\rho \sqrt{dt}} + P_U \, e^{\rho \sqrt{dt}}}$$

and

$$D' = \frac{e^{- \rho \sqrt{dt}}}{P_D \, e^{-\rho \sqrt{dt}} + P_U \, e^{\rho \sqrt{dt}}} .$$

Thus P_U, P_D, U' and D' are independent of n.

Using Eqs. (2.25a) and (2.25b) of Chapter 2 we find that the (annualized) growth rate for the $(n + 1)^{st}$ period is given by

$$\mu_n = r_n + \frac{\rho}{\sqrt{dt}} \cdot (P_U - P_D) - \frac{1}{dt} \ln\left[P_D \, e^{-\rho \sqrt{dt}} + P_U \, e^{\rho \sqrt{dt}} \right] . \tag{4.5a}$$

The average growth rate over the entire time period of T years, annualized, is therefore

$$\mu = \frac{1}{N} \sum_{n=0}^{N-1} r_n + \frac{\rho}{\sqrt{dt}} \cdot (P_U - P_D)$$

$$- \frac{1}{dt} \ln\left[P_D \, e^{-\rho \sqrt{dt}} + P_U \, e^{\rho \sqrt{dt}} \right] . \tag{4.5b}$$

REMARK 4.1   Notice that the trajectories followed by the price of the underlying asset form a *recombining* tree, in the sense that there are only n nodes at time n, just like the tree with constant U, D. Thus, a path that originates from a given point and goes first "up" and then "down" arrives at the same location after two time steps that a path that goes first "down" and then "up." In fact, we have

$$U_n \, D_{n+1} \;=\; e^{r_n \, dt} \, U' \cdot e^{r_{n+1} \, dt} \, D' \;=\; D_n \, U_{n+1} \, .$$

The only difference between the variable and the constant interest-rate models is that in the former the slopes of the trajectories vary according to the time period.   ∎

From Eq. (4.5b), it follows that the value of the underlying asset at node (n, j) is given by

$$S_n^j \;=\; S_0^0 \, e^{\sum_{k=0}^{n-1} r_k \, dt} \cdot (U')^j \, (D')^{N-j} \, . \tag{4.6a}$$

The recursion relation for the price of any derivative security contingent on the risky asset is

$$V_n^j \;=\; \frac{1}{1 + R_n} \left[P_U \, V_{n+1}^{j+1} + P_D \, V_{n+1}^{j} \right]$$

$$= e^{-r_n \, dt} \left[P_U \, V_{n+1}^{j+1} + P_D \, V_{n+1}^{j} \right] . \tag{4.6b}$$

Since the risk-neutral probabilities P_U and P_D are independent of the term-structure of interest rates, the valuation of derivative products using this model is particularly simple.

We observe from (4.5b) that

$$S_N^j = S_0^0 \, e^{\sum_{k=0}^{N-1} r_k \, dt} \cdot (U')^j \, (D')^{N-j}$$

$$= S_0^0 \, e^{\left(\frac{1}{N} \sum_{k=0}^{N-1} r_k \right) T} \cdot (U')^j \, (D')^{N-j} \, . \tag{4.7}$$

We conclude that the model prices European-style options and derivative securities expiring after N periods exactly like the binomial model of Chapter 2 with an effective, constant interest rate given by

$$\bar{r} \;=\; \frac{\sum_{n=0}^{N-1} r_n}{N} \, , \tag{4.8}$$

which represents the yield for risk-less lending over the duration of the contract.

> **Note.** It is important to notice that the values of path-dependent derivative securities and the corresponding hedge-ratios predicted by the recursion for *intermediate* time periods, i.e., for n between 1 and N, are different from the ones that would be obtained using a binomial tree with constant rate $\bar{r}$. This is because the recursion relation (4.6b) takes into account the fact that interest rates vary over the duration of the contract. Using the same average rate $\bar{r}$ of (4.8) in the recursion relation instead of r_n may give incorrect prices and hedge-ratios for $1 \leq n < N$.

For $dt \ll 1$, we can use the log-normal approximation to the binomial model. For this purpose, it is convenient to introduce a *function* $r(t)$ to model variability of interest rates. This function is connected to the discrete rates r_n through the formula

$$r_n = r(n\,dt)\,.$$

The effective average rate is then

$$\bar{r} \approx \frac{1}{T} \int_0^T r(s)\,ds\,.$$

It is easy to see from (4.6a) and (4.6b) that the expected (continuously compounded) return satisfies

$$\lim_{dt \to 0} \mu = \bar{r} - \frac{1}{2}\sigma^2\,.$$

Hence, under the risk-neutral probability, the price of the underlying asset satisfies

$$S_T = S_0\, e^{\sigma \sqrt{T}\, Z + \left(\frac{1}{T}\int_0^T r(s)\,ds - \frac{1}{2}\sigma^2\right)T}$$

$$= S_0\, e^{\sigma \sqrt{T}\, Z + \left(\bar{r} - \frac{1}{2}\sigma^2\right)T}\,,$$

where Z is normal with mean 0 and variance 1. In particular, *the Black–Scholes formula for European option pricing holds, with r replaced by $\bar{r}$.*

4.2 Constructing a Risk-Neutral Measure with Time-Dependent Volatility

We now apply the same philosophy to volatility. In Chapter 3 we defined the volatility parameter σ as the standard deviation of stock returns. Implicit in this definition was the fact that the standard deviation over single time-periods is constant. The annualized standard deviation of returns corresponding to future time intervals is what we call the *forward volatility*. We can generalize the binomial model by assuming that the forward volatility is time-dependent.

The forward volatilities can be specified as a sequence of parameters

$$\sigma_0\,,\ \sigma_1\,,\ \sigma_2\,,\ \ldots\ \sigma_{N-1}$$

corresponding to the annualized standard deviations of returns for each period. Thus,

$$\sigma_n^2 = \frac{1}{dt} \, \text{Var} \left[\ln \left(\frac{S_{n+1}}{S_n} \right) \right] \tag{4.9}$$

for $0 \leq n \leq N - 1$.

We want to construct a simple binomial model which is arbitrage-free and consistent with the given forward volatilities and interest rates. It is also convenient that the resulting trajectories for the price of the underlying asset form a "recombining" tree, so that derivative asset prices can be obtained by solving simple recursive relations. The problem consists in specifying parameters U_n, D_n, $P_U^{(n)}$, $P_D^{(n)}$, $n = 0, 1, \ldots, N - 1$ so that

$$1 + R_n = P_U^{(n)} \, U_n + P_D^{(n)} \, D_n \,, \tag{4.10}$$

with

$$P_U^{(n)} = \frac{1 + R_n - D_n}{U_n - D_n}$$

$$P_D^{(n)} = \frac{U_n - 1 - R_n}{U_n - D_n} \,, \tag{4.11}$$

and, in addition,

$$dt \, \sigma_n^2 = \left[\ln \left(\frac{U_n}{D_n} \right) \right]^2 \cdot P_U^{(n)} \cdot P_D^{(n)} \tag{4.12}$$

for all n. Finally, we need to impose the conditions

$$U_n \cdot D_{n+1} = D_n \cdot U_{n+1} \tag{4.13}$$

so as to have a recombining tree.

The no-arbitrage condition (4.10) is immediately satisfied if we set

$$U_n = e^{r_n \, dt} \, U_n' \,, \qquad D_n = e^{r_n \, dt} \, D_n' \,,$$

$$P_U^{(n)} = \frac{1 - D_n'}{U_n' - D_n'} \quad \text{and} \quad P_D^{(n)} = \frac{U_n' - 1}{U_n' - D_n'} \,. \tag{4.14}$$

Furthermore, condition (4.13) is equivalent to

$$\frac{U_n'}{D_n'} = \lambda \tag{4.15}$$

where $\lambda > 1$ is a constant. Therefore, matching the term-structure of volatilities requires finding λ, $P_U^{(n)}$, and $P_D^{(n)}$ such that

$$\Delta t \, \sigma_n^2 = (\ln \lambda)^2 \cdot P_U^{(n)} \cdot P_D^{(n)} \,, \qquad n = 1, \ldots, N \,. \tag{4.16}$$

Observe that the right-hand side of (4.16) reaches a maximum when the probabilities are equal. Thus, a necessary condition for the existence of solutions to the N equations corresponding to (4.16) is

$$dt \, \sigma_n^2 \leq \frac{1}{4} \, (\ln \lambda)^2$$

for all n, or

$$2 \sqrt{dt} \, \underset{n}{\text{Max}} \, \sigma_n \leq \ln \lambda \, .$$

Thus, if we set

$$\sigma_{max} = \underset{n}{\text{Max}} \, \sigma_n \, ,$$

an admissible λ should therefore have the form

$$\lambda = e^{\, 2\rho \, \sqrt{dt}}$$

with $\rho \geq \sigma_{max}$. Substituting into (4.16) and solving for the probabilities, we obtain

$$P_U^{(n)} = \frac{1}{2} \left[1 \pm \sqrt{1 - \frac{\sigma_n^2}{\rho^2}} \right]$$

$$P_D^{(n)} = \frac{1}{2} \left[1 \mp \sqrt{1 - \frac{\sigma_n^2}{\rho^2}} \right] . \tag{4.17}$$

The parameters U_n' and D_n' can then be obtained from (4.14). They are given by

$$U_n' = \frac{e^{\rho \sqrt{dt}}}{P_D^{(n)} \, e^{-\rho \sqrt{dt}} + P_U^{(n)} \, e^{\rho \sqrt{dt}}}$$

and

$$D_n' = \frac{e^{- \rho \sqrt{dt}}}{P_D^{(n)} \, e^{-\rho \sqrt{dt}} + P_U^{(n)} \, e^{\rho \sqrt{dt}}} . \tag{4.18}$$

Notice that the risk-neutral probabilities are now time-dependent. The prices of the underlying asset at the different nodes in the tree are, according to the model,

$$S_n^j = S_0^0 \cdot e^{\sum_{k=0}^{n-1} r_k \, dt} \cdot \frac{(e^{\rho \sqrt{dt}})^j \, (e^{-\rho \sqrt{dt}})^{n-j}}{\prod_{k=0}^{n-1} \left(P_D^{(k)} \, e^{-\rho \sqrt{dt}} + P_U^{(k)} \, e^{\rho \sqrt{dt}} \right)} . \tag{4.19}$$

Finally, the equation that gives the value of a European-style derivative security with payoff $F(S)$ maturing after N periods is

$$\begin{cases} V_n^j = e^{-r_n \, dt} \left[P_U^{(n)} \, V_{n+1}^{j+1} + P_D^{(n)} \, V_{n+1}^j \right] , & n = 0, \ldots, N - 1 \\[2mm] V_N^j = F(S_N^j) . \end{cases} \tag{4.20}$$

REMARK 4.2   The independence of increments in the binomial model implies that the variance of the logarithm of the price after n periods is

$$\text{Var.}(S_n/S_0) = \sum_{j=1}^{n} \sigma_j^2 \, dt \, .$$

This implies that the price of an option with expiration date t for $dt \ll 1$ be given by the Black–Scholes formula with volatility

$$\sigma \approx \sqrt{\frac{\sum_{j=1}^{n} \sigma_j^2}{n}} \, , \quad \text{for} \quad n\,dt \approx t \, .$$

∎

We will use this property heavily in the next section.

4.3   Deriving a Volatility Term-Structure from Option Market Data

Volatility is the most important variable in option pricing. Many methods have been proposed to "calibrate" the volatility variable in pricing models (this is an indication that there is no "correct" way of doing this!). Historically, there are two paradigms for volatility estimation: using *historical volatility* and using *implied volatility*.

DEFINITION 4.1   **Historical volatility** *is defined as an estimate of the variance of the logarithm of the price of the underlying asset, obtained from past data.*

DEFINITION 4.2   **Implied volatility** *is the numerical value of the volatility parameter that makes the market price of an option equal to the Black–Scholes value.*

The use of historical volatility estimates requires the construction of appropriate statistical estimators. One of the main problems in this regard is to select the sample size, or window of observations, that will be used to estimate σ (6 months of previous data, 3 months, 1 month, etc.). Different time-windows tend to give different volatility values.

The problem with using historical volatility is that it assumes that future market performance will be reflected in future option prices. Although this may be partially correct, such method will not survive a large "spike" in volatility, such as the one which occurred in the stock market crash of October 1987, for example. Another argument against historical volatility is that it does not incorporate arrivals of new information such as corporate mergers, sudden changes in exchange-rate policy (the Mexican peso devaluation of 1994), etc.

Implied volatility is a convenient way to quote option prices in terms of a single risk parameter. It is important to notice that implied volatility is a "forward-looking" parameter. Therefore, one can say that it incorporates the market's expectations about prices of derivative prices or, more concisely, about *risk*. Measuring risk through the construction of appropriate risk-neutral probability measures is the name of the game in Financial Mathematics.

Example 4.2

Consider the following situation: Stock XYZ is trading at $100.00. A 183-day call option with strike price $100 trades at $6.37. The interest rate is 7%. Assuming a 365-day year, the value of σ that makes $6.37 the Black–Scholes price is $\sigma_{implied} = 0.16$ or 16% annual volatility. (Check this with your Black–Scholes calculator.) □

It is important to realize that the implied volatilities of options on the same underlying asset are *not* constant for different strikes and maturities. At first, this seems like a serious "blow" to the theory, but what really happens is that the market assigns different risk-premia to different strikes and maturities. This does not mean necessarily that there exist arbitrages in the market, but instead that the way in which the market prices risk at different price levels and future dates is different.

The **term-structure of volatilities** is defined as the series of implied volatilities of at-the-money options with different maturities (most of the volume of traded options is in at-the-money options).

To illustrate how the term-structure of volatility arises consider the following example, drawn from the dollar/deutschemark options market on August 23, 1995. On that date, option market-makers (usually banks dealing over-the-counter) were trading options using the following volatilities for options with different maturities:

Maturity	Bid	Offer
1 month	.1390	.1425
2 months	.1365	.1390
3 months	.1330	.1360
6 months	.1300	.1320
9 months	.1290	.1310

The meaning of this table is the following: the bid/offer prices for at-the-money options on USD/DEM on that day were computed using the Black–Scholes formula with the above volatilities.

A simple way to incorporate information on the term-structure of volatilities is to use forward volatilities in the spirit of the previous section.

What is the relation between implied and forward volatilities? Let us represent the term-structure of volatilities as the sequence

$$\bar{\sigma}_1, \ \bar{\sigma}_2, \ \dots, \ \bar{\sigma}_N,$$

where $\bar{\sigma}_n$ is the implied volatility of an at-the-money option expiring after n periods. Given the fact that the variance of the logarithm of the price under the risk-neutral measure after n periods is

$$\sum_{j=1}^{n} \sigma_j^2 \, dt \ ,$$

the relationship between implied volatilities and forward volatilities that is consistent with no-arbitrage is

$$\bar{\sigma}^2 = \frac{1}{n\, dt} \sum_{j=1}^{n} \sigma_j^2 \, dt$$

$$= \frac{1}{n} \sum_{j=1}^{n} \sigma_j^2 \, .$$

In continuous time, these relations are expressed more simply as

$$\bar{\sigma}(t) = \sqrt{\int_0^t \sigma^2(s) \, ds}$$

and

$$\sigma(t) = \sqrt{\lim_{\Delta t \to 0} \frac{(t + \Delta) \bar{\sigma}^2(t + \Delta) - t \bar{\sigma}^2(t)}{\Delta t}} = \sqrt{\frac{\partial}{\partial t} \left(t \sigma^{-2}(t) \right)}.$$

In practice, we have a discrete term-structure of volatilities and therefore, the passage to the limit is replaced by a difference quotient, i.e.,

$$\sigma(s) = \sqrt{\frac{(t + \Delta t) \bar{\sigma}^2(t + \Delta) - t \bar{\sigma}^2(t)}{\Delta t}} \, , \qquad \text{for } t \le s < t + \Delta t \, .$$

This equation allows us to use market data to calculate local volatilities that can be used in the pricing model.

We outline the procedure using the US/DEM data. As a first step, we take the average between bid and offer implied volatilities. The result is

Maturity	Volatility
1 month	.14075
2 months	.13775
3 months	.13450
6 months	.13100
9 months	.13000

Notice that the data are not given over time intervals of the same duration. To calculate the "forward–forward" volatility from month 1 to month 2, we can use the above equation. Hence

$$2/12 \times (.13775)^2 = 1/12 \times (.14075)^2 + 1/12 \times (\sigma_{1,2})^2 \, .$$

Straightforward arithmetic gives $\sigma_{1,2} = .13468$. The next step is to compute the 2-to-3 months forward volatility. The corresponding equation is then

$$3/12 \times (.1345)^2 = 2/12 \times (.13775)^2 + 1/12 \times (\sigma_{2,3})^2 \, .$$

This gives $\sigma_{2,3} = 0.12775$. The 3-to-6 month volatility is found by solving

$$6/12 \times (.1310)^2 = 3/12 \times (.1345)^2 + 3/12 \times (\sigma_{3,6})^2 \, .$$

The result is $\sigma_{3,6} = 0.1274$. Finally, the equation for the 6-to-9 month volatility is

$$9/12 \times (.1300)^2 = 6/12 \times (.1310)^2 + 3/12 \times (\sigma_{6,9})^2 \, ,$$

Term Implied Volatilities		Forward Implied Volatility	
1 M	.14075	0-1 M	.14075
2 M	.13775	1-2 M	.13468
3 M	.13450	2-3 M	.12775
6 M	.13100	3-6 M	.12740
9 M	.13000	6-9 M	.12797

FIGURE 4.2a
Volatility: deriving forward-implied volatility from term-implied volatilities.

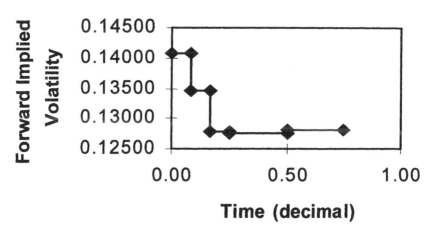

Forward Implied Volatility

FIGURE 4.2b
Graphical representation of forward-implied volatilities as a function of time.

which gives $\sigma_{6,9} = 0.1279$.

This calculation gives an approximate estimate of the forward volatilities. We can then set σ_n in the model equal to the appropriate value corresponding to the period under consideration.

4.4   Underlying Assets That Pay Dividends

We consider the valuation of European-style derivative securities that depend on a dividend-paying asset, such as the stock of a company. The binomial model must be slightly modified to account for this feature.

We assume that there are N trading periods and that dividend payments are always made at the end of a trading period. The values of the stock before dividends are paid are denoted by $\hat{S}_n$ or $\hat{S}_n^j$, for $n = 0, \ldots, N$. This is the *cum-dividend* value. We denote the value of the stock at the end of the nth period after a dividend payment—the *ex-dividend* value—by S_n or S_n^j.

A natural assumption regarding dividend payments is that the payment after the nth period is a fraction of the cum-dividend value, say

$$D_n = \alpha_{n-1} \hat{S}_n , \tag{4.21}$$

where $0 \le \alpha_n < 1$ for all $0 \le n \le N - 1$. The fractions α_n are assumed to be known in advance and are also allowed to depend on time (to model, for instance, periods without dividend payments).[5] Equation (4.21) gives a simple relation between the cum- and post-dividend values:

$$\hat{S}_n = S_n + D_n = S_n + \alpha_{n-1} \hat{S}_n ,$$

$$\hat{S}_n = \frac{1}{1 - \alpha_{n-1}} S_n \quad \text{or} \quad S_n = (1 - \alpha_{n-1}) \hat{S}_n . \tag{4.22}$$

We shall take as the basic variable in the model the ex-dividend price of the stock, assuming that, given the price history until the end of a period n, we have one of two possibilities for the price shock over the next period, namely

$$S_{n+1} = S_n U_n \quad \text{or} \quad S_n D_n .$$

We will also impose the condition $U_n D_{n+1} = D_n U_{n+1}$ $n = 0, \ldots, N - 1$, so as to have a recombining tree. We must determine a probability on the set of paths that makes the model arbitrage free. We know that this probability must be such that the present value of the stock is equal to its discounted future value, including dividends. In particular, from (4.22), we must have

$$S_n = \frac{1}{1 + R_n} \cdot E\{S_{n+1} + D_{n+1} | S_n\}$$

---

[5] As with the notations for interest rates and volatilities, we make the convention that α_n represents the fraction of the cum-dividend value paid after the $(n + 1)^{st}$ period. Modeling projected stock dividends is a very complex process, and the assumption of constant dividend yield is very rudimentary. On the other hand, this model is very useful for pricing currency options, where the "dividend rate" is interpreted as the interest rate in the foreign currency and the "term-structure of dividends" is the term-structure of foreign-currency deposits.

$$= \frac{1}{1 + R_n} \cdot E\{\hat{S}_{n+1} | S_n\}$$

$$= \frac{1}{1 + R_n} \cdot \frac{1}{1 - \alpha_n} \cdot E\{S_{n+1} | S_n\} . \tag{4.23}$$

Thus, the ex-dividend value is obtained by discounting the expectation of its future (post-dividend) values at a rate that depends on the risk-less interest rate and the fraction of dividends paid. The conditional probabilities $P_U^{(n)}$ and $P_D^{(n)}$ corresponding to the expectation in (4.23) should therefore satisfy

$$\begin{cases} P_U^{(n)} + P_D^{(n)} = 1 \\ \\ U_n P_U^{(n)} + D_n P_D^{(n)} = (1 + R_n)(1 - \alpha_n) . \end{cases} \tag{4.24}$$

To make a parallel with the no-dividend case, we introduce the term-structure of interest rates as in (4.1) and set

$$1 - \alpha_n = e^{-q_n \, dt} . \tag{4.25}$$

The constants q_n represent the annualized rate at which dividends accrue corresponding to the $(n+1)^{st}$ period. We can then rewrite the second equation in (4.24) as

$$U_n P_U^{(n)} + D_n P_D^{(n)} = e^{(r_n - q_n)dt} . \tag{4.26}$$

The calculation of the parameters U_n, D_n, $P_U^{(n)}$, and $P_D^{(n)}$ follows a route similar to the one of the two previous sections. We omit unnecessary details and state only the simplest result, corresponding to the case of constant local volatilities. Solving (4.24) and adjusting for volatility, we find that the post-dividend values of the stock at the different nodes are

$$S_n^j = S_0^0 \cdot e^{\sum_{k=0}^{n-1}(r_k - q_k) \, dt} \cdot \frac{(e^{\rho \sqrt{dt}})^j (e^{-\rho \sqrt{dt}})^{n-j}}{(P_D e^{-\rho \sqrt{dt}} + P_U e^{\rho \sqrt{dt}})^n} , \tag{4.27a}$$

where $\rho \geq \sigma$ is a parameter and where

$$P_U = \frac{1}{2}\left[1 \pm \sqrt{1 - \frac{\sigma^2}{\rho^2}}\right], \qquad P_D = 1 - P_U . \tag{4.27b}$$

The value of a European-style derivative security with payoff $F(S)$ maturing after N periods is then given by the familiar recursive relation

$$\begin{cases} V_n^j = e^{-r_n \, dt}\left[P_U V_{n+1}^{j+1} + P_D V_{n+1}^j\right] \\ \\ V_N^j = F(S_N^j) . \end{cases} \tag{4.28}$$

To get a better feeling for how dividends affect the pricing model, we observe that the expected (post-dividend) value of the stock under the no-arbitrage measure after the N periods is, from (4.27a) and (4.27b),

$$E\{ S_N \mid S_0 = S_0^0 \} = S_0^0 \cdot e^{\sum_{k=0}^{N-1} (r_k - q_k)\,\Delta t} \ .$$

Thus, the underlying variable (S_n) of the derivative security grows at a rate which is different from the one used to discount the value V_n in (4.28).

Dividend payments for the underlying asset can therefore be easily incorporated into the binomial pricing model. The pricing equations for European-style derivative securities are very similar to the case without dividend payments. However, if the derivative security can be exercised before its maturity date (as is the case for American-style options) the impact of dividend payments on the pricing equations is more substantial.

A *log-normal approximation* of the binomial model for dividend-paying underlying assets can be derived from the above considerations.[6] As with the case of interest rates, it is convenient to consider a *dividend function* $q(t)$ that interpolates between the discrete values, *viz.*,

$$q_n = q(n\,dt) \ , \quad n = 0 \ldots, N - 1 \ .$$

The asymptotic value of the mean annual yield is obtained from Eq. (4.27a). The key observation is that $r_k - q_k$ appears as the "effective" interest rate in the post-dividend price [compare with Eqs. (4.6a) and (4.6b)]. Therefore, in the log-normal approximation, the price of the underlying asset satisfies

$$S_T = S_0\, e^{\sigma\sqrt{T}\,Z + \left(\frac{1}{T}\int_0^T (r(s)-q(s))\,ds - \frac{1}{2}\sigma^2 \right) T}$$

$$= S_0\, e^{\sigma\sqrt{T}\,Z + \left(\bar{r} - \bar{q} - \frac{1}{2}\sigma^2 \right) T} \ ,$$

where

$$\bar{q} = \frac{1}{T} \int_0^T q(s)\,ds \ .$$

In particular, the Black–Scholes formula can be extended to dividend-paying assets. The general valuation formula for European-style derivative securities is

$$V(S, T) = e^{-rT}\, E \left\{ F \left(S\, e^{\sigma\sqrt{T}\,Z + (\bar{r}-\bar{q}-\frac{1}{2}\sigma^2) T} \right) \right\} \ ,$$

where Z is a standardized normal. Notice the difference with the previous results with respect to the level of the risk-neutral average yield. The Black–Scholes formula for the value of a European call option on an asset with continuous dividend yield q is

$$C(S, K; T) = S\, e^{-qT}\, N(d_1) - K\, e^{-rT}\, N(d_2) \ , \tag{4.29}$$

---

[6]This approximation assumes however that dividends are paid out continuously. The continuous dividend approximation is used in the case of options on indices such as the S&P 500 and options on foreign currencies. In the latter case, the dividend rate is simply the foreign exchange rate.

where

$$d_1 = \frac{1}{\sigma \sqrt{T}} \ln \left(\frac{S e^{(r-q) T}}{K} \right) + \frac{1}{2} \sigma \sqrt{T} , \qquad d_2 = d_1 - \sigma \sqrt{T} . \qquad (4.30)$$

The Delta of the option is now

$$\Delta(S, T) = e^{-qT} N(d_1).$$

Notice that the amount of shares of the underlying security is multiplied by the factor e^{-qT}. This is similar to what happens when hedging forward contracts with continuous dividend reinvestment.

4.5 Futures Contracts as the Underlying Security

Many exchange-traded and OTC derivative securities are based on futures contracts. Examples include options on Eurodollar futures and on Treasury bond futures contracts. In this section, we give a short introduction and consider the problem of pricing a European-style derivative security, assuming that the underlying asset is a "stylized" futures contract.[7] In constructing no-arbitrage models for derivatives based on futures contracts, the cash-flow structure of the futures contract must be taken into account. As we shall see, the situation bears a strong similarity with the case of assets with continuous dividend payments.

Let F_n, $n = 0, 1, \ldots, N$, represent the sequence of (random) futures prices after the different trading periods. We shall assume that the contract is marked-to-market after each trading period, so that an investor with a long (resp. short) position in one contract after the nth period will be credited/ debited the amount $F_{n+1} - F_n$ after the next period. We assume, as usual, that at each step the price follows a binomial model with

$$F_{n+1} = F_n U_n \quad \text{or} \quad F_n D_n ,$$

where U_n and D_n are parameters such that the trajectories will form a recombining binomial tree.

Positions can be taken at zero cost (according to our definition of "stylized" contract). Hence, the futures contract can be viewed as a security that has zero market value and obliges its holder to receive or pay out the cash-flows $F_{n+1} - F_n$, i.e., to mark-to-market after each trading period. To determine a possible risk-neutral probability measure for the random variables F_n, consider an investor who opens a long position in the futures contract after the nth period. Let us focus on the cash-flows associated with a single trading period. The risk-neutral probability measure on the set of paths $(F_n)_{n=0}^{N}$ should be such that the value of the futures contract (zero) is the discounted expectation of its cash-flows. Therefore,

$$0 = e^{-r_n \, dt} \, \text{E} \{ F_{n+1} - F_n \mid F_n \}$$

---

[7]By this we mean a contract that can be opened either going long or short at zero cost and that is marked-to-market daily. We do not take into account any cash flows resulting from maintenance or margin.

or simply

$$F_n = E\{ F_{n+1} \mid F_n \} . \tag{4.31}$$

This last equation states that $(F_n)_{n=0}^N$ must be a *martingale* under the no-arbitrage pricing measure, i.e., that any arbitrage free measure would make today's futures prices a "fair" bet on later futures prices.[8] As before, the conditional probabilities for the two states (up or down) that can occur in the binomial model can be completely determined from Eq. (4.31) and the parameters U_n and D_n. In fact, we have

$$\begin{cases} P_U^{(n)} + P_D^{(n)} = 1 \\ \\ U_n P_U^{(n)} + D_n P_D^{(n)} = 1 \end{cases}$$

so

$$P_U^{(n)} = \frac{1 - D_n}{U_n - D_n} \quad \text{and} \quad P_D^{(n)} = \frac{U_n - 1}{U_n - D_n} .$$

Since, these probabilities are independent of F_n, we conclude that any arbitrage-free probability measure is such that $(F_n)_{n=0}^N$ is a multiplicative random walk, i.e., that $\ln F_n$ follows a standard random walk with independent increments. The model can accommodate, if necessary, given term-structures of interest rates and/or volatilities using the methods shown above. In the simplest case of constant local volatilities, the (constant) probabilities are given by (4.27b) and the futures prices at the different nodes of the tree are

$$F_n^j = F_0^0 \cdot \frac{\left(e^{\rho \sqrt{dt}} \right)^j \left(e^{-\rho \sqrt{dt}} \right)^{n-j}}{\left(P_D e^{-\rho \sqrt{dt}} + P_U e^{\rho \sqrt{dt}} \right)^n} . \tag{4.32}$$

The pricing equation for a European-style contingent claim with value $G(F_N)$ at expiration is given by

$$\begin{cases} V_n^j = e^{-r \, dt} \left[P_U V_{n+1}^{j+1} + P_D V_{n+1}^j \right] \\ \\ V_N^j = G(F_N^j) . \end{cases} \tag{4.33}$$

Notice that the present scheme is formally equivalent to the one for pricing of a derivative security contingent on the price of an underlying asset with dividend yield $q_n = r_n$ [compare with Eqs. (4.27a) and (4.31)]. This equivalence can be seen more clearly by assuming that the futures contract is based on a traded underlying security with value S_n and that the futures price converges to the price of the underlying asset at the end of the Nth period, i.e., that $F_N = S_N$. (For simplicity, we assume that the traded security pays no dividends.) Since the cash-flow of the futures position after the Nth period coincides with that of a forward contract, the cost-of-carry formula implies that

$$F_n = S_n \, e^{\sum_{k=n}^{N-1} r_k \, \Delta t} . \tag{4.34}$$

Hence, the futures price is equal to the value of a certain number of shares of the traded security, with this number changing with time. Now, the value of this "equivalent portfolio" after the

---

[8]As mentioned earlier, this is not a statement about the statistics of future prices. The no-arbitrage probability measure is just a device for consistently pricing derivatives contingent on F_n.

next trading period is

$$\hat{F}_{n+1} \equiv S_{n+1} \, e^{\sum_{k=n}^{N-1} r_k \, \Delta t} = F_{n+1} \, e^{r_n \, \Delta t} \, .$$

This shows that the equivalent portfolio appreciates in value to more than F_{n+1}. The excess can be regarded as a dividend that is paid out after each period. The last equation can be viewed as giving the relation between the corresponding ex- and cum-dividend values of the equivalent portfolio, as in Eq. (4.22) with

$$1 - \alpha_n = e^{-r_n \, \Delta t} \, .$$

We now address the dynamic hedging of derivatives dependent on futures contracts. The hedging strategy consists of a money market account combined with an open position in futures. As in the case of equity, the number of contracts in the hedge is equal to the sensitivity of the theoretical value with respect to changes in the futures price. More precisely, suppose that the values V_n^j at all the nodes of the tree have been calculated. The replicating strategy corresponding to the node (n, j) will consist in investing V_n^j in a money-market account and opening Δ_n^j contracts. To find the "hedge-ratio" Δ_n^j, we must match the two possible cash-flows to the calculated values of the derivative security at the following nodes. Accordingly,

$$\begin{cases} \Delta_n^j \, (F_{n+1}^{j+1} - F_n^j) + (1 + R_n) \, V_n^j = V_{n+1}^{j+1} \\[2mm] \Delta_n^j \, (F_{n+1}^j - F_n^j) + (1 + R_n) \, V_n^j = V_{n+1}^j \, . \end{cases}$$

It is clear from these two equations that

$$\Delta_n^j = \frac{V_{n+1}^{j+1} - V_{n+1}^j}{F_{n+1}^{j+1} - F_{n+1}^j} \, , \tag{4.35}$$

as expected.

4.6 Valuation of a Stream of Uncertain Cash Flows

We conclude this chapter by writing down a general equation for pricing derivative securities that offer a series of cash flows that depend on the value of some underlying asset at different cash-flow dates. A simple example of such a security would be a *commodity-or equity-linked* debt instrument. These notes or bonds have coupon payments that are linked to the value of an index such as the price of copper or oil or the S&P 500 index. Such securities can be decomposed or "stripped" into a series of European-style derivative securities, in the same way that a coupon-paying bond can be viewed as a series of pure discount bonds. Moreover, the payoff for each maturity can be simple enough so that it can be regarded as a collection of simple options each of which could be valued separately. The latter point of view, which could be called "reverse-engineering," is useful in practice and will be discussed in detail in future chapters. Here we will show that in the "binomial world" with one risky asset, all cash flows can be incorporated into a single equation that can be solved recursively to price any stream of uncertain cash-flows.

Consider a derivative security maturing after N trading periods and paying a stream of cash flows after each period. These cash flows can be specified by means of N functions of S, namely[9] that there is a cash flow due on each trading date. This represents no loss of generality since some of these functions could be equal to zero for some values of

$$f_1(S), \ f_2(S), \ \ldots \quad \ldots \ f_N(S). \tag{4.36}$$

The cash flow at each node (n, j) in the tree is defined to be

$$f_n^j \equiv f_n(S_n^j).$$

Suppose that a risk-neutral probability for the underlying asset has been determined, consistently with a term-structure of interest rates, a term-structure of volatilities and the dividend payments. We have seen that such a model can be constructed in terms of a collection of probabilities $P_U^{(n)}$, $P_D^{(n)}$, and "up–down" parameters U_n and D_n, for $n = 0, \ldots, N-1$. The general recursion relation that we seek follows from the following observation: at any given time, the value of the derivative security is equal to the current "coupon" value plus the discounted expectation of future cash flows. Therefore, we have

$$\begin{cases} V_n^j = f_n^j + e^{-r_n \, \Delta t} \cdot \left[P_U^{(n)} \, V_{n+1}^{j+1} + P_D^{(n)} \, V_{n+1}^j \right] \\[2mm] V_N^j = f_N^j . \end{cases} \tag{4.37}$$

Thus, solving a single recursive relation, we can compute the theoretical value of such contingent claims.

References and Further Reading

[1] Boyle, P.P. (March 1988), A Lattice Framework for Option Pricing with Two State Variables, *Journal of Financial and Quantitative Analysis*, **23**, pp. 1–12.

[2] Cox, J., Ross, S., and Rubinstein, M. (October 1979), Option Pricing: A Simplified Approach, *Journal of Financial Economics*, **7**, pp. 229–264.

---

[9]Here, we assume that $f_n(S) = 0$ if there are no cash flows due on a particular date. For example, a European call with strike K and expiration date t_n has $f_j(S) = 0$ if $j \neq n$ and $f_n(S) = \max(S - K, 0)$.

Chapter 5

American-Style Options, Early Exercise, and Time-Optionality

European-style options can be exercised at specific dates. In this chapter, we study American options, which give the holder the right to exercise at an unknown date in the future.

The simplest derivative securities with time-optionality are American-style options. Exchange-traded stock options on the CBOT and AMEX are American-style. Other examples of securities with time-optionality are encountered in debt markets. U.S. Treasury bonds issued before the 1980s were *callable:* the government had the right to repay the principal after a certain period and retire the bond. This type of bond is called a *callable bond. Convertible bonds,* usually issued by corporations, are debt securities that can be converted into company stock. Corporate bonds usually contain a call provision as well as the possibility for the bondholder to convert the debt into stock. Callable and convertible bonds are said to have *embedded options.* Mortgages constitute another class of securities with time-optionality. The borrower has the right to increase the speed of payments (prepayment) or to pay the mortgage in full by refinancing the loan at a lower interest rate. Similarly, commercial loans (credit cards, lines of credit) have time-optionality since they can be paid off faster or refinanced.

5.1 American-Style Options

The holder of an American-style option has the right to exercise the option any time before the expiration date. The rationale for the early exercise of American options is simple: by exercising, the holder can claim the dividend income provided by the underlying security (call), or the interest derived by selling the underlying asset and investing the cash (puts).

To fix ideas, assume that the option is a call with expiration date T. If the option is exercised at time τ ($\tau \leq T$), the payoff is

$$\max\ (S_\tau - K,\ 0)\ ,\tag{5.1}$$

where S_τ is the price of the security and K is the strike price. According to arbitrage pricing theory, the fair value of a security that delivers this payoff at time τ is

$$\mathbf{E}\left\{ e^{-r\tau}\ Max\ (S_\tau - K,\ 0) \right\}\ .\tag{5.2}$$

Here, $\mathbf{E}$ represents the expectation with respect to a risk-neutral measure defined on the set of paths $\{\ S_t,\ 0 \leq t \leq T\ \}$. This formula is true irrespective of whether τ is deterministic or random, provided in the latter case that τ is a *stopping time*.

DEFINITION 5.1   *A stopping time τ is a function taking values in $[0, T]$ such that the event*
$\{\tau = t\}$, i.e., the decision to "stop at time t," is determined by the path $\{S_u, 0 \le u \le t\}$.

In the case of an American option, the exercise date τ is not known ahead of time. Nevertheless, since the holder has the right to exercise at *any* time, the theoretical value of this option should be

$$V_{am} = \sup_{0 \le \tau \le T} \mathbf{E}\left\{e^{-r\tau} Max\,(S_\tau - K, 0)\right\},\tag{5.3}$$

where the supremum is taken over all possible *stopping times*.[1]

How does one compute the optimal stopping time that realizes the supremum? This will be discussed in detail in Section 5.3 of this chapter. This stopping time is called the *optimal* or *rational* exercise time because it maximizes the present value of the cash flows received by the holder of the option.

5.2   Early-Exercise Premium

What additional value, if any, does an American-style option have compared to a European option with the same strike K and expiration date T? Let us denote the fair values at time $t \le T$ of an American option and a European option (put or call), respectively, by $V_{am}(S, t)$ and $V(S, t)$ and the intrinsic value by $F(S)$.[2] We observe that if the theoretical value of the European option satisfies

$$V(S, t) \ge F(S) \equiv \text{intrinsic value},\quad 0 \le t \le T\tag{5.4}$$

for all S, then the American option has no early-exercise premium (i.e., $V_{am} = V$). In fact, the theoretical value of the American option satisfies necessarily [cf. (5.3)]

$$V_{am}(S, t) \ge V(S, t)\tag{5.5}$$

and thus, from (5.4),
$$V_{am}(S, t) \ge F(S).$$

We conclude that the holder of the American option will achieve a higher "value" by not exercising the option at date t. Indeed, why exchange the option for something worth less? Equation (5.4) implies that $\tau = T$ and hence that $V_{am}(S, t) = V(S, t)$ for all $t \le T$.

Conversely, if $V_{am} = V$, then, since

$$V_{am}(S, t) \ge F(S),$$

we have $V(S, t) \ge F(S)$ for all t. From these considerations, we have

---

[1] Some derivative securities may be exercisable on a smaller set of dates, say, only after 1 year and then only on specific dates 6 months apart. In this case the supremum in (5.3) should be replaced by $\tau \in \Theta$, where Θ represents the subset of possible exercise dates.

[2] $F(S) = \max[S - K, 0]$ (calls) or $F(S) = \max[K - S, 0]$ (puts).

PROPOSITION 5.1

The statements
(i) $V(S, t) \geq F(S)$ *for all* (S, t)
and
(ii) $V_{am}(S, t) = V(S, t)$ *for all* (S, t)
are equivalent.

This implies

PROPOSITION 5.2

The following statements are equivalent:
(i) $V_{am}(S, t) > V(S, t)$ *for all S and* $t < T$
and
(ii) $F(S) > e^{-r(T-t)} F(e^{(r-q)(T-t)} S)$ *for some* (S, t).

PROOF Suppose that the option is a call. We denote its theoretical value by $C(S, K, t)$. Recall (say, from the Black–Scholes formula or from general principles) that the theoretical value of a call satisfies

$$C(S, K; t) \propto S e^{-q(T-t)} - K e^{-r(T-t)} \propto e^{-r(T-t)} F(e^{(r-q)(T-t)} S) \, , \quad S \gg K \, . \quad (5.6)$$

(The value of a deep-in-the-money call is asymptotic to the forward price.)

Suppose that condition (ii) holds for the pair $(\bar{S}, t)$. Then it is clear geometrically that the inequality in (ii) will hold for all values of S larger then $\bar{S}$. Therefore, by (5.6), we have $C(S, K; t) < F(S)$ for $S \gg K$ and some t, as illustrated by Figure 5.1.

Hence, by Proposition 5.1, the American option has an early-exercise premium.

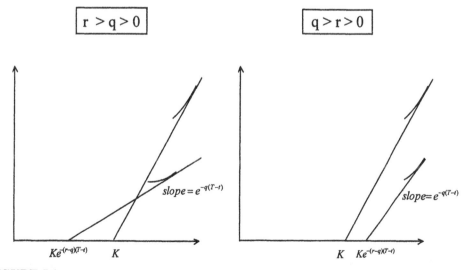

FIGURE 5.1
If the underlying security pays dividends, the slope of the intrinsic value of the call as a function of S is less than 1.

Conversely, assume that (i) holds. Then, by Proposition 5.1, $C(S, K; t) < F(S)$ for some S, t. From (5.6), we then conclude that $F(S) > e^{-r(T-t)} F(e^{(r-q)(T-t)} S)$. ∎

Puts can be analyzed similarly (this is left as an exercise).

Proposition 5.2 has the following useful consequence:

PROPOSITION 5.3

Assuming a risk-less rate r and a continuous dividend rate q,

 (i) Calls have early-exercise value if and only if q > 0,

 (ii) Puts have early-exercise value if and only if r > 0.

In particular, American-style call options on stocks that pay no dividends have no early-exercise premium. This is an important example, because it applies to exchange-listed stock options (assuming no dividend payouts during the lifetime of the option). Puts on stocks, on the other hand, have early-exercise value, since interest rates are nonzero. American options on futures always have an early-exercise premium (because $q = r$, as shown in Chapter 4).

5.3 Pricing American Options Using the Binomial Model: The Dynamic Programming Equation

In the simplest formulation, the risk-neutral probability can be taken to be either the binomial random walk (binomial tree) or its log-normal approximation. Even with such simple models, and assuming constant parameters (volatility and drift) the theoretical value of an American option cannot be expressed by a simple formula like the Black–Scholes formula for European-style options. The premium and optimal exercise time must be evaluated numerically.

Assume that the volatility, dividend rate and interest rate are given. Consider a binomial tree with parameters

$$U = \frac{e^{\sigma \sqrt{dt} + r\,dt - q\,dt}}{\cosh(\sigma \sqrt{dt})} \, ,$$

$$D = \frac{e^{-\sigma \sqrt{dt} + r\,dt - q\,dt}}{\cosh(\sigma \sqrt{dt})}$$

and

$$P_U = P_D = \frac{1}{2} \, .$$

Recall that the value of the underlying index at the node (n, j) is given by

$$S_n^j = S_0^0 \, U^j \, D^{n-j} \quad , \ 0 \le j \le n \le N \, .$$

Using this model, we shall calculate the value of an American option recursively. Let V_n^j represent the value of the option at the node (n, j), and assume that V_{n+1}^{j+1} and V_{n+1}^{j}, the values at the two "offspring" of node (n, j), have been determined. From the analysis of the binomial model, we know that, at time n, there exists a portfolio of stocks and cash that will replicate the two "cash-flows" V_{n+1}^{j+1} and V_{n+1}^{j} at time $n + 1$. The value of this replicating portfolio is

$$e^{-r\,dt} \left\{ P_U \, V_{n+1}^{j+1} + P_D \, V_{n+1}^{j} \right\} . \tag{5.7}$$

Hence, the theoretical value of the option, *conditional on it not being exercised at time t_n,* is (5.7). However, the holder can exercise at time t_n and collect the intrinsic value $F(S_n^j)$. Clearly, the rational decision to exercise should be made according to whether the value of holding the option, given by (5.7), exceeds or does not exceed the intrinsic value $F(S_n^j)$. Therefore, the theoretical value of the American option at time t_n if $S_n = S_n^j$ should satisfy

$$V_n^j = Max \left\{ F(S_n^j), \; e^{-r\,dt} \left[P_U V_{n+1}^{j+1} + P_D V_{n+1}^j \right] \right\}.$$ (5.8)

This equation is known as a **dynamic programming equation.**[3] Once the values of the option at the expiration date T are specified, *viz.,*

$$V_N^j = F(S_N^j),$$ (5.9)

Eq. (5.8) can be solved recursively, as in the case of the linear recursion relation used for pricing of European options.

The dynamical programming (DP) equation can be used to find out whether the option should be exercised, given the spot price and the time to expiration. More precisely, the nodes of the tree can be divided into two classes, according to whether

$$V_n^j > F(S_n^j), \qquad \text{i.e.,}$$

$$Max \left\{ F(S_n^j), \; e^{-r\,dt} \left[P_U V_{n+1}^{j+1} + P_D V_{n+1}^j \right] \right\}$$

$$= e^{-r\,dt} \left[P_U V_{n+1}^{j+1} + P_D V_{n+1}^j \right],$$ (5.10a)

or

$$V_n^j = F(S_n^j), \qquad \text{i.e.,}$$

$$Max \left\{ F(S_n^j), \; e^{-r\,dt} \left[P_U V_{n+1}^{j+1} + P_D V_{n+1}^j \right] \right\} = F(S_n^j).$$ (5.10b)

Nodes satisfying (5.10a) correspond to spot levels S_n^j where the option should not be exercised. Conversely, if (5.10b) is satisfied at a node, the option should be exercised. In particular, the optimal stopping time is given by

$$\tau^* = Min \left\{ n\,dt : \; V_n^j = F(S_n^j) \right\}.$$

The *exercise region,* i.e., the set of pairs $(S_n^j, n\,dt)$ such that $V_n^j = F(S_n^j)$, has a simple geometry, as indicated in the above equations. For calls, the option will be exercised if the price of the underlying asset is sufficiently high. Puts are exercised if the price of the underlying asset is low enough.

---

[3]Because the optimal exercise decision is determined dynamically.

5.4 Hedging

The Delta of an American option is given by the difference-quotient

$$
\Delta_n^j = \frac{V_{n+1}^{j+1} - V_{n+1}^j}{S_n^{j+1} - S_{n+1}^j} \,,
\tag{5.11}
$$

where V_n^j is the solution of (5.8)–(5.9). Suppose that a trader writes (sells) an American option and decides to hedge his exposure using a self-financing portfolio of stocks and cash with Delta given by (5.11). More precisely, assume that

- At time $t = 0$ an individual sells an American-style option,

- He implements thereafter a dynamic hedge that consists in holding $\Delta_n^{\bullet}$ shares and $B_n^{\bullet} = V_n^{\bullet} - \Delta_n^{\bullet} \cdot S_n^{\bullet}$ in bonds at time $t_n = n\,dt, n \geq 0$.[4]

- The buyer exercises the option at some time $\tau \leq T$.

We claim that this strategy is riskless for the seller of the option, regardless of whether the option is exercised. To see this, notice that the dynamical programming equation ensures that the value of the replicating portfolio at time t_n, *before the holder exercises*, is equal to $V_n^{\bullet}$. Since the solution of the dynamical programming equation satisfies

$$
V_n^j \geq F(S_n^j) \,, \quad \forall j \,, \quad \forall n \,,
$$

the value of the replicating portfolio will always be at least equal to the intrinsic value of the option (the liability faced by the seller of the option). Therefore, the strategy is effectively riskless.

5.5 Characterization of the Solution for $dt \ll 1$: Free-Boundary Problem for the Black–Scholes Equation

In practice, stock options can be exercised anytime between the settlement date and the expiration date. Therefore, it is natural to consider the "lognormal" model as the one to be used in this case. Computationally, we consider a binomial lattice with very refined mesh. To illustrate the issue of convergence as $dt \to 0$, consider the case of a 1-year American put. Assuming that $r = 8\%$ and $\sigma = 20\%$ and zero dividends, we find the following dependence for the value of the option at $S = K = 100$:

---

[4]The superscript [•] represents an arbitrary level $0 \leq j \leq n$.

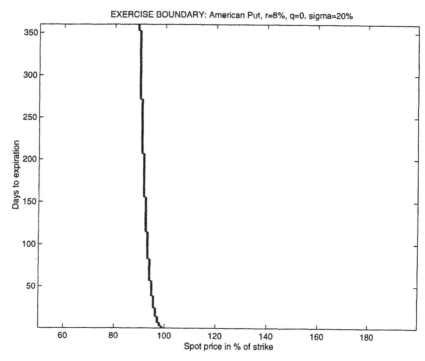

FIGURE 5.2
The jagged line represents the exercise boundary for an American put option calculated with a binomial lattice with 25 periods per day. In reality, the exercise boundary is a smooth curve that is well approximated by the curve shown here.

dt	V(100)
1/12	5.20
1/24	5.24
1/48	5.257
1/120	5.268
1/360	5.272
1/1520	5.274

We see that a considerable mesh refinement (small dt) must be taken to achieve convergence. The first value in the table corresponds roughly to a 1-month interval approximation. The last term corresponds to exercising up to 20 times in a day, which is essentially the continuous limit.

In the limit $dt \ll 1$, the solution of the recursion relation (5.8)–(5.9) can be expressed as a partial differential equation with a *free boundary condition*. This free boundary represents the geometric boundary of the set where the value of the option is equal to the intrinsic value (exercise region).

The same caveat regarding the approximation of the value of the option applies to the free boundary. Unless advanced PDE techniques are used, the free boundary obtained by numerical integration will be only an approximation to the "exact" exercise region. The value of the option varies very slowly in a neighborhood of the free boundary, as we shall establish below. In Figure 5.2, we present the exercise boundary of the put option of the previous example.

This section provides a heuristic discussion of the limit $dt \ll 1$ of the binomial pricing of American options. A characterization in terms of partial differential equations is useful for constructing more advanced numerical schemes for pricing and hedging American options. Moreover, the PDE description is useful to understand the Gamma-risk at the exercise boundary.[5]

From the previous analysis, we know that the (S, t)-plane can be decomposed into an *exercise region*, where $V_n^j = F(S_n^j)$ and its complement, where the dynamic programming equation reduces to the linear relation

$$V_n^j = e^{-r\,dt}\left[P_U\, V_{n+1}^{j+1} + P_D\, V_{n+1}^j \right]. \tag{5.12}$$

This relation is identical to the one satisfied by European-style derivative securities. It is therefore not surprising that the value of the American option should satisfy the Black–Scholes differential equation in the complement of the exercise region for $dt \ll 1$. More precisely, let $V(S, t)$ represent the limiting value of the American option outside the exercise region, i.e.,

$$V(S, t) = \lim_{S_n^j = S,\ (N-n)\,dt = t,\ dt \to 0} V_n^j.$$

Here, t represents the time-to-expiration of the option. Then, according to the analysis of European contingent claims, we know that $V(S, t)$ satisfies the equation

$$\frac{\partial V(S, t)}{\partial t} = \frac{1}{2}\sigma^2 S^2 \frac{\partial^2 V(S, t)}{\partial S^2} + (r - q) S \frac{\partial V(S, t)}{\partial S} - r\, V(S, t) \tag{5.13}$$

for (S, t) in the complement of the exercise region. (The correspondence between the binomial model (5.13) and the Black–Scholes PDE (5.14) was shown at the end of Chapter 3.) At expiration, we have $V(S, 0) = F(S)$ [the analogue of Eq. (5.8)]. However, the valuation of the American option requires that we specify the location of the boundary of the exercise region and appropriate boundary conditions.

PROPOSITION 5.4
Let $\mathcal{B}$ represent the boundary of the exercise region, i.e., the region where $V(S, t) = F(S)$. Then

$$V(S, t) = F(S) \ \text{for}\ S \in \mathcal{B}, \tag{5.14}$$

$$\frac{\partial V(S, t)}{\partial S} = F'(S) \ \text{for}\ S \in \mathcal{B}. \tag{5.15}$$

Also,

$$\Theta(S, t) = -\frac{\partial V(S, t)}{\partial t} = 0 \ \text{for}\ S \in \mathcal{B}. \tag{5.16}$$

REMARK 5.1   Equation (5.14) is obvious. For call options, (5.15) reads $\frac{\partial V(S,t)}{\partial S} = +1$. For puts, the boundary condition is $\frac{\partial V(S,t)}{\partial S} = -1$. In words, the graph of the function

---

[5]PDE solvers, usually based on finite-difference schemes, may give a more accurate characterization of the exercise boundary (see Wilmott et al. [1995]).

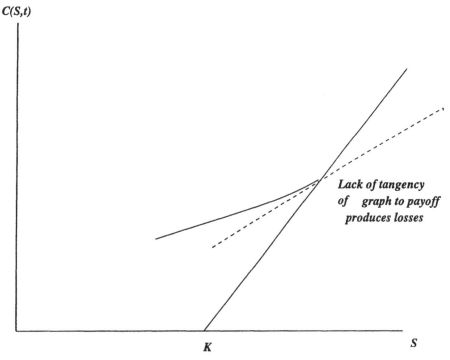

$C(S,t)$

*Lack of tangency
of   graph to payoff
produces losses*

K S

FIGURE 5.3
**A lack of tangency at the contact point with the graph of the intrinsic value of the call
would lead to unbounded losses.**

$V(S, t)$ is *tangent to the graph of the intrinsic value* along the exercise boundary, as illustrated
by Figures 5.3 and 5.4.   ∎

Proof of Proposition 5.4   For simplicity of presentation we consider the case of the simple
payoff $F_{call}(S) = \max(S - K, 0)$, and $F_{Put} = \max(K - S, 0)$ so that $F'_{Call}(S) = +1$ and
$F'_{Put} = -1$.
   For each t, let $S_f(t)$ represent the contact point

$$V(S_f(t), t) = F(S_f(t)) . \tag{5.17}$$

(Hence, $(S_f(t), t) \in \mathcal{B}$.) Since the value of the option cannot be less than the intrinsic value
we have, along $\mathcal{B}$,

$$\frac{\partial V(S, t)}{\partial S}\Big|_{S=S_f(t)} = \begin{cases} \leq -1 \text{ for puts} \\ \\ \geq +1 \text{ for calls} . \end{cases} \tag{5.18}$$

We claim that the strict equality holds in (5.18) for both puts and calls. To see this, consider
the problem from the point of view of a trader that is replicating the option by Delta-hedging.
In Section 5.4, we demonstrated that this is possible. In other words the existence of a riskless
hedge implies that the graph of $V(S, t)$ must, for fixed t, be *tangent* to the graph of the intrinsic
value $F(S)$ at the point of contact $S_f(t)$. To see this, consider for example the case of a call.
Clearly, as we have seen in Sections 5.1 and 5.2, $V(S, t) \geq F_{Call}(S)$, the graph of $V(S, t)$ (as

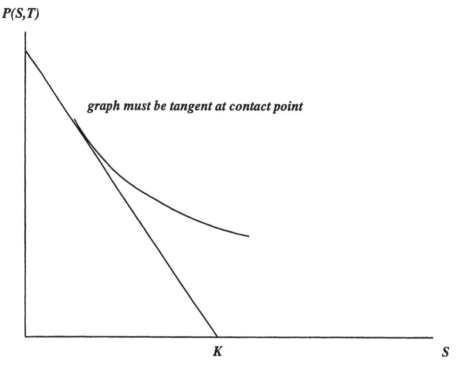

FIGURE 5.4
The graph of an American put is tangent to the graph of the intrinsic value.

a function of S for any fixed t) lies above the graph of $F_{Call}(S)$. This implies that the slope of the tangent to the graph of $V(S, t)$ at the point $S = S(t)$ is *less than* or equal to that of $F_{Call}(S)$ that has slope 1 for $S > K$. So we need to show that strict inequality $\frac{\partial V}{\partial S}|_{S=S_f^-} < 1$ does not hold, where we denote by a superscript $[-]$ the derivative from the left. Here we give a financial argument, based on the stochastic calculus methods developed in Part II, and in the appendix to this chapter we give an argument using PDE's. The argument shows that if $\frac{\partial V}{\partial S}|_{S=S_f^-} < 1$, the writer of the call option will be unable to Delta hedge to protect himself and may suffer unbounded losses.

Indeed, at a point $S = S_f(t)$ on the free boundary, if, in time dt the stock moves up to a value $S > S_f(t)$, it remains inside the exercise region (recall the remarks at the end of Section 5.3) and the value of the call option is equal to its intrinsic value. Thus

$$dV = dS ,$$

and this dictates that the writer of the option must buy $\Delta = 1$ shares to protect himself against an upward tick in the stock price.

Consider, however, his situation if the stock moves down. It has then moved back into the no-exercise region. The change in the stock's value in a small time dt is to leading order

$$dS = \sigma S_f(t) \, dZ \sim \sigma \, S_f \, \epsilon \sqrt{dt} ,$$

where ϵ is the standard normal random variable. The change in the options value in a small time dt, is to leading order

$$dV = \frac{\partial V}{\partial S}\big|_{S=S_f^-(t)} dS$$

$$= \frac{\partial V}{\partial S}\big|_{S=S_f^-(t)} \sigma S_f \epsilon \sqrt{dt} .$$

Since the writer has chosen to hold $\Delta = 1$ unit of stock, the change dP in the value of his portfolio will be $-dC + 1dS$ given by

$$dP = \begin{cases} 0 & \text{for } dS > 0 \\ \sigma S_f \left(\frac{\partial V}{\partial S}\big|_{S=S_{f(t)}^-} - 1 \right) \epsilon \sqrt{dt} & \text{for } dS < 0 . \end{cases}$$

Thus the expected value of the change of his portfolio will be (the mean being taken with respect to the standard Gaussian)

$$\mathbf{E}(dP) = \frac{\sigma S_f^-(t)}{\sqrt{2\pi}} \left(\frac{\partial V}{\partial S}\big|_{S=S_f^-(t)} - 1 \right) \sqrt{dt} , \qquad (5.19)$$

and it is clear that if $\frac{\partial V}{\partial S}\big|_{S=S_f^-(t)}$ is strictly smaller than 1, this loss will be negative on average.

More precisely if, during a time δt, with $dt < \delta t \ll T$, the stock performs a random walk starting at the value $S = S_f(t)$ on the free boundary, the writer will suffer a loss of order

$$O \left(\sigma S_f \sqrt{dt} \frac{\delta t}{dt} \left(\frac{\partial V}{\partial S}\big|_{S=S_f} - 1 \right) \right) ,$$

since the number of transactions possible during such a period is $O(\frac{\delta t}{dt})$. The writer's losses therefore become unbounded as $dt \to 0$. The only possibility, therefore, is that the derivative with respect to S be continuous across $\mathcal{B}$.

The fact that the option has no time-decay across the free-boundary can be deduced immediately from this property and Eq. (5.17). Consider first the case of a put. Differentiating (5.17) with respect to t, we find that

$$\frac{\partial V(S(t), t)}{\partial S} \dot{S}(t) + \frac{\partial V(S(t), t)}{\partial t} = - \dot{S}(t)$$

where $\dot{S}(t)$ represents the derivative of $S(t)$ with respect to t and where we used (5.14). Therefore, using (5.15) and $F'(S_t) = -1$ we obtain

$$\frac{\partial V(S(t), t)}{\partial t} = 0 .$$

The proof of (5.16) for an American call is similar.

The vanishing of Θ along the free boundary means that the agent who sells the American option and hedges using Delta will be *Gamma-neutral* along the free-surface. In fact, the Γ of the option satisfies, from the Black–Scholes equation,

$$0 = \frac{\partial V(S(t), t)}{\partial t}$$

$$= \frac{1}{2} \sigma^2 S(t)^2 \Gamma(S(t), t) + (r - q) S(t) \Delta(S(t), t) - r V(S(t), t) . \qquad (5.20)$$

Hence, if the dynamic hedge is such that the portfolio Delta and the total value of the portfolio are both zero, it follows from (5.19) that $\Gamma = 0$ along the exercise boundary.

The financial interpretation of this result is that if an American option is far from its expiration date, then the hedger faces very little "slippage risk" across the exercise boundary. Of course, the risk increases as $t \rightarrow 0$ due to the fact that $S(t) \rightarrow K$, which is a point of discontinuity of Δ. (This is the usual "pin risk" problem for at-the-money options near expiration.) ∎

References and Further Reading

[1] Barone-Adesi, G. and Whaley, R. (June 1987), Efficient Analytic Approximation of American Option Values, *Journal of Finance*, **42**, pp. 301–20.

[2] Barone-Adesi, G. and Whaley, R. (1987), On the Valuation of American Put Options on Dividend Paying Stocks, *Advances in Futures Options Research*, **3**, pp. 1–14.

[3] Bensoussan, A. (1984), On the Theory of Option Pricing, *Acta Applied Mathematics*, **2**, pp. 139–158.

[4] Black, F. and Scholes, M. (May/June 1973), The Pricing of Options and Corporate Liabilities, *Journal of Political Economy*, **81**, pp. 637–54.

[5] Brennan, M. and Schwartz, E. (1977), The Valuation of American Options, *Journal of Finance*, **32**, pp. 449–462.

[6] Cox, J., Ross, S., and Rubinstein, M. (1979), Option Pricing: A Simplified Approach, *Journal of Finance Economics*, **7**, pp. 229–263.

[7] Geske, R. and Johnson, H.E. (December 1984), The American Put Option Valued Analytically, *Journal of Finance*, **39**, pp. 1511–24.

[8] Harrison, J.M. and Kreps, D. (1979), Martingales and Arbitrage in Multiperiod Security Markets, *Journal of Economic Theory*, **20**, pp. 381–408.

[9] Johnson, H. (1983), An Analytic Approximation of the American Put Price, *Journal of Financial Quantitative Analysis*, **18**, pp. 141–148.

[10] Karatzas, I. (1988), On the Pricing of American Options, *Applied Math and Optimization*, **17**, pp. 37–60.

[11] Lamberton, D. and Lapeyre, B. (1996), *Introduction to Stochastic Calculus Applied to Finance*, Chapman & Hall, London.

[12] MacMillan, L. (1986), Analytic Valuation for the American Put Option, *Advances Futures and Options Research,* **1**, pp. 119–139.

[13] Omberg, E. (1987), The Valuation of American Puts with Exponential Exercise Policies, *Advances in Futures Options Research,* **2**, pp. 117–142.

[14] Parkinson, M. (January 1997), Option Pricing: The American Put, *Journal of Business,* **50**, pp. 21–36.

[15] Samuelson, P. and McKean, H.P. (Spring 1966), Rational Theory of Warrant Pricing: A Free Boundary Problem for the Heat Equation Arising from a Problem in Mathematical Economics, *Industrial Management Review,* **6, no. 2**.

[16] Van Moerbeke, P. (1976), On Optimal Stopping and Free Boundary Problems, *Archive for Rational Mechanics and Analysis,* **60**, pp. 101–148.

[17] Wilmott, P., Dewynne, J.N., and Howison, S.D. (1995), *Option Pricing: Mathematical Methods and Computation,* Oxford Financial Press.

Appendix: A PDE Approach to the Free-Boundary Condition

Condition (5.3) at the beginning of this chapter relative to an arbitrary initial time t, reads

$$V_{am} = \sup_{0 \leq \tau \leq T} \mathbf{E}_t \left\{ e^{-r\tau} Max\,(S_\tau - K,\, 0) \right\}, \qquad (A.1)$$

or equivalently that

$$V_{am} \geq \mathbf{E}_t \left\{ e^{-r\tau} Max\,(S_\tau - K,\, 0) \right\} \text{ for all } \tau : t \leq \tau \leq T \qquad (A.2)$$

where $\mathbf{E}_t$ is the conditional expectation given the information at time t. This inequality can be shown to be equivalent to the partial differential inequality

$$\left\{ \frac{\sigma^2}{2} S^2 \frac{\partial^2}{\partial S^2} + rS \frac{\partial}{\partial S} + \frac{\partial}{\partial t} \right\} \tilde{P} \leq 0 \qquad (A.3)$$

where $\tilde{P}$ is the value of the put discounted to time t

$$\tilde{P} = e^{-rt} P .$$

The equivalence of (A.3) and (A.2) is not surprising, since it is analogous to the equivalence, demonstrated in Chapter 3, of relation (A.2) when there is equality and the Black–Scholes equation, i.e., (A.3) with equality. For a more detailed discussion of this point we refer to the article of Myeni.[6]

---

[6]Actually a precise derivation of (A.3) establishes first that it holds in the sense of distributions or weak solutions.

A.1   A Proof of the Free Boundary Condition

In this appendix we show, in the case of an American put, the equivalence of the PDE (5.2) with the free boundary condition (5.15). In this case the free boundary condition

$$\lim_{S \to S_f(t)} \frac{\partial P}{\partial S}(S, t) = -1$$

is satisfied.

Let $\mathcal{L}$ correspond to the operator

$$\mathcal{L} = \frac{\sigma^2}{2} x^2 \frac{\partial^2}{\partial x^2} + rx \frac{\partial}{\partial x} + \frac{\partial}{\partial t}$$

where we have replaced S by x for simplicity. In this notation (A.3) reads

$$\mathcal{L}[e^{-rt} P(S, t)] \le 0 \qquad\qquad (S, t) \in \Re^+ \times [0, T] . \tag{A.4}$$

To facilitate the analysis, we introduce the new variable $\xi = \ln S$, whose effect is to transform the differential (A.4) into one with constant coefficients.

$$\hat{P}(\xi, t) = P(\xi(x), t) .$$

We get

$$\frac{\sigma^2}{2} \hat{P}_{\xi\xi} \le - \left(r - \frac{\sigma^2}{2} \right) \hat{P}_\xi - \hat{P}_t + r \hat{P} . \tag{A.5}$$

Also note that in the new variables the free boundary condition $P_x = -1$ reads $\hat{P}_\xi = -e^\xi$.

Now we consider a region $\Sigma_\epsilon^{t_1, t_2}$ that is a small strip of width ϵ around the free boundary $S_\xi = \partial\{(\xi, t) : P(x, t) \ge 0\}$, which is the image of the free boundary $S = \partial\{(S, t) : P(S, t) \ge 0\}$ in the (ξ, t) plane. We will integrate the expression (A.5) over $\Sigma_\epsilon^{t_1, t_2}$ and derive the free boundary condition by passing to the limit as $\epsilon \to 0$ and as $t_2 \to t_1$. See Figure A.5. Indicating by $\xi^*(t)$ a generic point on $\Sigma_\epsilon^{t_1, t_2} \cap S_\xi$, this gives

$$\int_{t_1}^{t_2} \frac{\sigma^2}{2} \left[\hat{P}_\xi(\xi^*(t) + \epsilon, t) - \hat{P}_\xi(\xi^*(t) - \epsilon, t) \right] dt$$

$$\le - \int_{t_1}^{t_2} \left(r - \frac{\sigma^2}{2} \right) \left[\hat{P}(\xi^*(t) + \epsilon, t) - \hat{P}(\xi^*(t) - \epsilon, t) \right] dt$$

$$- \int_{\Sigma_\epsilon^{t_1, t_2}} \left(\hat{P}_t + r \hat{P} \right) dt . \tag{A.6}$$

Now to deal with the integral over $\Sigma_\epsilon^{t_1, t_2}$ of P_t we express the latter as an iterated integral. If we integrate P_t first with respect to t along horizontal strips Σ_ξ, denote the vertical strips

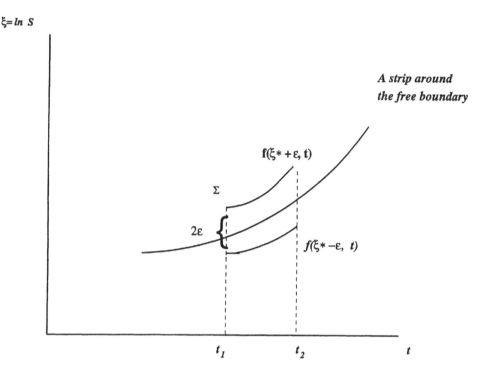

FIGURE A.5
Strip around the free boundary used in the derivation of Equation (A.7).

by Σ_t, and the intersection of a horizontal strip with the free boundary by $t^*(\xi)$ (the inverse function of $\xi^*(t)$), we obtain

$$\int_{\Sigma_\epsilon^{t_1,t_2}} \hat{P}_t = \int_{\Sigma_{t_2}} \hat{P}(\xi, t_2)\, d\xi - \int_{\Sigma_{t_1}} \hat{P}(\xi, t_1)\, d\xi \,.$$

We have used that $P = 0$ on the free boundary to cancel, for each ξ, the contribution at $t^*(\xi)$. When ϵ gets vanishingly small the vertical strips given by the intersection Σ_{t_i}, $i = 1, 2$ are of vanishingly small length. Thus rewriting the inequality (A.6) in the form

$$\int_{t_1}^{t_2} \frac{\sigma^2}{2} \left[\hat{P}_\xi(\xi^*(t) + \epsilon, t) - \hat{P}_\xi(\xi^* - \epsilon, t) \right] dt$$

$$\leq \int_{t_1}^{t_2} \left(r - \frac{\sigma^2}{2} \right) \left[\hat{P}(\xi^* + \epsilon, t) - \hat{P}(\xi^* - \epsilon, t) \right] dt$$

$$+ \int_{\Sigma_{t_2}} \hat{P}(\xi, t_2)\, d\xi - \int_{\Sigma_{t_1}} \hat{P}(\xi, t_1)\, d\xi - \int_{\Sigma_\epsilon^{t_1,t_2}} r \hat{P} d\xi dt \,, \tag{A.7}$$

and letting ϵ tend to zero, we see that the right-hand side of (A.7) vanishes. Now since $\hat{P}_\xi$ is identically equal to $-e^\xi$ on the stopping region, the left-hand side can be written

$$\int_{t_1}^{t_2} \left(\lim_{\xi \to \xi *} \hat{P}_\xi + e^{\xi *} \right) dt \; .$$

(A.8)

Now divide (A.8) by $t_2 - t_1$ and let $t_2 \to t_1$ to recover the pointwise relation

$$\hat{P}_\xi \le -e^{-\xi *}$$

which is equivalent to $P_{S_f} \le -1$.

Also, since $P(S, t) \ge (K - S)^+$, we clearly have that $P_x(S_f(t), t) \ge -1$ and thus $P_x(S_f(t), t) = -1$. This establishes the "smooth-pasting" condition.

Chapter 6

Trinomial Model and Finite-Difference Schemes

6.1 Trinomial Model

This chapter discusses the implementation of multi-period trinomial trees and of numerical techniques associated with these models. Trinomial trees present more flexibility than binomial trees. For example, they allow us to implement dynamic models for asset prices in which the volatility and drift are functions of time and the current index level. Trinomial trees can also be viewed as *finite-difference approximations* to the Black–Scholes partial differential equation and more general PDEs. This will lead to the issue of implementation of numerical schemes, stability conditions, far-field boundary conditions, and implicit schemes.

The one-period trinomial tree is shown in Figure 6.1. This structure is reproduced at each time-step to generate a tree in which each vertex has three "offspring." A necessary and sufficient condition for the tree to recombine is that the parameters U, M, and D satisfy

$$U \cdot D = M^2 . \tag{6.1}$$

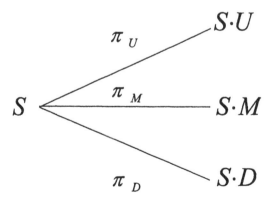

FIGURE 6.1
The one-period trinomial tree.

We assume, without loss of generality, that these parameters are given by the formulas

$$U = e^{\bar{\sigma}\sqrt{dt} + \bar{v}\,dt}$$

$$M = e^{\bar{v}\,dt}$$

$$D = e^{-\bar{\sigma}\sqrt{dt}\,+\,\bar{v}\,dt} \ . \tag{6.2}$$

Here $\bar{\sigma}$ and $\bar{v}$ are constants (with $\bar{\sigma} > 0$) that will be determined later and dt represents a (small) interval of time between successive shocks (measured in years).

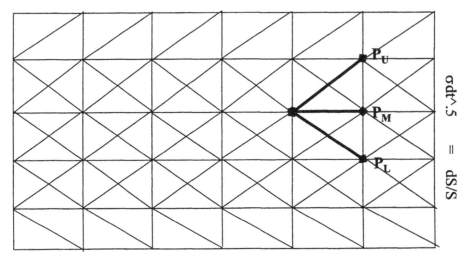

FIGURE 6.2
The multi-period trinomial lattice.

We assign conditional probabilities, P_U, P_M, and P_D, to each of the three outcomes (up, middle, down). Let us denote by $v\,dt$ and $\sigma^2\,dt$ the mean and variance of the logarithm of the price change over one period. Then, we have,

$$v\,dt \ = \ \bar{\sigma}\sqrt{dt} \cdot (P_U - P_D) + \bar{v}\,dt \tag{6.3}$$

and

$$\sigma^2\,dt = \bar{\sigma}^2\,dt \cdot (P_U + P_D) - (v\,dt)^2$$

$$= \bar{\sigma}^2\,dt \cdot (P_U + P_D) + O(dt^{3/2}) \ . \tag{6.4}$$

We will always assume that $dt \ll 1$ and that the parameters σ and v are $O(1)$. Under these conditions, we are only interested in keeping terms of order dt in the calculation of the mean and variance.[1] In particular, we will replace Eq. (6.4) by the simpler equation

$$\sigma^2\,dt \ = \ \bar{\sigma}^2\,dt \cdot (P_U + P_D) \ . \tag{6.5}$$

Equations (6.3) and (6.5) can be used to calculate the probabilities P_U, P_M, and P_D in terms of the mean and the variance of the logarithm of the shock over a period. In fact, we can easily

---

[1]Lower-order terms of order $dt^{3/2}$ give a negligible contribution to the sum of the local variances.

deduce from (6.3) and (6.5) that

$$P_U = \frac{1}{2}\left[\frac{\sigma^2}{\bar{\sigma}^2} + \left(\frac{\nu - \bar{\nu}}{\bar{\sigma}}\right)\sqrt{dt}\right],$$ (6.6)

$$P_D = \frac{1}{2}\left[\frac{\sigma^2}{\bar{\sigma}^2} - \left(\frac{\nu - \bar{\nu}}{\bar{\sigma}}\right)\sqrt{dt}\right],$$ (6.7)

and

$$P_M = 1 - \frac{\sigma^2}{\bar{\sigma}^2}.$$ (6.8)

Financial considerations typically require adjusting the drift $\mu(S, t)$ of the process in (6.1) to satisfy no-arbitrage conditions. For instance, in the world of currency derivatives, $\mu = r_d - r_f =$ the difference between the domestic and foreign interest rates. Thus, in practice, we would like to treat μ (i.e., the annualized expected return) as the input parameter rather than ν. Of course, we know from the results of Chapter 3,

$$\nu = \frac{d\mathbf{E}\{\log S_t\}}{dt} = \mu - \frac{1}{2}\sigma^2.$$ (6.9)

Let us rewrite Eqs. (6.6) and (6.7) in terms of μ. The result is

$$P_U = \frac{1}{2}\frac{\sigma^2}{\bar{\sigma}^2}\left(1 - \frac{\bar{\sigma}\sqrt{dt}}{2}\right) + \left(\frac{\mu - \bar{\nu}}{2\bar{\sigma}}\right)\sqrt{dt}$$ (6.10)

and

$$P_D = \frac{1}{2}\frac{\sigma^2}{\bar{\sigma}^2}\left(1 + \frac{\bar{\sigma}\sqrt{dt}}{2}\right) - \left(\frac{\mu - \bar{\nu}}{2\bar{\sigma}}\right)\sqrt{dt}.$$ (6.11)

6.2   Stability Analysis

We would like to derive conditions on the parameters so that (P_U, P_M, P_D) are probabilities, i.e., are nonnegative and have sum 1. Introducing the parameters

$$p \equiv \frac{\sigma^2}{\bar{\sigma}^2}, \quad q \equiv \frac{\mu - \bar{\nu}}{\bar{\sigma}^2},$$ (6.12)

the equations for the probabilities become

$$P_U = \frac{p}{2}\left(1 - \frac{\bar{\sigma}\sqrt{dt}}{2}\right) + q\frac{\bar{\sigma}\sqrt{dt}}{2},$$

$$= \frac{p}{2} - \left(\frac{q}{2} - \frac{p}{4}\right)\bar{\sigma}\sqrt{dt}$$ (6.13)

$$P_D = \frac{p}{2} \left(1 + \frac{\overline{\sigma} \sqrt{dt}}{2} \right) - q \frac{\overline{\sigma} \sqrt{dt}}{2} ,$$

$$= \frac{p}{2} + \left(\frac{q}{2} - \frac{p}{4} \right) \overline{\sigma} \sqrt{dt} \tag{6.14}$$

and

$$P_M = 1 - p . \tag{6.15}$$

These expressions are useful to determine whether P_U, P_M, and P_D are probabilities. We have the following result:

PROPOSITION 6.1
("CFL condition") *Given the parameters* σ, μ, $\overline{\sigma}$, $\overline{\nu}$, *and* dt, *we have* $P_U \geq 0$, $P_D \geq 0$, $P_M \geq 0$ *if and only if*

$$\sigma \leq \overline{\sigma} \tag{6.16}$$

and

$$\overline{\sigma} \sqrt{dt} < \frac{\sigma^2}{|\mu - \overline{\nu} - \frac{1}{2}\sigma^2|} . \tag{6.17}$$

PROOF   Condition (6.16) follows from (6.15) and the fact that $P_M \geq 0$ must hold. It is then seen immediately from (6.13) and (6.14) that the condition ensuring that (P_U, P_M, P_D) are probabilities reduces to

$$\overline{\sigma} \sqrt{dt} < \frac{p}{|q - \frac{p}{2}|} . \tag{6.18}$$

Inequality (6.17) is obtained by substituting the values of p and q from Eq. (6.12).[2]   ∎

REMARK 6.1   We interpret the constants $\overline{\sigma}$, $\overline{\nu}$, and dt as "lattice parameters" that define the tree structure. The volatility σ and drift μ at a given node are determined by the parameters p and q and vice-versa. One observes, therefore, that a trinomial tree allows us to specify an entire range of parameters σ and μ at any give node, without changing the underlying tree structure. This range corresponds to those values of p and q such that the CLF condition is satisfied. In contrast, as we saw in previous chapters, the binomial tree does not have this flexibility.   ∎

6.3   Calibration of the Model

For $\sigma = \sigma(S, t)$ and $\mu = \mu(S, t)$, we derive a criterion for choosing $\overline{\sigma}$, $\overline{\nu}$, and dt consistent with the CFL condition. For this, we consider the four numbers

$$\sigma_{min} \equiv \inf_{S,t} \sigma(S, t) , \quad \sigma_{max} \equiv \sup_{S,t} \sigma(S, t) \tag{6.19a}$$

---

[2]This last proposition is a version of the famous Courant–Friedrichs–Lewy (CFL) stability condition for finite-difference schemes.

$$\mu_{min} \equiv \inf_{S,t} \mu(S, t) \text{ and } \mu_{max} \equiv \sup_{S,t} \mu(S, t). \qquad (6.19b)$$

We shall assume that $\sigma_{min} > 0$.[3] We shall derive conditions on the parameters $\overline{\sigma}$, $\overline{v}$, and dt that lead to positive probabilities.

We first determine $\overline{\sigma}$. Since Eq. (6.5) tells us that

$$\sigma^2 = p \cdot \overline{\sigma}^2, \quad 0 \le p \le 1, \qquad (6.20)$$

we must have $\sigma^2 \le \overline{\sigma}^2$. One possible choice for $\overline{\sigma}$ is

$$\overline{\sigma} = \sigma_{max},$$

which is the smallest value of $\overline{\sigma}$ compatible with the range of $\sigma(S, t)$. Of course, any choice of $\overline{\sigma}$ greater than σ_{max} is also acceptable.[4]

Let us turn next to condition (6.17).

PROPOSITION 6.2

Suppose that $\sigma(S, t)$ and $\mu(S, t)$ are given and let μ_{min}, μ_{max}, σ_{min}, σ_{max} be defined as in (6.19a), (6.19b). A sufficient condition that ensures that probabilities are positive is

$$\begin{cases} dt \le \left(\dfrac{1}{\overline{\sigma}^2} \dfrac{2\sigma_{min}^2}{\mu_{max} - \mu_{min}} \right)^2 & \text{if } \sigma_{min}^2 < \mu_{max} - \mu_{min} \\[2em] dt \le \dfrac{1}{\overline{\sigma}^2} \left(\dfrac{1 + \frac{\sigma_{max}^2}{\sigma_{min}^2}}{\frac{\mu_{max} - \mu_{min}}{\sigma_{min}^2} + \frac{1}{2}\frac{\sigma_{max}^2}{\sigma_{min}^2}} \right)^2 & \text{if } \sigma_{min}^2 \ge \mu_{max} - \mu_{min} \end{cases} \qquad (6.21a)$$

with

$$\overline{v} = \begin{cases} \dfrac{\mu_{max} + \mu_{min} - \sigma_{min}^2}{2} & \text{if } \sigma_{min}^2 < \mu_{max} - \mu_{min} \\[2em] \dfrac{\frac{\mu_{min}}{\sigma_{max}^2} + \frac{\mu_{min}}{\sigma_{max}^2} \frac{1}{\sigma_{max}^2}}{\frac{1}{\sigma_{max}^2} + \frac{2}{\sigma_{max}^2}} & \text{if } \sigma_{min}^2 \ge \mu_{max} - \mu_{min}. \end{cases} \qquad (6.21b)$$

PROOF The proof follows easily from (6.17) written in the form

$$\sqrt{dt} \le \frac{1}{\overline{\sigma}} \frac{1}{\left| \frac{\mu - \overline{v}}{\sigma^2} - \frac{1}{2} \right|}.$$

---

[3]This last requirement is not necessary—in fact, it is possible to model using the trinomial tree processes with a local volatility that vanishes at certain levels of spot prices.

[4]We note that the greater the value of $\overline{\sigma}$, the smaller we must choose dt in order to ensure that (6.17) holds. Thus, larger $\overline{\sigma}$ implies longer computing times. This would suggest taking $\overline{\sigma}$ as small as possible. On the other hand, given a value of the volatility σ at a node, the use of $\overline{\sigma} > \sigma$ may give a more accurate numerical approximation to the PDE and hence be more desirable.

Indeed we write the denominator as

$$\max\left(\left(\frac{\mu - \bar{\nu}}{\sigma^2} - \frac{1}{2}\right)^+, \left(\frac{\bar{\nu} - \mu}{\sigma^2} + \frac{1}{2}\right)^+\right)$$

and note that

$$\left(\frac{\mu - \bar{\nu}}{\sigma^2} - \frac{1}{2}\right)^+ \leq \frac{\mu_{max} - \bar{\nu}}{\sigma_{min}^2} - \frac{1}{2} \text{ if } \bar{\nu} < \mu_{max}$$

and

$$\left(\frac{\mu - \bar{\nu}}{\sigma^2} - \frac{1}{2}\right)^+ \leq \begin{cases} \frac{\bar{\nu} - \mu_{min}}{\sigma_{max}^2} + \frac{1}{2} & \text{ if } \bar{\nu} < \mu_{min} \\[2ex] \frac{\bar{\nu} - \mu_{min}}{\sigma_{min}^2} + \frac{1}{2} & \text{ if } \bar{\nu} \geq \mu_{min} . \end{cases}$$

To proceed we now distinguish, as indicated in Proposition 6.2, the two cases $\mu_{min} < \mu_{max} - \sigma_{min}^2$ and $\mu_{min} > \mu_{max} - \sigma_{min}^2$. (The two additional cases $\mu_{min} < \mu_{max} - \frac{\sigma_{min}^2}{2}$ and $\mu_{min} \geq \mu_{max} - \frac{\sigma_{min}^2}{2}$ also require consideration but, as it turns out, do not affect the optimal value of $\bar{\nu}$.) We then consider the maximum of the three piecewise linear functions as a function of $\bar{\nu}$ and then take the minimum height of the graph over all possible values of the variable $\bar{\nu}$. ∎

In practice, we often consider trinomial trees with $\bar{\nu} = 0$. These models correspond to the symmetric case $U = D^{-1}$. They are important when considering barrier options, for example (see Chapter 8).

PROPOSITION 6.3
Suppose that σ, ν, and $\bar{\sigma}$ are given and that $\bar{\nu} = 0$. Then a sufficient condition for having positive probabilities is

$$dt \leq \frac{1}{\bar{\sigma}^2}\left(\frac{1}{\max\left(\frac{\mu_{max}}{\sigma_{min}^2} - \frac{1}{2}, \frac{1}{2} - \frac{\mu_{min}}{\sigma_{max}^2}\right)}\right)^2 .$$

This condition is equivalent to

$$dt \leq \begin{cases} \frac{1}{\bar{\sigma}^2}\left(\frac{1}{\frac{\mu_{max}}{\sigma_{min}^2} - \frac{1}{2}}\right)^2 & \text{if } 1 < \frac{\mu_{max}}{\sigma_{min}^2} + \frac{\mu_{min}}{\sigma_{max}^2} \\[3ex] \frac{1}{\bar{\sigma}^2}\left(\frac{1}{\frac{1}{2} - \frac{\mu_{min}}{\sigma_{max}^2}}\right)^2 & \text{if } 1 \geq \frac{\mu_{max}}{\sigma_{min}^2} + \frac{\mu_{min}}{\sigma_{max}^2} . \end{cases} \tag{6.22}$$

To construct a trinomial tree with volatility and drift parameters $\sigma = \sigma(S, t)$ and $\mu = \mu(S, t)$, we calculate σ_{min}, σ_{max}, μ_{min}, and μ_{max} and define the parameters $\bar{\sigma}$ and $\bar{\nu}$ so that the resulting probabilities are positive. This is done by choosing $\bar{\sigma} \geq \sigma_{max}$ and dt so that conditions (6.21a) and (6.21b) or (6.22) hold (according to whether we optimize over $\bar{\nu}$ or take

$\bar{\nu} = 0$). Once this is done, we obtain a trinomial tree, or lattice, that describes the values of the index S in discrete increments. These values can be denoted by S_n^j, where n represents the time variable and j the "height" on the tree.

Next, we discretize the drift and volatility surfaces, setting

$$\sigma_n^j \equiv \sigma(S_n^j, t_n) , \quad \mu_n^j \equiv \mu(S_n^j, t_n) .$$

Finally, we set

$$p_n^j = \frac{\left(\sigma_n^j\right)^2}{\bar{\sigma}^2} ,$$

$$q_n^j = \frac{\mu_n^j - \bar{\nu}}{\bar{\sigma}^2}$$

and define the probabilities at the node (n, j) according to Eqs. (6.13), (6.14), and (6.15), substituting p_n^j for p and q_n^j for q. In this way, we have specified a discrete approximation for the diffusion process (6.1) on a trinomial tree.

Finally, we write down the pricing equation for general (European-style) contingent claims. For this purpose, let r_n^j represent the riskless rate for the time interval (t_n, t_{n+1}) conditional on $S = S_n^j$, and let F_n^j represent the cash-flow that this security pays at time t_n given that $S = S_n^j$. Then, the value function satisfies

$$V_n^j = F_n^j + e^{-r_n^j dt} \left[(P_U)_n^j V_{n+1}^{j+1} + (P_M)_n^j V_{n+1}^j + (P_D)_n^j V_{n+1}^{j-1} \right]$$

$$= F_n^j + e^{-r_n^j dt} \cdot \left[\frac{1}{2} p_n^j \left(1 - \frac{\bar{\sigma}\sqrt{dt}}{2}\right) V_{n+1}^{j+1} + \frac{1}{2} p_n^j \left(1 + \frac{\bar{\sigma}\sqrt{dt}}{2}\right) V_{n+1}^{j-1} \right.$$

$$\left. + (1 - p_n^j) V_{n+1}^j + \frac{1}{2} q_n^j \bar{\sigma} \sqrt{dt} \left(V_{n+1}^{j+1} - V_{n+1}^{j-1}\right) \right] \qquad (6.23)$$

(cf. Section 5.5, Chapter 5). This equation can be viewed as a *finite-difference scheme* for solving the Black–Scholes equation. The construction of the previous sections guarantees the stability and consistency of the scheme.

Of course, dynamic programming equations derived from (6.23) can be used for pricing American options and other contingent claims that involve stopping times. In particular, it is noteworthy that the trinomial tree provides a more accurate alternative than the binomial model for the pricing of barrier options, because the barrier can be made to coincide with a particular level in the tree. This eliminates to some extent numerical roundoff error.[5]

---

[5]Other applications involve dynamical programming equations that are used in worst-case scenario analysis of portfolios (Avellaneda et al. [1995] and Avellaneda and Parás [1996]).

6.4 "Tree-Trimming" and Far-Field Boundary Conditions

The number of nodes in the trinomial tree at date $t_n = n \, dt$ is $2n + 1$ (this can be verified easily by induction). Therefore, the total number of nodes is

$$1 + 3 + 5 + \ldots + 2N + 1 = (N + 1)^2 \, .$$

In this section, we show that the trinomial scheme can be implemented in a rectangular region (as opposed to a "tree"), which has a number of nodes of order $N^{3/2}$ rather than N^2, without sacrificing accuracy. We will make use of the fact that the probability that the price S_{t_n} reaches extremely high or extremely low values is negligible.

Recall that for $dt \ll 1$ the distribution of $\ln S$ is approximately normal. Let us assume for simplicity that $v = 0$, so that $\ln S$ has approximately a centered normal distribution. In this case, we have

$$\text{Prob.} \left\{ |\ln S_{t_N}| > a \right\} \propto \exp\left\{ -\frac{a^2}{2\sigma \, t_N} \right\} = \exp\left\{ -\frac{a^2}{\sigma \, N \, dt} \right\}$$

Hence, for any real number x,

$$\text{Prob.} \left\{ |\ln S_{t_N}| > x \overline{\sigma} \sqrt{N \, dt} \right\} \propto \exp\left\{ -\frac{x^2 \overline{\sigma}^2 \, N \, dt}{2\sigma^2 \, N \, dt} \right\}$$

$$\leq \exp\left\{ -\frac{x^2}{2} \right\} \, .$$

If we choose $x = 3.5$, for example,

$$\text{Prob.} \left\{ |\ln S_{t_N}| > x \sqrt{N \, dt} \, \overline{\sigma} \sqrt{dt} \right\} \approx \exp\left\{ -\frac{3.5^2}{2} \right\} \approx 0.0022 \, .$$

With $x = 5$, the probability is approximately equal to $3.7 \cdot 10^{-6}$.

Based on these considerations, we consider a rectangular domain corresponding to the range of spot prices

$$S_n^j \, , \quad -M \leq j \leq +M \, , \quad M = \text{integer part of } x \sqrt{N} \, .$$

This domain has therefore $(N + 1) \times (2M + 1)$ nodes, i.e., $O(N^{1.5})$ as opposed as $O(N^2)$. For a tree with 1000 time steps, trimmed with 3.5 standard deviations, the number of nodes decreases from approximately 10^6 to approximately $2 \cdot 10^5$, a gain of a factor of 5 (Figure 6.3).

If the logarithm of the underlying price has a nonzero mean, trimming should be done relative to the "center" of the distribution. Assuming a constant v for simplicity, we find that the "far-field" boundaries should correspond roughly to

$$v T - x \overline{\sigma} \sqrt{T} < S < v T + x \overline{\sigma} \sqrt{T} \, ,$$

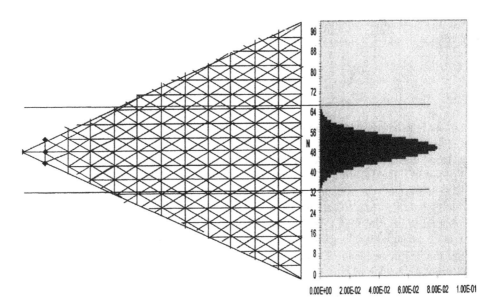

FIGURE 6.3
The tree-trimming technique saves computational time. Only nodes that are at a distance less than 3.5 standard deviations from the mean of the distribution are used in the computation.

where T is the final time. Recalling that $T = N\,dt$ and dividing by $\overline{\sigma}\,\sqrt{dt}$, we obtain a possible range for j, namely

$$\left(\frac{\nu\sqrt{T}}{\overline{\sigma}} - x\right)\sqrt{N} < j < \left(\frac{\nu\sqrt{T}}{\overline{\sigma}} + x\right)\sqrt{N}.$$

If the contribution from the drift is significant, the computational domain has to be constructed taking into account the fact that the center of the distribution drifts away from the level $j = 0$ as t increases.

Trimming should be done in such a way that the probability of exiting the domain is negligible. For example, if $\nu = .10$, $\overline{\sigma} = .20$, and $T = 2$, we have $\frac{\nu\sqrt{T}}{\overline{\sigma}} = \frac{\sqrt{2}}{2} = 0.74$. If we take the trimming constant to be $x = 5$, we obtain the bounds

$$-4.26\,\sqrt{N} < j < 5.74\,\sqrt{N},$$

which means that the contribution due to the drift is not very important. However, if T is large, e.g., $T = 25$, we have $\frac{\nu\sqrt{T}}{\overline{\sigma}} = 2.5$, so one must take care to take the domain wide enough to avoid boundary effects. The effect of the drift is also important in situations when interest rates are very high.

One possible way to avoid the "drift effect" when ν is large is to use $\overline{\nu} \approx \nu$. This is equivalent to a change of variables on the price process, in which one considers the evolution of $S_t\,e^{-\overline{\nu}t}$ rather than S_t.

We turn next to the solution of the pricing equation (6.23). We interpret the backward-induction equation as a linear operator, i.e.,

$$\mathbf{V}_n = A_n \mathbf{V}_{n+1} , \qquad \mathbf{V}_n = \left(v_n^j \right)_{j=-M}^{M} , \qquad n = 0, 1, \ldots , N \qquad (6.24)$$

where A_n is a matrix. Notice, however, that Eq. (6.23) does not give the values of $\mathbf{V}_n$ at the extremal nodes $(n, -M)$ and (n, M), V_n^{-M} and V_n^M. In fact, the backward-induction equation (6.23) only determines the entries of the vector $\mathbf{V}_n$ at those nodes (n, j) for which the values of $\mathbf{V}_{n+1}$ at its three offspring are known. These nodes are such that $-M < j < M$.

To remedy this, we need to specify *boundary conditions* for $\mathbf{V}_n$ at the extreme nodes $(n, -M)$ and $(n, +M)$. Recall that the boundaries were originally chosen so that they would correspond to the "far-field." One possible approach is to identify the boundary conditions with these asymptotic values of the true solution for $S \ll 1$ or $S \gg 1$. In many cases, these values can be guessed without solving the equation. For example, in the case of a call option with strike K on an asset with dividend yield d, assuming a risk less rate r, we have

$$C(S, K, t, T) \approx \begin{cases} S e^{-d(T-t)} - K e^{-r(T-t)} & S \gg 1 \\ 0 & S \ll 1 , \end{cases}$$

a relation that follows from the fact that the value of a deep-in-the money call is asymptotically equal to the forward price. Accordingly, the boundary conditions could be

$$V_n^M = e^{-d \cdot dt (N-n)} S_n^M - e^{-r \cdot dt (N-n)} K$$

and

$$V_n^{-M} = 0 .$$

We call these boundary conditions **asymptotic** boundary conditions.

A different approach, which we call the **zero-Gamma** boundary condition, is based on the idea that the BS equation is sensitive to convexity. Accordingly, we would like to choose the boundary values in such way that there are no spurious sources of convexity at the boundaries. This is tantamount to assuming that

$$\frac{V_n^M - V_n^{M-1}}{S_n^M - S_n^{M-1}} = \frac{V_n^{M-1} - V_n^{M-2}}{S_n^{M-1} - S_n^{M-2}} ,$$

and

$$\frac{V_n^{-M} - V_n^{-M+1}}{S_n^{-M} - S_n^{-M+1}} = \frac{V_n^{-M+1} - V_n^{-M+2}}{S_n^{-M+1} - S_n^{-M+2}} .$$

Solving for V_n^M and V_n^{-M} in the latter equations, we obtain the boundary conditions

$$V_n^M = \left(\frac{U - D}{1 - D} \right) V_n^{M-1} - \left(\frac{U - 1}{1 - D} \right) V_n^{M-2} , \qquad (6.25a)$$

$$V_n^{-M} = \left(\frac{U - D}{U - 1} \right) V_n^{-M+1} - \left(\frac{U - 1}{U - 1} \right) V_n^{-M+2} . \qquad (6.25b)$$

The advantage of zero-Gamma boundary conditions is that they do not require us to determine (and program) the asymptotic values of the option or of the portfolio of options under consideration (Figure 6.4). At the same time, they give excellent numerical approximations to solutions of the Black–Scholes-type equations in unbounded domains, often better than the ones derived with asymptotic boundary conditions.[6]

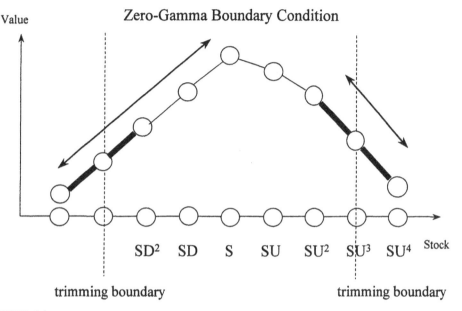

FIGURE 6.4
The "zero-Gamma" boundary condition is such that the numerical derivative of the function does not change across the boundary. More precisely, the values of the function at external boundary nodes are obtained by linear extrapolation.

6.5 Implicit Schemes

The backward-induction relation (6.23) with the appropriate boundary conditions [e.g., (6.25a)–(6.25b)] constitutes an **explicit** scheme for solving the Black–Scholes differential equation. The term "explicit" stems from the fact that the values are derived inductively from one step to the next, as symbolized in (6.24). The quantities P_U, P_M, P_D are conditional probabilities and the scheme computes the discounted expected value of cash-flows recursively.

Another, more general, point of view is to consider P_U, P_M, P_D as "weights" such that the matrix

$$(L\,V)^j \;=\; \frac{P_U\,V^{j+1} + P_M\,V^j + P_D\,V^{j-1} - V^j}{dt}$$

approximates the partial differential operator

$$\mathcal{L}\,V \;=\; \frac{\sigma^2\,S^2}{2} V_{SS} + \mu\,S\,V_S \,.$$

---

[6]See, for instance, Robert Buff's *Virtual option pricer* at the URL http://marco-pc.cims.nyu.edu.

Notice that we can write the solution of the final-value problem for the Black–Scholes PDE as

$$V(\cdot, t) = e^{-r(T-t)} e^{(T-t)\mathcal{L}} V(\cdot, T) .$$

(6.26)

The explicit backward-induction scheme

$$\mathbf{V}_n = e^{-rdt} (\mathbf{I} + dt\, L)\, \mathbf{V}_{n+1} ,$$

(6.27)

gives, after iteration,

$$\mathbf{V}_n = e^{-(N-n)rdt} (\mathbf{I} + dt\, L)^{N-n}\, \mathbf{V}_N$$

$$= e^{-r(T-t)} (\mathbf{I} + dt\, L)^{(T-t)/dt}\, \mathbf{V}_N ,$$

(6.28)

which is an approximation to (6.26).

At first sight, the latter viewpoint does not require that the weights be positive and less than one – only that they be consistent with the PDE for $dt \ll 1$. However, an important consideration to avoid exponential divergence of $(\mathbf{I} + dt\, L)^{N-n}$ is that the eigenvalues of $\lambda_k, k = 1, 2, \ldots, 2M+1$ of the matrix $\mathbf{I} + dt\, L$ satisfy the condition

$$|\lambda_k|^2 = Re(\lambda_k)^2 + Im(\lambda_k)^2 \leq 1 .$$

(6.29)

If this condition is not satisfied, the powers of $\mathbf{I} + dt\, L$ increase exponentially and the scheme is *unstable*.

We must therefore ascertain under which conditions (6.29) holds. The analysis of the eigenvalues of the matrix $\mathbf{I} + dt\, L$ is straightforward if we assume periodic boundary conditions.[7] In this case, we obtain

$$\lambda_k = P_U\, e^{-ix} + P_M + P_D\, e^{ix}$$

$$= 1 - (P_U + P_D)(1 - \cos(x)) + i(P_D - P_U)\sin(x) ,$$

$$= 1 - p(1 - \cos(x)) - i\left(q - \frac{p}{2}\right)\sin(x)\,\overline{\sigma}\,\sqrt{dt}$$

(6.30)

where

$$x = \frac{2\pi k\,\overline{\sigma}\,\sqrt{dt}}{2M + 1} \quad -M \leq k \leq M .$$

The reader can easily verify from the first line of (6.30) that the eigenvalue condition (6.29) is satisfied if P_u, P_M, P_D are nonnegative. Conversely, for any $x \neq 0$, the spectral condition (6.29) will be violated if one of the probabilities is negative. Therefore, *for the one-dimensional trinomial model, stability of the explicit scheme is equivalent to the positivity of the probabilities P_u, P_M, and P_D.*[8]

---

[7]See, for instance, Bellman [1970]. The consideration of periodic boundary conditions simplifies matters and is sufficient for our purposes, as we shall see.

[8]For two-dimensional explicit finite-difference schemes, the notions of stability (CFL condition) and positive conditional probabilities are not equivalent.

As shown in the previous section, the positivity condition and hence the spectral condition are always satisfied by taking dt sufficiently small. In practice, however, this may lead to unnecessarily small mesh sizes and long computational times. To solve the PDE numerically *without* satisfying the spectral condition (6.29), we use an **implicit scheme**. A typical implicit scheme has the form

$$\mathbf{V}_n = e^{-r\,dt}\,(\mathbf{I} - \theta\,dt\,L\,)^{-1} \cdot (\mathbf{I} + (1-\theta)\,dt\,L\,)\,\mathbf{V}_{n+1}, \tag{6.31}$$

where $0 < \theta < 1$. Observe that the matrix $\mathbf{I} - \theta\,dt\,L$ is invertible because, from (6.30), the eigenvalues λ_k satisfy $Re(\lambda_k) \le 1$, which implies that the eigenvalues of $\mathbf{I} - \theta\,dt\,L$ are

$$1 - \theta\,(\lambda_k - 1) = (1+\theta) - \theta\,\lambda_k$$

and cannot vanish if $\theta > 0$.

Notice that the scheme (6.31) is consistent with the underlying partial differential equation (6.30) for $dt \ll 1$, since we have

$$(\mathbf{I} - \theta\,dt\,L\,)^{-1} = \mathbf{I} + \theta\,dt\,L + \theta^2\,dt^2\,M^2 + o(dt^2)$$

for $dt \ll 1$. It follows that

$$(\mathbf{I} + \theta\,dt\,L\,)^{-1} \cdot (\mathbf{I} + (1-\theta)\,dt\,L\,) = \mathbf{I} + dt\,L + \theta\,dt^2\,L^2 + o(dt^2). \tag{6.32}$$

Thus, the implicit scheme (6.31) and the explicit scheme (6.27) agree to first order in dt and represent the solution of the same PDE [cf. Eq. (6.26)] as dt tends to zero. At the same time, the eigenvalues of the matrix on the left-hand side of (6.31) are

$$\Lambda_k = \frac{\theta + (1-\theta)\,\lambda_k}{1 + \theta - \theta\,\lambda_k}.$$

The scheme will be stable if the eigenvalues Λ_k satisfy the spectral condition $|\Lambda_k| \le 1$. We can therefore choose θ to our advantage. The key observation here is that the rational function

$$\lambda \to \Lambda = \frac{\theta + (1-\theta)\,\lambda}{(1+\theta) - \theta\lambda}$$

maps the half-plane $\{\lambda :\ Re(\lambda) < 1\}$ onto the circle of center $\left(\frac{2\theta - 1}{\theta},\ 0\right)$ passing through the point $(1, 0)$. The radius of this circle is $\frac{1}{2\theta}$.

We conclude from this that the spectrum of the matrix corresponding to the scheme (6.31) is contained in the unit disk $\{\Lambda :\ |\Lambda| \le 1\}$ provided that $\theta \ge 1/2$. The value $\theta = \frac{1}{2}$ is particularly significant because it leads to a stable scheme and, furthermore, the right-hand side of (6.32) agrees with $e^{L\,dt}$ to second order in dt, approximating the operator $e^{dt\,L}$ more accurately than if $\theta \ne 1/2$. The case $\theta = 1/2$ is the **Crank–Nicholson scheme**. It gives excellent approximations for the solutions of the Black–Scholes PDE without need to enforce the CFL condition.

References and Further Reading

[1] Avellaneda, M., Levy, A., and Parás, A. (1995), Pricing and Hedging Derivative Securities in Markets with Uncertain Volatilities, *Applied Mathematical Finance,* **2**, pp. 73–88.

[2] Avellaneda, M. and Parás, A. (1996), Dynamic Hedging with Transaction Costs: Lattices Models, Non-Linear Volatility and Free-Boundary Problems, *Communications in Pure and Applied Math.*

[3] Bellman, R. (1970), *Introduction to Matrix Analysis,* Mc-Graw Hill, New York.

[4] Dewynne, J., Howison, S. and Wilmott, P. (1993), *Option Pricing: Mathematical Models and Computation,* Oxford Financial Press, Oxford.

[5] Parás, A. and Avellaneda, M. (1996), Managing the Volatility Risk of Portfolios of Derivative Securities: The Lagrangian Uncertain Volatility Model, *Applied Mathematical Finance,* **3**, pp. 21–52.

[6] Press, W. (1988), *Numerical Recipes in C: The Art of Scientific Computing,* University Press, Cambridge.

Chapter 7

Brownian Motion and Ito Calculus

Brownian motion is the mathematical model of a *continuous random walk*. K. Ito's stochastic calculus is a collection of tools that permit us to perform operations—such as composition, integration and differentiation—on functions of Brownian paths and more general random functions known as *Ito processes*. As we shall see, Ito calculus and Ito processes are extremely useful in the formulation of financial risk-management techniques. These notes are intended to introduce the reader to Brownian motion and stochastic calculus in a straightforward, intuitive way. For rigorous treatments of this rich subject, the reader can consult Ikeda and Watanabe [6], Varadhan [14] or Karatzas and Shreve [8].

7.1 Brownian Motion

Intuitively, Brownian motion corresponds to the concept of a *homogeneous, continuous-time, continuous random walk*. One way to visualize Brownian paths is to consider a simple random walk on the real line, in which the walker starts at position $X_0 = 0$ and moves up or down by an amount $\sqrt{dt}$ after each time interval of duration dt. If X_n denotes the position of the walker after the nth jump, we have

$$X_n = X_{n-1} \pm \sqrt{dt} , \quad n = 1, 2, \ldots \tag{7.1}$$

where the [+] and [−] signs occur with probability $1/2$. This process is called a simple random walk. Note that the magnitude of the jump and the lag between successive jumps are chosen so that the variance of the displacement of the walker after time T (where T is an integer multiple of dt) is exactly T.

A continuous path can be built from the variables X_n by interpolating linearly between the different points:

$$\overline{X}(t) = X_n + (t - n\,dt) \cdot (X_{n+1} - X_n) , \quad \text{for } n\,dt \le t \le (n+1)\,dt . \tag{7.2}$$

These paths have the following properties:

1. If $t = n\,dt$ and $a > 0$, the increment $\overline{X}(t + a) - \overline{X}(t)$ is independent of the "past" $\{\overline{X}(s) , \; s \le t\}$;

2. $\mathbf{E}\left(\overline{X}(t)\right) = 0$;

3. $\mathbf{E}\left(\overline{X}(t)^2\right) = t$.

For small time increments ($dt \ll 1$), these paths have nearly *independent increments* (neglecting the small persistence effect due to the linear interpolation). Moreover, the mean and variance of the walkers displacement are independent of dt.

For $dt \ll 1$, each increment of $\overline{X}(t)$ is a sum of many independent binomial random variables with mean zero and finite variance. Therefore, by the Central Limit Theorem, the limiting probability distribution of the increments of $\overline{X}(t)$ as $dt \to 0$ is Gaussian (or normal). More precisely, we have

$$\lim_{dt \to 0} \mathbf{P}\left\{\overline{X}(t+a) - \overline{X}(t) \geq x\right\} = \frac{1}{\sqrt{2\pi a}} \int_x^{+\infty} e^{-\frac{y^2}{2a}}\, dy$$

for all x. This property, together with the independence of the increments, characterizes the statistics of the paths $\overline{X}(t)$ in the limit.

This discussion motivates the following:

DEFINITION 7.1   *A one-dimensional Brownian motion is a real-valued stochastic process $Z(t), 0 \leq t < \infty$, with the following properties:*

1.   *$Z(0) = 0$ with probability 1;*

2.   *For all $t > 0$ and $a > 0$, the increments $Z(t+a) - Z(t)$ are Gaussian with mean zero and variance a; and*

3.   *$Z(t+a) - Z(t)$ is independent of $\{ Z(s), 0 \leq s \leq t \}$.*

Items 1 – 3 completely specify the probability distribution of any n-tuple

$$(Z(t_1), \ Z(t_2), \ Z(t_3), \ \ldots \ Z(t_n))$$

where $t_1 < t_2 < \cdots < t_n$ are arbitrary times. This distribution is a (centered) multivariate Gaussian[1] with covariance

$$\mathbf{E}\left\{ Z(t_i)\, Z(t_j) \right\} = \operatorname{Min}(t_i, \ t_j).$$

Given that the increments are independent, we have for ($t_i < t_j$)

$$\mathbf{E}\left\{ Z(t_i) Z(t_j) \right\} = \mathbf{E}\left\{ \big(Z(t_j) - Z(t_i) + Z(t_i) - Z(0)\big)\ (Z(t_i) - Z(0)) \right\}$$

$$= \mathbf{E}\left\{ (Z(t_i) - Z(0))^2 \right\}$$

$$= t_i.$$

---

[1]Notice that the statistical distribution of the paths for finite dt is more complicated, because increments have multinomial distributions with different parameters, according to the number of elementary jumps between the times t and $t + a$. In this respect, Brownian motion is a simpler object than a random walk with a finite jump size.

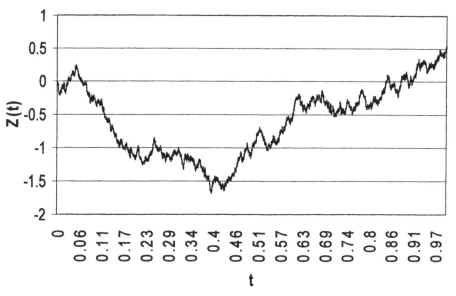

FIGURE 7.1
Random walk with Gaussian increments having mean zero and standard deviation $\sqrt{\Delta t}$, where $\Delta t = 0.0001$. This random path behaves essentially like a Brownian motion for time scales greater than 10^{-4}.

7.2 Elementary Properties of Brownian Paths

The first important property is the **continuity of Brownian paths,** namely

$$\mathbf{P}\left\{Z(\cdot) \text{ is a continuous function}\right\} = 1 .$$

The proof of this fact is mathematically nontrivial (see Ikeda and Watanabe [6] or Varadhan [13]).

Brownian paths are continuous but also very *irregular.* Any student of random walks (or, for the matter, of financial time series) has noticed that the sample paths of random walks are nonsmooth and appear to have an infinite slope (making trends difficult to predict). One way to see that Brownian paths are not differentiable is to consider their **quadratic variation.**

PROPOSITION 7.1

Let $0 = t_0 < t_1 < \cdots < t_n = T$ represent a partition of the time interval $[0, T]$ and let $dt = \max\limits_{j} \left(t_j - t_{j-1}\right)$. Set

$$\Delta Z_j = Z(t_j) - Z(t_{j-1}) .$$

With probability 1, Brownian paths satisfy

$$\lim_{dt \to 0} \left\{ \sum_{j=1}^{N} (\Delta Z_j)^2 \right\} = T .$$

(The right-hand side of this last equation is known as the quadratic variation of the path $Z(\cdot)$ on the interval $[0, T]$.) This result is a direct consequence of the Law of Large Numbers of probability theory. In fact, the random variables $(\Delta Z_j)^2$ are independent and have means $\mathbf{E}\{(\Delta Z_j)^2\} = t_j - t_{j-1}$. Therefore, as $dt \to 0$, the sum converges to its expected value, $\sum_j (t_j - t_{j-1}) = T$.

This result implies that Brownian paths are not differentiable. To see this, we note that if the function $f(t)$ is differentiable, then we have[2]

$$\lim_{dt \to 0} \left\{ \sum_{j=1}^{N} |\Delta f_j| \right\} \approx \int_0^T |f'(s)| \, ds .$$

But then, in view of the inequality

$$\sum_{j=1}^{N} (\Delta f_j)^2 \leq \max_j |\Delta f_j| \cdot \sum_{j=1}^{N} |\Delta f_j| ,$$

and the fact that

$$\max_j |\Delta f_j| = O(dt) ,$$

we conclude that differentiable functions f satisfy

$$\sum_{j=1}^{N} (\Delta f_j)^2 \overset{<}{\approx} \left(\text{constant} \cdot \int_0^T |f'(s)| \, ds \right) \cdot dt \to 0 \text{ as } t \to 0 .$$

In other words, differentiability implies that the quadratic variation must vanish. The finiteness of the quadratic variation of Brownian motion implies that its paths are not differentiable.

Another remarkable property of Brownian motion is statistical **self-similarity**. For any parameter $\lambda > 0$, the transformation

$$Z(t) \mapsto \lambda^{-\frac{1}{2}} Z(\lambda t)$$

maps Brownian paths into Brownian paths. This means that if one considers, for instance, the ensemble of Brownian paths on the interval $[0, 1]$ and "stretches" them according to the above transformation with $\lambda = 1/2$, the result is Brownian motion on the interval $[0, 2]$. This self-similarity is consistent with the fact that Brownian paths are "fractal" objects.

---

[2]Using the Mean Value theorem.

7.3   Stochastic Integrals

DEFINITION 7.2   *A function $f(t)$ is said to be nonanticipative with respect to the Brownian motion $Z(t)$ if, for all $t > 0$,*

$$f(t) = \tilde{f}(\{Z(s); s \le t\}, t),$$

i.e., if the value of the function at time t is determined by the values taken by the history of the path $Z(\cdot)$ up to time t.

We also include in this definition deterministic functions—these have a "trivial" dependence on the random paths. The point of introducing the concept of a nonanticipative function is to distinguish between general functions of Brownian paths and those that are determined by the natural "flow of information" associated with the path $Z(\cdot)$ as time progresses.

Examples. The function

$$f_1(t) = \begin{cases} 0 & \text{if} & \max_{0 \le s \le t} Z(s) < 5 \\ 1 & \text{if} & \max_{0 \le s \le t} Z(s) \ge 5 \end{cases}$$

is nonanticipative. (This function is equal to zero at time t if the walk has not reached the value 5 by time t and is equal to 1 otherwise.) On the other hand,

$$f_2(t) = \begin{cases} 0 & \text{if} & \max_{0 \le s \le 1} Z(s) < 5 \\ 1 & \text{if} & \max_{0 \le s \le 1} Z(s) \ge 5 \end{cases}$$

is not. The reason is that the value of the latter function at any time $t < 1$ is determined by the realization of the path $Z(\cdot)$ over the entire interval $[0, 1]$—the information gained by knowing $Z(s)$ for $s \le t$ is insufficient to determine $f_2(t)$.

Nonanticipative functions are the "natural" objects to perform integration with respect to Brownian increments.

PROPOSITION 7.2

Let $f(\cdot)$ be a continuous, nonanticipative function such that

$$\mathbf{E}\left\{\int_0^T |f(t)|^2 \, dt\right\} < \infty.$$

Then, given any sequence of partitions of the interval $[0, T]$ with mesh size $\Delta t \to 0$,

$$\lim_{\Delta t \to 0} \sum_{j=1}^{N} f(t_{j-1}) \cdot \left(Z(t_j) - Z(t_{j-1}) \right)$$

$$= \lim_{\Delta t \to 0} \sum_{j=1}^{N} f(t_{j-1}) \cdot \Delta Z(t_j)$$

exists and is independent of the sequence used to take the limit. This limit is, by definition, the **Ito integral of** f. *It is denoted by*

$$\int_{0}^{T} f(t) \, dZ(t) \, .$$

Stochastic integration is a natural operation associated with Brownian paths: a path is "sliced" into consecutive Gaussian increments, each increment is multiplied by a random variable, and these numbers are then added together again to reconstruct the stochastic integral. Thus, the stochastic integral can be viewed as a random walk with increments that have different amplitudes, or *conditional variances*—a sort of "inhomogeneous random walk." In this respect, it is important to emphasize the role of the nonanticipative assumption. Consider the jth increment after multiplication by the random variable $f(t_{j-1})$:

$$f\left(t_{j-1}\right) dZ_j = f\left(t_{j-1}\right) \cdot \left(Z\left(t_j\right) - Z\left(t_{j-1}\right) \right) \, . \tag{7.3}$$

Once the history of the path up to time t_{j-1} is revealed, the value of $f(t_{j-1})$ is also known. Therefore, the increment of the stochastic integral over the next period *conditional on the past up to time t* is Gaussian with mean zero and variance $f(t_{j-1})^2 \cdot \left(t_j - t_{j-1} \right)$. If the function $f(t)$ was anticipative (for lack of a better word), the two factors in (7.3) need not be conditionally independent. In the latter case, the mean of the increment may not be zero necessarily and the variance cannot be calculated in explicit form. Thus, nonanticipative functions are the correct functions to define a continuous, inhomogeneous random walk through the stochastic integral.

Some basic properties of the stochastic integral are given in

PROPOSITION 7.3
Under the above assumptions, we have

$$\mathbf{E} \left\{ \int_{0}^{T} f(t) \, dZ(t) \right\} = 0 \, , \tag{7.4}$$

$$\mathbf{E} \left\{ \left(\int_{0}^{T} f(t) \, dZ(t) \right)^2 \right\} = \mathbf{E} \left\{ \int_{0}^{T} |f(t)|^2 \, dt \right\} \, . \tag{7.5}$$

Moreover, $t \mapsto \int_0^t f(s) \, dZ(s)$ *is a nonanticipative process that is continuous with probability 1.*

Proofs of Propositions 7.2 and 7.3 are sketched in the appendix to this chapter. The main idea behind the definition of the stochastic integral is the fact that Brownian motion increments dZ "point to the future of f" at each discretization time.

Example 7.1

Consider a function $f(t) = \sigma(t)$ that is deterministic (i.e., independent of the Brownian motion).[3] In this case, the stochastic integral

$$X(t) = \int_0^t \sigma(s) \, dZ(s) \ , \quad t \geq 0 \ , \tag{7.6}$$

is a Gaussian random process, in the sense that $(X(t_1), X(t_2), \ldots, X(t_N))$ is a multivariate normal vector for any set of times $t_1 < t_2 < \cdots < t_N$. The reason is that the stochastic integral is a limit (in the sense of the mean-square norm) of sums of independent Gaussian random variables. Thus, if the integrand $\sigma(t)$ is deterministic, the stochastic integral is a Gaussian random walk with *time-dependent local variance*. The variance of any increment is given by the formula

$$\mathbf{E} \left\{ (X(t+a) - X(t))^2 \right\} = \int_t^{t+a} \sigma^2(s) \, ds \ .$$

In Options Theory, stochastic processes of type (7.6) are often used to model the term structure of volatility (Figure 7.2).

Mathematicians have a different way of thinking about $X(t)$: in fact, suppose that we define a *new time scale* $\theta(t)$ by the equation

$$\theta(t) = \int_0^t \sigma^2(s) \, ds \ .$$

The reader may think of θ as the time that a special clock would give you whenever the "real" time was t. Defining a new process $\tilde{X}(\theta) \equiv X(t)$ (the displacement with respect to the "rubber clock"), we have

$$\mathbf{E} \left\{ \left(\tilde{X}(\theta) \right)^2 \right\} = \mathbf{E} \left\{ (X(t))^2 \right\}$$

$$= \int_0^t \sigma^2(s) \, ds$$

$$= \theta \ .$$

---

[3] We use the notation $\sigma(t)$ to suggest the notion of a "local variance."

Sample Path for Ito Integral

sigma(t)=1 if t<1/2 and sigma(t)=5 if t>=1/2

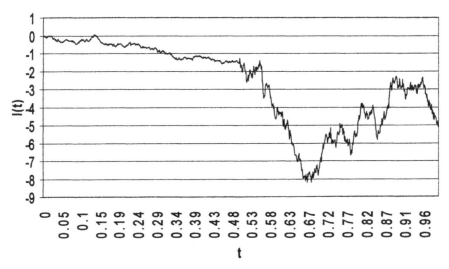

FIGURE 7.2
Diagram for the value of a stochastic integral of the form (7.6) where $\sigma(t) = 1$ for $t < 0.5$
and $\sigma(t) = 5$ for $t \geq 0.5$.

Thus the process induced by the stochastic integral can be regarded as a Brownian motion with respect to the new clock. This is a nice result, because it characterizes the probability distribution of Ito integrals in the case of deterministic integrands.[4]   ∏

Example 7.2
 Suppose that $f(t) = f(Z(t))$, i.e., that f depends on the current value of the Brownian path. This is, of course, a nonanticipative function so the Ito integral can be defined. Intuitively, the stochastic integral corresponds to a walk in which the local variance of increments is a function of the auxiliary Brownian path. The stochastic integral

$$\int_0^t f(Z(s))\, dZ(s)$$

will be non-Gaussian in general. We should now make an important remark, which is related to the results of the previous section and to what lies ahead. If Brownian paths were smooth, or rather, if one ignored the fact that they are nonsmooth, then one might be tempted to consider

---

[4]This trick of changing time has also made its way into financial modeling. Some exchanges close while others remain open, and interest accrual on deposits may take into account time at which there is no trading. Because of this, the local volatility of certain assets over holidays and quiet periods can be lowered to reflect the lack of strong trading activity. Similarly, the short-term volatility parameter might be increased on the date of an important economic or political announcement that will have a large impact on prices or rates. To take into account these effects in valuation models it is useful to introduce the notion of "nonlinear" time.

the antiderivative (primitive) of f and to write

$$dF(Z(t)) = f(Z(t)) \cdot dZ(t)$$

whereby

$$\int_0^t f(Z(s)) \, dZ(s) = F(Z(t)) - F(0) . \tag{7.7}$$

This is wrong!! (Unless f is constant.) For instance, if the primitive $F(Z)$ is positive and vanishes at $Z = 0$, we would conclude from (7.7) that the stochastic integral is positive. This cannot be, since we know from Proposition 7.3 that stochastic integrals have mean zero. Consider, for instance, the elementary case $f(Z) = Z$ (and hence $F(Z) = Z^2/2$). Using the definition of the stochastic integral and standard notation, we have

$$\int_0^T Z(s) \, dZ(s) \approx \sum_j Z_{j-1} \, dZ_j$$

$$= \sum_j Z_{j-1} \left(Z_j - Z_{j-1} \right)$$

$$= -\sum_j \left(Z_j - Z_{j-1} \right)^2 + \sum_j Z_j \left(Z_j - Z_{j-1} \right)$$

$$= -\sum_j \left(Z_j - Z_{j-1} \right)^2 + \sum_j \left(Z_j^2 - Z_{j-1}^2 \right)$$

$$- \sum_j Z_{j-1} \left(Z_j - Z_{j-1} \right)$$

$$\approx -T + Z^2(T) - \int_0^T Z(s) \, dZ(s) , \tag{7.8}$$

where we used the result on the quadratic variation of Brownian motion (Proposition 7.1). Thus, we conclude that

$$\int_0^T Z(s) \, dZ(S) = \frac{Z(T)^2}{2} - \frac{T}{2} . \tag{7.9}$$

The right-hand side now has expectation zero (at it should). The additional term $\frac{T}{2}$, which is missing in the "naive" formula (7.7) comes from the quadratic variation $\sum (dZ_j)^2$. $\square$

Example 7.3

Suppose that one wants to implement the idea of a random walk with conditionally Gaussian increments (i.e., a stochastic integral) in which *the local variance depends on the position of the walk*. This means that the "elementary increment" of the stochastic integral should have the form

$$\Delta X(t) = \sigma \left(X(t) \right) \cdot \Delta Z(t) .$$

More formally, we would like to define a process that satisfies

$$X(t) = \int_0^t \sigma(X(s)) \, dZ(s) \, . \tag{7.10}$$

As opposed to Example 7.3, the stochastic integral on the right-hand side depends on the left-hand side of the equation. This is therefore an **integral equation** for $X(t)$, which is often written in differential form

$$dX(t) = \sigma(X(t)) \, dZ(t) \, , \tag{7.11}$$

in which case it is termed a **stochastic differential equation**. If $Z(t)$ were smooth, then (7.11) has a clear meaning and can be solved by standard methods of ordinary differential equations. In the case of Brownian differentials, (7.11) should be interpreted as the integral equation (7.10) (Figure 7.3). The existence and uniqueness of solutions of stochastic differential or integral equations such as (7.11) and (7.10) is treated in most books on stochastic calculus (Ikeda and Watanabe [6], Varadhan [13], and Karatzas and Shreve [8], among many others).   ▯

Ito Integral:

sigma(X)=1+2*(1+sign(|X|-0.5))

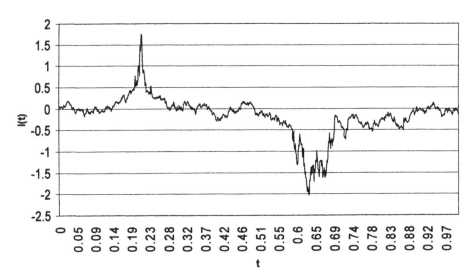

FIGURE 7.3
Diagram for the value of a stochastic integral of the form (7.10) where $\sigma(x) = 1$ for $|x| < 0.5$ and $\sigma(x) = 5$ for $|x| \geq 0.5$.

The Ito integral, or stochastic integral, is a powerful analytic tool for constructing stochastic processes that are similar to Brownian motion but have local characteristics that depend on time, the value of the process itself, or more general nonanticipative factors. The implications for financial modeling are very interesting: stochastic integrals and stochastic differential equations can be used to model heterogeneity of the local price volatility, which is a key theme in chapters to follow.

7.4   Ito's Lemma

We now discuss a systematic approach for evaluating Ito integrals and functions of Brownian motion. This result can be viewed as the analogue of the Fundamental Theorem of Calculus for functions of Brownian motion.

PROPOSITION 7.4

Let $F(Z, t)$ be a smooth function of two real variables Z and t with bounded derivatives of all orders and let $\{Z(t),\ t > 0\}$ represent a Brownian path. Then

$$F(Z(T),\ T) = F(0,\ 0) + \int_0^T \frac{\partial F}{\partial Z}(Z(s),\ s)\, dZ(s)$$

$$+ \int_0^T \left\{ \frac{\partial F}{\partial t}(Z(s),\ s) + \frac{1}{2}\frac{\partial^2 F}{\partial Z^2}(Z(s),\ s) \right\} ds\ . \tag{7.12}$$

This proposition is known as **Ito's Lemma**. It provides the correction to formula (7.7) that would result from a naive application of standard calculus. The additional term is

$$\int_0^T \frac{1}{2}\frac{\partial^2 F}{\partial Z^2}(Z(s),\ s)\, ds \tag{7.13}$$

[compare with (7.9)]. The Fundamental Theorem of Calculus [cf. (7.7)] does not involve second derivatives, the reason for this being that the contribution to the integral due to quadratic and higher-order terms is negligible. (The quadratic variation of a smooth function is zero.) In contrast, the $(dZ)^2$-terms contribute to the differential in the case of Brownian motion because the quadratic variation of the paths is non-trivial. The beauty of Ito's Lemma is that it provides a "closed-form" expression for $F(Z(t),\ t)$ for any reasonable smooth function F, elucidating the effect of the quadratic variation. This avoids having to go through manipulations of sums like the ones done in the Example 7.2 of the previous section.

Sketch of the Proof of Ito's Lemma.[5]

We consider the Taylor expansion of F about some point (Z, t). Formally, we have

$$\Delta F = F_Z \Delta Z + F_t \Delta t + \frac{1}{2}F_{ZZ}(\Delta Z)^2$$

$$+ F_{Zt}\Delta Z \Delta t + \frac{1}{2}F_{tt}(\Delta t)^2 + \cdots \tag{7.14}$$

Now, assume that we have a partition of the interval $[0,\ T]$, $\{t_j\}$, and that

$$Z = Z(t_{j-1}),\ t = t_{j-1},$$

$$\Delta t = t_j - t_{j-1},$$

---

[5]For a rigorous proof, see Ikeda and Watanabe [6].

and

$$\Delta Z_j = Z_{t_j} - Z_{t_{j-1}} .$$

Considering the Taylor expansion (7.14) and adding up the successive increments, we find that the first term in the right-hand side of (7.14) gives the contribution $\int F_Z(Z(s), s) \, dZ(s)$ to (7.12). Similarly, the second term in the Taylor expansion (7.14) contributes to the integral $\int F_t \, dt$ in (7.12). These are the two terms that we expect from standard differential calculus.

Let us turn to the higher-order terms in the Taylor expansion. Since we have

$$\mathbf{E} \left\{ |\Delta Z|^p \, (\Delta t)^q \right\} \propto (\Delta t)^{p/2 + q} ,$$

the contribution to $\sum dF_j$ that arises from adding the ($N = T/\Delta t$) terms proportional to $(\Delta Z_j)^p (\Delta t)^q$ has order ($\Delta t)^{p/2 + q - 1}$. The conclusion is that the only terms with $p + q \geq 2$ that contribute are those with $p = 2$ and $q = 0$, i.e., the $(\Delta Z)^2$ terms. All other terms vanish asymptotically as $\Delta t \to 0$.

This means that we should study the asymptotic behavior of the sums

$$\frac{1}{2} \sum_{j=1}^{N} F_{ZZ} \left(Z(t_{j-1}), t_{j-1} \right) (\Delta Z_j)^2 , \tag{7.15}$$

recalling that each increment ΔZ_j points to the future of t_{j-1}. Heuristically, replacing $(\Delta Z_j)^2$ by its expectaion Δt_j, this sum becomes

$$\frac{1}{2} \sum_{j=1}^{N} F_{ZZ} \left(Z(t_{j-1}), t_{j-1} \right) \Delta t ,$$

which is a Riemann sum approximating the integral in (7.13). Thus, to establish Ito's Lemma, it suffices to prove that the sums

$$\frac{1}{2} \sum_{j=1}^{N} F_{ZZ} \left(Z(t_{j-1}, t_{j-1}) \right) \left[\Delta Z_j)^2 - \Delta t \right] \tag{7.16}$$

converge to zero as $\Delta t \to 0$ in a suitable sense. To see this, we analyze the mean and the variance of the sum. This analysis is similar to the one of the proof of Propositions 7.2 and 7.3 given in the appendix of this chapter. The crucial point is that the increments in (7.16) have mean zero and point toward the future. Because of the latter property, we have

$$\mathbf{E} \left\{ F_{ZZ} \left(Z(t_{j-1}, t_{j-1}) \right) \left[\Delta Z_j)^2 - \Delta t \right] \right\} = 0 .$$

The expected value of the sums (7.16) is therefore zero. Consider now the variance of (7.16),

$$\frac{1}{4} \mathbf{E} \left\{ \left(\sum_{j=1}^{N} F_{ZZ} \left(Z(t_{j-1}), t_{j-1} \right) \left[\Delta Z_j)^2 - \Delta t \right] \right)^2 \right\} .$$

We observe two things: first, the variables $F_{ZZ}\left(Z(t_{j-1}),\, t_{j-1}\right)\left[\,(\Delta Z_j)^2 - \Delta t\,\right]$ and $F_{ZZ}\left(Z(t_{k-1}),\, t_{k-1}\right)\left[\,(\Delta Z_k)^2 - \Delta t\,\right]$ are uncorrelated for $j \neq k$. The reason is that if, say, $t_j < t_k$,

$$F_{ZZ}\left(Z(t_{j-1}),\, t_{j-1}\right)\left[\,(\Delta Z_j)^2 - \Delta t\,\right] \cdot F_{ZZ}\left(Z(t_{k-1}),\, t_{k-1}\right)$$

and

$$(\Delta Z_k)^2 - \Delta t$$

are independent (increments point to the future!). Second, the latter random variable has expectation zero. This guarantees that the expectation of their product is zero. Consequently, the variance is given, to leading order, by

$$\frac{1}{4}\sum_{j=1}^{N} \mathbf{E}\left\{\left(F_{ZZ}\left(Z(t_{j-1}),\, t_{j-1}\right)\right)^2\right\} \mathbf{E}\left\{\left((\Delta Z_j)^2 - \Delta t\right)^2\right\}$$

$$\approx \frac{1}{2}\,\Delta t\,\cdot\left(\int_0^t \mathbf{E}\left\{\left(F_{ZZ}\left(Z(t),\, t\right)\right)^2\right\}\,dt\right),$$

a negligible quantity as $\Delta t \to 0$. Here, we used the fact that

$$\mathbf{E}\left\{\left(\Delta Z^2 - \Delta t\right)^2\right\} = (\Delta t)^2\,\mathbf{E}\left\{(N^2 - 1)^2\right\} = 2\,(\Delta t)^2,$$

where N is a standard normal.[6] We have thus shown that the sums in (7.15) converge to the integral (7.13) (the "Ito correction term") in the sense that the difference has mean zero and variance converging to zero as $\Delta t \to 0$.

Example 7.4

Consider the following function of Brownian motion:

$$S(t) = S_0\, e^{\sigma\, Z(t) + \mu t}\,,\quad t \geq 0,$$

where σ and μ are constants. This random process is sometimes called **geometric Brownian motion**. Let us apply Ito's Lemma to $S(t)$, with the object of finding the integral equation satisfied by it. Accordingly, applying (7.12) to the function $F(Z, t) = S_0 \exp\{\sigma Z + \mu t\}$, we find that

$$S(t) - S(0) = \int_0^t S(\tau)\,\sigma\,dZ(\tau) + \int_0^t S(\tau)\left(\mu + \frac{\sigma^2}{2}\right)d\tau.\qquad(7.17)$$

If we use differential notation, we obtain

$$dS(t) = S(t)\,\sigma\,dZ(t) + S(t)\left(\mu + \frac{\sigma^2}{2}\right)dt.\qquad(7.18)$$

---

[6]From the explicit form of the moment-generating function for the standard normal distribution, $\mathbf{E}\left\{e^{\lambda N}\right\} = e^{\frac{\lambda^2}{2}}$, it follows that $\mathbf{E}\left\{N^{2k}\right\} = (2k)!/\,2^k k!$ for all integers $k > 0$.

This stochastic differential equation states that geometric Brownian motion has the property that the infinitesimal *relative increments*

$$\frac{dS(t)}{S(t)} = \sigma \, dZ(t) + \left(\mu + \frac{\sigma^2}{2} \right) dt$$

are normal with mean $\mu + \frac{\sigma^2}{2}$ and variance σ^2. This is the Black & Scholes model for stock price returns, which we will discuss shortly using ideas from stochastic calculus.   ☐

7.5 Ito Processes and Ito Calculus

What is the most general class of random processes that can be described in terms of sums of stochastic integrals and standard integrals? The answer is the class of **Ito processes**.

DEFINITION 7.3   *We say that a random process $X(t), t \geq 0$, is an Ito process if there exist a Brownian motion measure and two nonanticipative functions $\sigma(t)$ and $b(t), t \geq 0$, such that*

$$X(t) = X(0) + \int_0^t \sigma(s) \, dZ(s) + \int_0^t b(s) \, ds \, , \quad t > 0 \, , \tag{7.19}$$

or, in terms of differentials,

$$dX(t) = \sigma(t) \, dZ(t) + b(t) \, dt \, . \tag{7.20}$$

Intuitively speaking, Ito processes are continuous random functions with infinitesimal increments which, conditionally on the past until time t, are Gaussian with mean $b(t)$ and variance $\sigma^2(t)$.[7]

An important class of Ito processes that we already encountered in this chapter are functions of Brownian motion, i.e.,

$$X(t) = F(Z(t), t) \, . \tag{7.21}$$

In fact, Ito's Lemma shows that the **local parameters** associated to this process are

$$\sigma(t) = \frac{\partial F(Z(t), t)}{\partial Z}$$

and

$$b(t) = \frac{\partial F(Z(t), t)}{\partial t} + \frac{1}{2} \frac{\partial^2 F(Z(t), t)}{\partial Z^2} \, .$$

The next result is very important for applications. It states that any smooth function of an Ito process is also an Ito process and gives a formula for the local parameters.

---

[7]It is important to observe that, in general, the increments will not be Gaussian over a *finite* time interval.

PROPOSITION 7.5

(Generalized Ito Lemma) *Suppose that $X(t)$ is an Ito process with local parameters $\sigma(t)$ and $b(t)$, i.e., that (7.20) or (7.21) hold. Let $F(X, t)$ be smooth function of (X, t) with bounded derivatives of all orders. Then*

$$Y(t) = F(X(t), t)$$

is an Ito process with local parameters $\Sigma(t)$ and $B(t)$ given by

$$\Sigma(t) = \frac{\partial F(X(t), t)}{\partial X} \cdot \sigma(t) \tag{7.22}$$

and

$$B(t) = \frac{\partial F(X(t), t)}{\partial t} + \frac{1}{2}\sigma^2(t)\frac{\partial^2 F(X(t), t)}{\partial X^2} + b(t)\frac{\partial F(X(t), t)}{\partial X} . \tag{7.23}$$

In other terms, the Generalized Ito Lemma states that, for all $t > 0$, the integral equation

$$F(X(t), t) = F(X(0), 0) + \int_0^t \frac{\partial F(X(s), s)}{\partial X} \cdot \sigma(s)\, dZ(s) \tag{7.24}$$

$$+ \int_0^t \left\{ \frac{\partial F(X(s), s)}{\partial t} + \frac{1}{2}\sigma^2(s)\frac{\partial^2 F(X(s), s)}{\partial X^2} + b(s)\frac{\partial F(X(s), s)}{\partial X} \right\} ds$$

holds for all sufficiently smooth functions $F(X, t)$. It is convenient to express equation (7.24) in differential form, like we did in the case of Brownian motion. Accordingly,

$$dF(X(t), t) = \frac{\partial F(X(t), t)}{\partial X}\sigma(t)\, dZ(t) + \frac{\partial F(X(t), t)}{\partial X} b(t)\, dt$$

$$+ \frac{\partial F(X(t), t)}{\partial t} dt + \frac{1}{2}\sigma^2(t)\frac{\partial^2 F(X(t), t)}{\partial X^2} dt$$

$$= \frac{\partial F(X(t), t)}{\partial X} \left(\sigma(t)\, dZ(t) + b(t)\, dt \right)$$

$$\frac{\partial F(X(t), t)}{\partial t} dt + \frac{1}{2}\sigma^2(t)\frac{\partial^2 F(X(t), t)}{\partial X^2} dt$$

$$= \frac{\partial F(X(t), t)}{\partial X} \cdot dX(t) + \frac{\partial F(X(t), t)}{\partial t} dt + \frac{1}{2}\sigma^2(t)\frac{\partial^2 F(X(t), t)}{\partial X^2} dt$$

$$\equiv \frac{\partial F(X(t), t)}{\partial X} \cdot dX(t) + \frac{\partial F(X(t), t)}{\partial t} dt$$

$$+ \frac{1}{2}\frac{\partial^2 F(X(t), t)}{\partial X^2} (dX(t))^2 , \tag{7.25}$$

where we used Eq. (7.20) and the formal relation

$$(dX(t))^2 \equiv \sigma^2(t)\,dt .$$ (7.26)

The latter equation is a convention: it expresses in synthetic form the contributions from quadratic terms $((dX(t))^2$ or $(dZ(t))^2$-terms). Thus, the Generalized Ito Lemma can be written concisely in the form

$$dF(X(t),\,t) = = \frac{\partial F(X(t),\,t)}{\partial X} \cdot dX(t) + \frac{\partial F(X(t),\,t)}{\partial t}\,dt$$

$$+ \frac{1}{2}\frac{\partial^2 F(X(t),\,t)}{\partial X^2}\,(dX(t))^2 .$$

The result is simple to remember: *the infinitesimal increment of a smooth function of an Ito process is obtained by making a Taylor expansion of order 1 in dt, of order 2 in dX, and using the convention (7.26)*. This convention is often referred to as the **Ito multiplication rule**. Note that if $\sigma(t) \equiv 1$ and $b(t) \equiv 0$, we recover the "little" Ito Lemma (Proposition 7.4) of the previous section.

In summary, *stochastic differential calculus is an extension of standard calculus in which we use the Ito multiplication rule (7.26) to account for the effect of nontrivial quadratic variations.*

The proof of the Generalized Ito Lemma is very similar to the proof of Proposition 7.4. We omit it. The interested reader should consult Ikeda and Watanabe [6], Varadhan [13], or Karatzas and Shreve [8].

References and Further Reading

[1] Bachelier, L., Théorie de la spéculation, *Ann. Ecole Norm. Sup.*, (S. 3), **17**, pp. 21–86.

[2] Brown, R. (June and July 1827), A Brief Description of Microscopical Observations (German transl.), *Ann. Phys.*, 14, pp. 294–313.

[3] Chung, K.I. and Williams, R.J. (1990), *Introduction to Stochastic Integration*, 2nd ed., Birkauser.

[4] Cootner, P.H. *The Random Character of Stock Prices*, MIT Press, Cambridge, MA.

[5] Einstein, A. (1956), *Investigations on the Theory of Brownian Movement*, Dover Publications, London.

[6] Ikeda, N. and Watanabe, S. (1989), *Stochastic Differential Equations and Diffusion Processes*, 2nd ed., North Holland-Kodansha, Amsterdam.

[7] Ito, K. (1944), Stochastic Integral, *Proc. Imp. Acad. Tokyo* 20, pp. 648–665.

[8] Karatzas, I. and Shreve, S. (1988), *Brownian Motion and Stochastic Calculus*, Springer-Verlag, New York.

[9] Lévy, P. (1948), Sur certains processus stochastiques homogènes, *Composition Math.* 7, pp. 283–339.

[10] Lévy, P. (1948), *Processus Stochastiques et Mouvement Brownien,* Gauthier Villars, Paris.

[11] Oksendaal, B. (1995), *Stochastic Differential Equations, An Introduction with Applications,* 4th ed., Springer Verlag, Berlin.

[12] Paley, R.E.A.C. and Wiener, N.Z.A. (1933), Notes on Random Functions, *Math. Z.,* 37, pp. 47–668.

[13] Varadhan, S.R.S. (1980), *Diffusion Processes and Partial Differential Equations,* Tata Institute Lectures, Springer Verlag.

[14] Varadhan, S.R.S. (1980), *Lectures on Brownian Motion and Stochastic Differential Equations,* Tata Institute of Fundamental Research, Bombay.

Appendix: Properties of the Ito Integral

The aim of this appendix is to provide additional technical elements to understand the basic properties of the stochastic integral stated in Propositions 7.1 and 7.2.[8]

Let us consider Proposition 7.2. Notice that any "approximating sum"

$$s_N = \sum_{j=1}^{N} f(t_{j-1}) \left(Z(t_j) - Z(t_{j-1}) \right) \equiv \sum_{j=1}^{N} f_{j-1} \, dZ_j \tag{A.1}$$

has mean zero and uniformly bounded variance. In fact, since $f(t)$ is nonanticipative, we have

$$\mathbf{E} \left\{ f_{j-1} \, dZ_j \right\} = \mathbf{E} \left\{ \mathbf{E} \left\{ f_{j-1} \, dZ_j \, | Z(s), \ s \le t_{j-1} \right\} \right\}$$

$$= \mathbf{E} \left\{ f_{j-1} \, \mathbf{E} \left\{ dZ_j \, | Z(s), \ s \le t_{j-1} \right\} \right\}$$

$$= 0 \,,$$

where $\mathbf{E} \left\{ \bullet \, | Z(s), \ s \le t_{j-1} \right\}$ represents the conditional expectation operator given the history of the path up to time t_{j-1}. The conditional expectation of dZ_j vanishes because the increments of $Z(t)$ are independent of their past. We conclude that the sum in (A.1) has mean zero. To compute its variance, we must evaluate the sum of terms of the form

$$\mathbf{E} \left\{ f_{j-1} \, dZ_j \cdot f_{k-1} \, dZ_k \right\} \,,$$

---

[8]These are in lieu of formal proofs, which would take us too far from the subject of these chapters into Measure Theory. We encourage the interested reader to consult the aforementioned references to obtain mathematically precise hypotheses on $f(t)$, on the types of convergence of random variables used in the proofs, etc.

where $1 \leq j, k \leq N$. The key observation is that this expectation vanishes if $j \neq k$. In fact, if $j < k$, we have

$$\mathbf{E}\left\{f_{j-1}\, dZ_j \cdot f_{k-1}\, dZ_k\right\} = \mathbf{E}\left\{\mathbf{E}\left\{f_{j-1}\, dZ_j \cdot f_{k-1}\, dZ_k \mid Z(s),\ s \leq t_{k-1}\right\}\right\}$$

$$= \mathbf{E}\left\{f_{j-1}\, dZ_j\, f_{k-1}\, \mathbf{E}\left\{dZ_k \mid Z(s),\ s \leq t_{k-1}\right\}\right\}$$

$$= 0.$$

The variance on (A.1) is therefore equal to the sum of the "diagonal" terms

$$\sum_{j=1}^{N} \mathbf{E}\left\{\left(f_{j-1}\, dZ_j\right)^2\right\} = \sum_{j=1}^{N} \mathbf{E}\left\{\left(f(t_{j-1})\right)^2\right\} (t_j - t_{j-1})$$

$$\approx \int_0^T \mathbf{E}\left\{(f(t))^2\right\} dt.$$

(This assumes implicitly that the sum can be approximated by the integral, i.e., that $\mathbf{E}\left\{(f(t))^2\right\}$ is continuous.)

We have established that the random variables s_N have bounded mean and variance. Next, we show that the sums *converge to a limit* as the partitions become finer and finer.

Consider a sequence of partitions $\left\{t_j^{(\nu)}\right\}_{j=1}^{N_\nu}$, where

$$\max_{1 \leq j \leq N_\nu}\left(t_j^{(\nu)} - t_{j-1}^{(\nu)}\right) \to 0 \text{ as } \nu \to \infty,$$

and the corresponding sequence of sums s_{N_ν}. We claim that

$$\lim_{\nu,\, \nu' \to \infty} \mathbf{E}\left\{\left(s_{N_\nu} - s_{N_{\nu'}}\right)^2\right\} = 0.$$

Consider two such sums, s_{N_ν} and $s_{N_{\nu'}}$, corresponding to different partitions. The idea is to "merge" the two partitions to form a finer one that combines the nodes of each of them. With respect to this refined partition (which we call $\{t_j\}$), both sums can be written in the form

$$s_{N_\nu} = \sum_{j=1}^{N} f(t_{j-1}^*)\left(Z(t_j) - Z(t_{j-1})\right)$$

and

$$s_{N_{\nu'}} = \sum_{j=1}^{N} f(t_{j-1}^{**})\left(Z(t_j) - Z(t_{j-1})\right).$$

Note that t_j^* s and the t_j^{**} are times (of the original partitions) that satisfy

$$t_j^* \leq t_j, \quad t_j^{**} \leq t_j. \tag{A.2}$$

Moreover, the differences $t_j^{**} - t_j^*$ converge to zero as $\nu \to \infty$. The difference of the sums can therefore be expressed in the form

$$\sum_{j=1}^{N} \left(f(t_{j-1}^*) - f(t_{j-1}^{**}) \right) \cdot \left(Z(t_j) - Z(t_{j-1}) \right)$$

Taking into account the fact that increments point to the future, it is easy to show that this sum has mean zero and variance

$$\sum_{j=1}^{N} \mathbf{E} \left\{ \left(f(t_{j-1}^*) - f(t_{j-1}^{**}) \right)^2 \right\} \cdot \left(t_j - t_{j-1} \right) . \tag{A.3}$$

If $f(t)$ satisfies suitable boundedness and continuity assumptions, e.g., if $f(t)$ is bounded and

$$\lim_{h \to 0} \mathbf{E} \left\{ \left(f(t + h) - f(t) \right)^2 \right\} = 0 ,$$

the sum in (A.3) tends to zero with the mesh size.

This argument shows the existence of the Ito stochastic integral. The properties stated in Proposition 7.3 also follow from these calculations.

Chapter 8

Introduction to Exotic Options: Digital and Barrier Options

Exotic options are a generic name given to derivative securities that have more complex cash-flow structures than standard puts and calls. The principal motivation for trading exotic options is that they permit a much more precise articulation of views on future market behavior than those offered by "vanilla" options. Like standard options, exotics can be used as part of a risk-management strategy or for speculative purposes. From the investor's perspective, some exotics provide high leverage because they can focus the payoff structure very precisely (this is the case of barrier options discussed below). Exotics are usually traded over the counter and are marketed to sophisticated corporate investors or hedge funds. Dealers are generally banks or investment houses. They manage their risk-exposure by

- Making two-way markets and attempting to be market-neutral as much as possible,

- Hedging with "vanilla" options as well as cash instruments.

The risk-management of exotics is more delicate than that of standard options because exotics are less liquid. This means that the seller of an exotic may not be able to buy it back, if his theoretical hedging strategy failed, without having to pay a large premium. Therefore, making a market in exotic options requires acute timing skills in hedging and the use of options to manage volatility risk. Roughly speaking, we can say that "exotic options are to standard options what options are to the cash market." By this we mean that exotic options are very sensitive to higher-order derivatives of option prices such as Gamma and Vega. Some exotics can be seen essentially as bets on the future behavior of higher-order "Greeks." Another important issue is the notion of **pin risk:** since some exotics have discontinuous payoffs, they can have huge Deltas and Gammas near expiration that make them very difficult, if not impossible, to Delta-hedge.

This chapter gives an introduction to the most important types of exotics that are used in financial markets, namely digital and barrier options. We will discuss various aspects of these contracts, namely (i) binomial tree pricing, (ii) closed-form solutions assuming the underlying index follows a geometric Brownian motion, and (iii) price sensitivity and hedging. The second point will require that we develop several mathematical results on the distribution of first-passage times and of the supremum of Brownian motion with drift over a given time-interval.

Aside from providing an introduction to these instruments, this study is interesting because it gives us a better perspective on the risks associated with hedging derivative products in general, including the risk-management of portfolios of standard options.[1]

---

[1] The material for this lecture was taken from various research publications and our own notes. Recommended reading: Hull [3], chapter on exotics; Rubinstein [4], (a compilation of his articles in *RISK Magazine*); and *From*

8.1   Digital Options

A **digital,** or **binary** option is a contingent claim on some underlying asset or commodity that has an "all-or-nothing" payoff. A digital call has a payoff (Figure 8.1)

$$F(S_T) = \begin{cases} 1 \text{ if } S_T \geq K \\ \\ 0 \text{ if } S_T < K \end{cases} \tag{8.1}$$

and a digital put has payoff $1 - F(S_T)$. Like standard options, digitals can be classified as **European** or **American** style. A European digital call provides a payoff of \$1 if the asset ends up above the strike price at the option's maturity date, and \$0 otherwise (Figure 8.2). The American digital call has a payoff of \$1 if the underlying asset reaches the value K before or at the expiration date T (Figure 8.3).

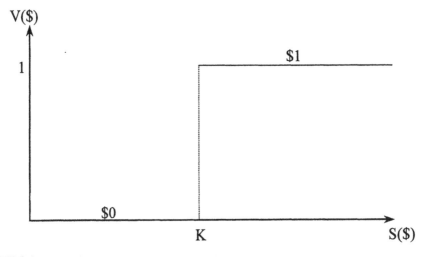

FIGURE 8.1
The payoff of a digital call option is a Heaviside step function. The holder of the option receives a fixed dollar amount if the underlying asset price is above K and nothing otherwise. Notice the jump discontinuity at $S = K$.

European Digitals

The fair value of a digital call with payoff (8.1) can be derived easily under the assumptions of log-normal prices (the "Black–Scholes world"). In fact, it is given by

$$V(S, T) = e^{-rT} \mathbf{E} \{ H(S_T - K) \} = e^{-rT} \mathbf{P} \{ S_T \geq K \}, \tag{8.2}$$

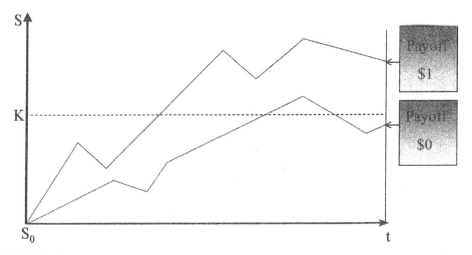

FIGURE 8.2
European-style digitals pay if and only if the price of the underlying asset at the *final date* is greater than K.

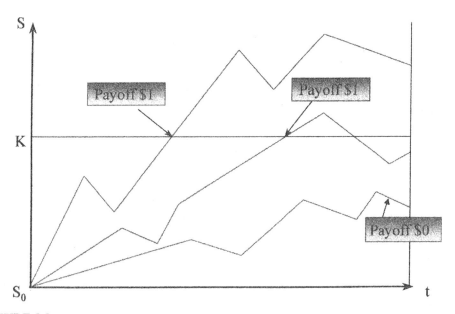

FIGURE 8.3
The American-style option pays if the path crosses the strike value any time between the inception date and the expiration date. This option is also known as a "one-touch" option.

where r is the interest rate (assumed constant) and the expectation is taken with respect to a risk-neutral probability. Here,

$$
H(X) = \begin{cases} 1 & \text{if } X \geq 0 \\ 0 & \text{if } X < 0 \end{cases}
$$

is the **Heaviside step function**. The calculation of the last probability in (8.2) is straightforward. Since the terminal price of the underlying asset (dividend yield= q) satisfies

$$S_T = S e^{\sigma Z \sqrt{T} + (r - q - \frac{1}{2}\sigma^2) T} ,$$

where Z is normal with mean zero and variance 1, we have

$$\mathbf{P}\{S_T \geq K\} = \int_{Z_K}^{+\infty} \frac{1}{\sqrt{2\pi}} e^{-\frac{z^2}{2}} dz$$

$$= N(-Z_K) .$$

Here $N(\cdot)$ is the cumulative distribution function of the standard normal and Z_K is defined by the equation

$$S e^{\sigma Z_K \sqrt{T} + (r - q - \frac{1}{2}\sigma^2) T} = K ,$$

i.e.,

$$Z_K = \frac{1}{\sigma \sqrt{T}} \ln\left(\frac{K e^{-rT}}{S e^{-qT}}\right) + \frac{1}{2}\sigma \sqrt{T} .$$

Therefore, defining

$$d_2 \equiv -Z_K = \frac{1}{\sigma \sqrt{T}} \ln\left(\frac{S e^{-qT}}{K e^{-rT}}\right) - \frac{1}{2}\sigma \sqrt{T} , \qquad (8.3)$$

we conclude that the fair value of the European digital call is given by

$$V(S, T) = e^{-rT} N(d_2) \qquad (8.4)$$

(Figure 8.4). This formula resembles the expression derived previously for the amount of cash held in the equivalent hedging portfolio for a *vanilla call*, namely

$$- K e^{-rT} N(d_2) .$$

This is no coincidence. In fact, the holder of a European call is, by equivalence of the final cash-flows, long a contingent claim that delivers one share if $S_T \geq K$ or nothing if $S_T < K$, and short K European digital calls. In other words, the standard call can be viewed as a "portfolio" of two digital options (one digital call with payoff consisting of one share and $-K$ digital calls with payoff of $1). Of course, this is consistent with the Black–Scholes formula for the value of a call

$$S \cdot e^{-qT} N(d_1) - K \cdot e^{-rT} N(d_2) , \qquad (8.5)$$

where

$$d_1 = \frac{1}{\sigma \sqrt{T}} \ln\left(\frac{S e^{(r-q)T}}{K}\right) + \frac{1}{2}\sigma \sqrt{T} .$$

We can interpret the two terms in this formula as the values of the two digital payoffs that "make up" the standard option (see, if necessary, the derivation of the Black–Scholes formula in Chapter 3.)

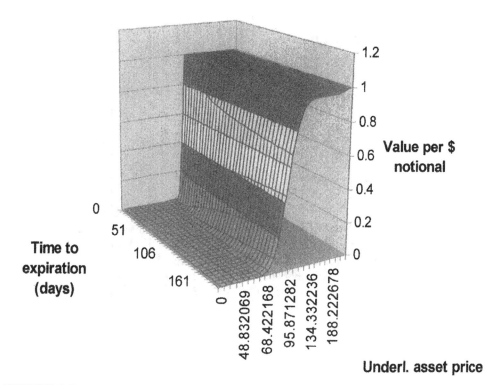

FIGURE 8.4
Graph of the price of a European digital option according to formula (8.4) generated using a trinomial tree.

Mathematically, European digital options are simple to price. On the other hand, they are harder to hedge. In fact, there are two fundamental differences between digital options and standard options:

- Digital options have **mixed convexity**

- Digitals have **discontinuous payoffs**.

Mixed convexity is an important feature. Recall that if a trader is **short Gamma** (at some price level) he is vulnerable to large moves in the underlying asset, whereas if he is **long Gamma** he is vulnerable to small moves, i.e., to time decay.[2] Digital options present both features: for the short position, an in-the-money digital is Gamma-positive and an out-of-the-money digital is Gamma-negative.

Another consequence of mixed-convexity is that the exposure to changes in volatility is nontrivial: for out-of-the money digitals the value of the option increases as the volatility increases. However, the value decreases if the option is in-the-money.[3]

---

[2]Gamma is the second derivative of the value function with respect to the price of the underlying asset. The simplest long Gamma position is achieved by buying options, and the simplest short-Gamma position by selling options.

[3]Market-makers beware! Suppose that we take the implied volatility of an at-the-money vanilla option with the same expiration to price an out-of-the money digital option. The exposure to a change in implied volatility is not as simple as for a vanilla option: its value increases if volatility increases and the market remains near the spot price. However, its value decreases if the market rallies and the volatility increases.

Let us examine these issues more closely by looking at the Greeks of the European digital. Differentiation with respect to S in (8.4) gives

$$\Delta_{digital} = \frac{e^{-rT} e^{-\frac{d_2^2}{2}}}{S \sqrt{2\pi \sigma^2 T}} \tag{8.6}$$

and

$$\Gamma_{digital} = -\frac{e^{-rT} e^{-\frac{d_2^2}{2}}}{S^2 \sigma^2 T \sqrt{2\pi}} \cdot d_1 . \tag{8.7}$$

These sensitivities become large as $T \to 0$ for $S \approx e^{-(r-q)T} K$. As T converges to zero, the Delta of the digital option approaches the Dirac delta function. This has two consequences: first, when expiration is far away the value of the digital option is small compared to \$1 and the Deltas and Gammas are small. As the expiration date approaches, hedging the digital becomes much more complicated due to the unbounded Deltas and Gammas. The second consequence is **pin risk:** if the price of the underlying asset oscillates around the strike price near expiration, the dynamic hedger will have to buy and sell large numbers of shares very quickly to replicate the option. At some point, the amount of shares bought or sold can be so large that the potential loss due to a small (but finite) move in the stock price exceeds the losses that would result by not hedging! At this point, Delta-hedging becomes extremely risky.[4]

The Gamma of the option vanishes for

$$S = S^*(T) = K \cdot e^{-(r-q)T} \cdot e^{\frac{1}{2}\sigma^2 T} . \tag{8.8}$$

(this is the value of spot for which the Delta of the standard call is exactly 1/2). For $S < S^*(T)$, Gamma is large and negative. This means that the hedger is exposed to significant risk near expiration if there is a big move in the spot price. If $S > S^*(T)$, the hedger is subject to time-decay risk: he must rebalance his position frequently in order to offset time decay. When $S = S^*(T)$ the hedger is subject to risk from both large and small moves!

The sensitivity of the price of the digital with respect to the volatility parameter is

$$Vega_{digital} = \frac{\partial V(S, T)}{\partial \sigma}$$

$$= -\frac{e^{-rT} e^{-\frac{d_2^2}{2}} \sigma}{\sqrt{2\pi}} \cdot d_1 . \tag{8.9}$$

Vega also changes sign at $S = S^*(T)$. In particular, the seller of the option is vulnerable to an increase or a decrease in market volatility according to whether S is smaller or greater than $S^*(T)$.

---

[4]To better understand pin risk, we should recall that prices do not change continuously—the log-normal approximation is just a convenient device for generating simple pricing formulas. The *discrete nature* of price movements can make continuous-time hedging techniques very risky when Delta changes rapidly as the spot moves. This observation applies also to highly leveraged option portfolios. A seasoned foreign-exchange trader once told one of the authors that the best solution in the case of a severe pin for digital options near the expiration date was "to go out and have lunch." By this he was making the point that adjusting the delta to disproportionately high levels would be equally as costly as not hedging at all.

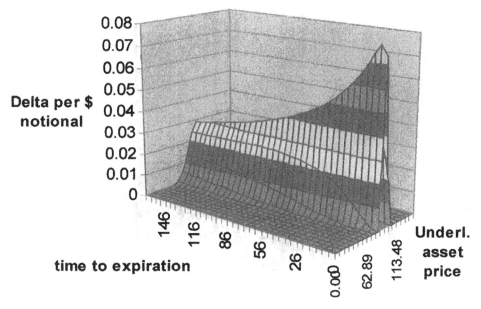

FIGURE 8.5
Delta of the European digital option in Figure 8.4.

One way to understand the European digital option in terms of standard options is with **call spreads**. Recall that a call spread is a position that consists of being long one call with a given strike and short another call with a different strike. Let ϵ denote a small number. Then, the position

- Long $1/\epsilon$ calls with strike $K - \epsilon$ and

- Short $1/\epsilon$ calls with strike K,

where the calls have the same expiration date as the digital, has a final payoff

$$\frac{1}{\epsilon} Max\,[\,S_T - (K - \epsilon),\,0\,] - \frac{1}{\epsilon} Max\,[\,S_T - K),\,0\,]\,. \qquad (8.10)$$

This function is greater than $H(S_T - K)$ for all S_T. This implies that the value of the digital is less than that of $1/\epsilon$ $(K - \epsilon, K)$ call spreads. We say that the call spreads **dominate** the digital. This observation suggests that a good hedging strategy for digital options would be to use call spreads instead of Delta-hedging, that is, to adopt a **static** hedging strategy. If ϵ is large, this will require relatively few options, but the fair value of the spread may be much greater than that of the digital. On the other hand, diminishing ϵ will make the difference in prices arbitrarily small but the hedge requires many options. This may be difficult and costly to execute. Moreover, options with strikes very close to K may not exist in the market. Nevertheless, the idea of using call spreads is a useful one. In fact, hedging with an option spread that *approximates* the binary payoff (but not necessarily dominates it or replicates it exactly) can help offset pin risk by diminishing the magnitude of the jump. In other words, a portion of the risk can be diversified by hedging with options and the **residual**, i.e., the difference between the payoffs of the digital and the option spread, can be hedged in the cash market at a lesser risk.

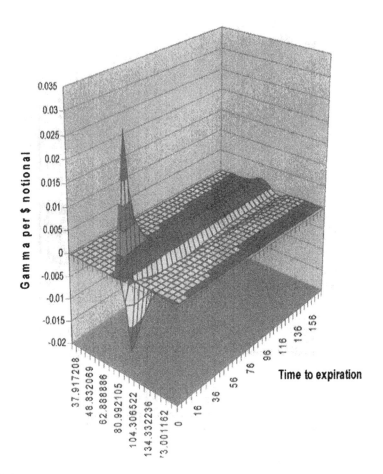

FIGURE 8.6
Gamma of the European digital option in Figure 8.4.

Example 8.1

An important example where digitals appear in finance is in the pricing of **contingent premium options**. These are derivative securities that are structured as standard European options, except for the fact that the holder pays the premium *at maturity if and only if the option is in the money.* A contingent premium option can be viewed as a portfolio consisting of

- Long one standard option with strike K and maturity T,

- Short V binary calls with strike K and maturity T,

where V represents the premium, to be paid if and only if $S_T > K$. The intrinsic value of a contingent-premium call option is therefore zero for $S_T \leq K$ and $S_T - K - V$ for $S_T > K$. Notice, in particular, that the investor makes money only if $S_T > K + V$ and will actually lose money if $K < S_T < K + V$. (The situation is analogous to that of a person who has "free" medical insurance but with a large "deductible.") If the option is structured so that no down-payment is required, then, according to Equations (8.4) and (8.5), V should satisfy

$$e^{-rT} \, \mathbf{E} \, \{ Max(\, S_T - K \, , 0 \,) \} \; - \; e^{-rT} \, V \, \mathbf{P} \, \{ S_T > K \, \} \; = \; 0. \qquad (8.11)$$

Therefore, from (8.4) and (8.5), the "fair" deferred premium should be

$$V = S e^{(r-q)T} \frac{N(d_1)}{N(d_2)} - K .$$
(8.12)

More generally, such options can be structured so that a portion of the premium is paid up front and another is contingent on the option being in-the-money at maturity . All such options have "embedded" European binaries. The idea was implemented in recent years for designing debt securities known as **structured notes**. A simple example of a structured note would consist of a note with coupons indexed to LIBOR containing the following characteristics:

- Coupon payment = max (LIBOR, 5%)

- Contingent premium = 0.25% if LIBOR < 5% on the coupon date.

This note guarantees the holder a "floor" of 5% on the interest rate income. The structure resembles that of a floating-rate note with an interest-rate floor (series of puts on interest rates.) However, the investor does not pay for the floor when he buys the structured note. Instead, he can (and also must) take advantage of the interest-rate floor by paying 0.25% (25 basis points) if LIBOR goes below 5% on any given coupon date. This derivative security could be desirable to investors that believe that if interest rates go below 5% then they will be significantly *below* 5% for some period of time. Also, the structure may be desirable to investors seeking protection if interest rates drop but who are unwilling to finance the interest-rate insurance up front. ▢

American Digitals

The payoff of the American digital option is similar, but now the holder receives $1 *if and when the underlying asset trades above* $K *for the first time*. This introduces additional time-optionality to the problem. The "fair value" of the American digital is, according to the general principles,

$$V(S, T) = \mathbf{E}\{e^{-r\tau} ; \tau \leq T\} ,$$
(8.13)

where expectation is taken with respect to a risk-neutral probability measure and τ represents the **first hitting time** of the strike price $S = K$.

It is worthwhile to consider the two valuation methods that we are familiar with: numerical methods using the binomial and trinomial pricing models and closed-form solutions using log-normal approximation. Numerical schemes are extremely easy to implement. The log-normal approximation is very useful to have a first approximation to the fair value and for "benchmarking" numerical schemes.

To implement the binomial/trinomial approach, we proceed as follows: by definition, the value of the American binary is $1 if the option is in the money. Therefore, with the usual notation, we have the boundary conditions

$$V_n^j = 1 \quad \text{if} \quad S_n^j \geq K$$
(8.14a)

and

$$V_N^j = 0 \quad \text{if} \quad S_N^j < K .$$
(8.14b)

The portion of the tree or lattice that needs to be determined by "roll-back" corresponds to the nodes (n, j) such that

$$S_n^j < K \quad \text{with } n < N . \tag{8.15}$$

The value of the binary option at these nodes is calculated using the familiar recursive relations

$$V_n^j = e^{-r_n \, dt} \cdot \left\{ P_U^{(n)} V_{n+1}^{j+1} + P_D^{(n)} V_{n+1}^{j} \right\} \quad \text{(binomial)} \tag{8.16a}$$

or

$$V_n^j = e^{-r_n \, dt} \left[P_U^{(n)} V_{n+1}^{j+1} + P_M^{(n)} V_{n+1}^{j} + P_D^{(n)} V_{n+1}^{j-1} \right] \quad \text{(trinomial scheme)}$$

$$\text{if } S_n^j < K . \tag{8.16b}$$

with $V_n^j = 1$ if $S_n^j \geq K$.

The only difference, from a numerical point of view, between American and European binaries is that the value at the nodes that are one step away from the boundary $\{S = K\}$ are computed using Eq. (8.14a) for the value of V_{n+1}^{j+1} (i.e., setting $V_{n+1}^{j+1} = 1$). This type of problem, which involves a lateral boundary condition, is known as a **Dirichlet problem** or **boundary-value problem** in the theory of partial differential equations.

In practice, the trinomial lattice scheme is better than the binomial tree, especially for the numerical evaluation of digital options and barrier options. One of the great advantages of the trinomial lattice is that nodes can be aligned with barriers and boundaries. This reduces computational roundoff errors near the boundaries.

Notice that formulas (8.16a) and (8.16b) allow for term-structures of volatility and interest rates, i.e., time-dependent parameters. This point is important for pricing American digitals, because the hitting time of the barrier is unknown. Therefore, the effect on pricing due to a time-varying volatility may be significant.[5] From the point of view of interest-rate and volatility term structures, the main difference between European and American binaries is due to the fact that the volatility parameter that is relevant for pricing European binaries is the annualized standard deviation of the change of price between now and the maturity date, i.e., the *term volatility*. In contrast, American binaries are sensitive to the entire volatility "path," or if you prefer, to the prices of options with maturities varying between the inception and the expiration of the digital.

Closed-form expressions for American binary options can be obtained under the assumption that the volatility and interest rates remain constant. This requires introducing new mathematical tools. Let $f(\theta)$ represent the probability density function of the random variable τ, i.e.,

$$\mathbf{P}\{\tau < T\} = \int_0^T f(\theta) \, d\theta . \tag{8.17}$$

Then, the value of the American digital in Eq. (8.13) can be written as

$$V(S, T) = \int_0^T e^{-r\theta} f(\theta) \, d\theta . \tag{8.18}$$

---

[5]In contrast, European binary options can be priced in a time-dependent volatility environment like standard options, using an "effective" mean-square volatility $\overline{\sigma}_T$ such that $\overline{\sigma}_T^2 = T^{-1} \int_0^T \sigma_t^2 \, dt$.

An explicit expression for the function $f(\theta)$ time can be obtained from:

LEMMA 8.1
Let Z_t^μ represent a Brownian motion with drift μ, i.e.,

$$Z_t^\mu = Z_t + \mu t$$

where Z_t is a Brownian motion and μ is a constant. Let τ_A represent the first time the path Z_t^μ hits A, where $A > 0$. Then,

$$P\{\tau_A < T\} = N\left(\frac{-A + \mu T}{\sqrt{T}}\right) + e^{2A\mu} N\left(\frac{-A - \mu T}{\sqrt{T}}\right). \tag{8.19}$$

We defer the proof of this lemma for the moment. To apply this result to the case of a log-normal random walk of the form

$$S_t = S e^{\sigma Z_t + \left(r - q - \frac{1}{2}\sigma^2\right)t} \tag{8.20}$$

we set

$$\mu \equiv \frac{r - q}{\sigma} - \frac{1}{2}\sigma$$

and

$$A \equiv \frac{1}{\sigma} \ln\left(\frac{K}{S}\right).$$

The reader will verify easily that if τ represents the first hitting time of $S = K$, then

$$P\{\tau < T\} = P\{\tau_A < T\}. \tag{8.21}$$

Using Eq. (8.19), we conclude that

$$P\{\tau < T\} = N(d_2) + \left(\frac{K}{S}\right)^{\left(\frac{2(r-q)}{\sigma^2} - 1\right)} \cdot N(d_3) \tag{8.22}$$

where

$$d_3 \equiv \frac{1}{\sigma\sqrt{T}}\left[\ln\left(\frac{S}{K}\right) - \left(r - q - \frac{1}{2}\sigma^2\right)T\right]. \tag{8.23}$$

Equation (8.22) gives a closed-form expression for the probability distribution of the first-exit probability of the set $\{S < K\}$. To compute the value of an American digital call, we can use the fact that

$$f(\theta) = \frac{d}{d\theta} P\{\tau < \theta\}.$$

Accordingly, we have

$$V(S, T) = \int_0^T e^{-r\theta} f(\theta)\, d\theta$$

$$= \int_0^T e^{-r\theta} \frac{d}{d\theta} \mathbf{P}\{\tau < \theta\} d\theta$$

$$= \left[e^{-r\theta} \mathbf{P}\{\tau < \theta\} \right]_{\theta=0}^{\theta=T} + r \int_0^T e^{-r\theta} \mathbf{P}\{\tau < \theta\} d\theta$$

$$= e^{-rT} \mathbf{P}\{\tau < T\} + r \int_0^T e^{-r\theta} \mathbf{P}\{\tau < \theta\} d\theta . \qquad (8.24)$$

The final expression for $V(S, T)$ is, from (8.22),

$$V(S, T) = e^{-rT} \cdot N(d_2) + e^{-rT} \cdot \left(\frac{K}{S} \right)^{\left(\frac{2(r-q)}{\sigma^2} - 1 \right)} N(d_3)$$

$$+ r \int_0^T e^{-r\theta} \mathbf{P}\{\tau < \theta\} d\theta , \qquad (8.25)$$

where the probability inside the integral is given by (8.22) (with T replaced by θ). This last integral can be computed numerically by quadrature. Notice that in the special case $r = 0$, the formula simplifies further. The same is true if the option is modified so that the holder collects \$1 *at time T* if the price *ever* touches K (as opposed to collecting K the *first time, τ, S_t* hits K). In both cases, the value of the American digital is

$$V(S, T) = e^{-rT} \cdot N(d_2) + e^{-rT} \cdot \left(\frac{K}{S} \right)^{\left(\frac{2(r-q)}{\sigma^2} - 1 \right)} N(d_3)$$

$$= e^{-rT} \mathbf{P}\{\tau < T\} . \qquad (8.26)$$

Next, we discuss the option's sensitivities to changes in spot price or market volatility.

Both the Delta and Gamma of the American digital are monotone increasing for $0 < S < K$ and become unbounded as $T \to 0$ in the neighborhood of $S = K$. Therefore, the option has significant pin risk. The "worst-case scenario" for the hedger, which is short, would be a market rallying slowly toward the strike level that collapses immediately before the option's maturity. In this event, Delta-hedging builds up a large spot position in the rally (long the market). If the market falls suddenly, the hedger may incur a significant loss, defeating the purpose of Delta-hedging. There is, however, an important difference with respect to European binaries: the hedger does not have to worry about market "whipping" around the strike price, since the option expires after K is hit for the first time. The exposure to Gamma and pin risk are simpler than for the European counterpart.

The Vega of the American digital option is positive at all values of spot below K because the graph of V as a function of S is convex for $S < K$. This is intuitively clear, since the risk-neutral probability of hitting the level K increases with Sigma. Thus, the risk-exposure due to an incorrect estimate of the volatility is only one-sided.[6] Just like with standard options,

---

[6]Recall that the European digital has two-sided volatility risk. The difference stems from the fact that in-the-money European digitals may go out-of-the-money, whereas American digitals pay off once the price of the underlying asset crosses the strike level for the first time.

the seller fears an increase in volatility and the buyer a decrease in volatility.

Near the "barrier" $S = K$, we can make a straightforward analysis of the sensitivity of the Delta of the option to volatility (what some traders call "D-Delta-D-Vol"). In fact, we know that $V(S, T)$ is nondecreasing as a function of the volatility and that $V(K, T) = 1$. This implies that the difference quotient

$$\frac{V(K, T) - V(K - \epsilon, T)}{\epsilon} \approx \Delta(K, T)$$

decreases as σ increases. Hence, Δ varies inversely to σ in a neighborhood of K. Therefore, increasing the volatility parameter (with respect to, say, the implied volatility of vanilla options traded in the market) will provide protection against slippage when the spot is away from K and improve the exposure to pin risk by decreasing the Delta at the barrier.[7]

8.2 Barrier Options

Barrier options are a generic name given to derivative securities with payoffs that are contingent on the spot price reaching a given level, or barrier, over the lifetime of the option. The most common types of barrier options are

- **Knock-out options**. These are contingent claims that *expire* automatically when the spot price touches one or more predetermined barriers.

- **Knock-in options**. These contingent claims are *activated* when the spot price touches one or more predetermined barriers.

The most common barrier options are structured as standard European puts and calls with one knock-in or knock-out barrier. For instance,

- A **down-and-out call** with strike K, barrier H, and maturity T is an option to buy the underlying asset for $\$K$ at time T, provided the spot price never goes *below* $\$H$ between now and the maturity date.

- An **up-and-out call** with strike K, barrier H, and maturity T is an option to buy the underlying asset for $\$K$ at time T, provided the spot price never goes *above* $\$H$ between now and the maturity date.[8]

- A **down-and-in call** with strike K, barrier H, and maturity T is an option to buy the underlying asset for $\$K$ at time T, provided that the spot price goes *below* $\$H$ between now and the maturity date.

- An **up-and-in call** with strike K, barrier H, and maturity T is an option to buy the underlying asset for $\$K$ at time T, provided that the spot price goes *above* $\$H$ between now and the maturity date.

---

[7]Of course, since increasing the volatility increases the premium, the seller will have to charge more if he wishes to follow this augmented-volatility strategy. He must therefore charge above the market volatility or else "set aside" some of his other funds to finance the strategy.

[8]In this case, we must have $K < H$, since otherwise the option has value 0. A similar remark applies to down-and-out-puts.

Similar definitions apply to puts. Barrier options are especially used in foreign-exchange derivatives markets. The *London Financial Times* of November 16, 1996 reported that exotic options now constitute around 10% of the currency option business. Barrier options, which are relatively simple variations on the European put and call, enjoy a great popularity.

A first observation regarding barrier options is that they are much cheaper than standard options. The optionality feature can be targeted more precisely by introducing a barrier. This is illustrated in the following example, described to one of us by a trader.

Example 8.2

A large multinational corporation based in Europe must convert its U.S. business revenue into DEM periodically. Given the weakness of the dollar with respect to the deutschemark in the past years and drop of the dollar at the beginning of 1995, the company fears a decrease in revenues in DEM terms. Its treasury department could have anticipated the problem by purchasing standard options but did not do this. Ideally, the company would like to have an at-the-money DEM call/dollar put with 6 months to expiration. If the spot exchange rate is 1.4225 DEM/USD, the value of a dollar put with strike 1.42 expiring in 180 days is $0.0391 per dollar notional. On a $100 million notional, the cost of this option is therefore approximately $3,910,000.[9] On the other hand, suppose that the company purchases now a down-an-out dollar put (or, equivalently, an up-and-out mark call) with a knock-out barrier at 1.27 DEM/USD. The value of this option is instead $0.01181 per dollar notional, or $1,181,000 to the nearest $1,000. (We will derive below a pricing formula for knock-out options.) Thus, the knock-out option with a 1.27 barrier is nearly four times cheaper than the vanilla. Therefore, if the treasurer believes that the dollar will not drop below $1.27 over the next 6 months, the knock-out option provides a cheaper alternative with the "same" terminal payoff.   ▯

The option described in the above example, which knocks out when the option is *in-the-money*, is often called a **reverse knock-out**. The difference between in-the-money and out-of-the-money barriers is significant, because the former have discontinuous payoffs at expiration. Thus, Delta-hedging reverse knock-in and knock-out options may lead to significant pin risk, similar to the one encountered in digitals. In contrast, options with out-of-the-money barriers do not seem to be very interesting from a hedging perspective. We will therefore discuss primarily barrier options that knock in or out when the option is in-the-money.

Knock-in and knock-out options are related by the simple formula

$$\text{KI} + \text{KO} = \text{Vanilla}. \tag{8.27}$$

This formula is self-evident: the holder of a portfolio consisting of one knock-in call and one knock-out call with the same strike, barrier and maturity will effectively hold a call at maturity regardless of whether the barrier was crossed. We can therefore reduce the question of pricing barrier options to the pricing of knock-outs.

We note that in some cases, the structure of barrier options is more complicated. We note two cases that were mentioned to us by professional traders:

- **Double knock-in** or **double knock-out** options, which have two barriers.

- **Knock-out options with rebate**. The holder receives a "consolation prize" in the form of a cash rebate on the premium paid if the option knocks out.[10]

Pricing Barrier Options Using Trees or Lattices

Barrier options are priced by solving a boundary-value problem similar to the one for American digitals. In the case of an up-and-out call, the value of this derivative security is determined recursively by solving the problem

$$V_n^j = e^{-r_n \, dt} \left[P_U^{(n)} V_{n+1}^{j+1} + P_D^{(n)} V_{n+1}^j \right]$$

$$\text{if } S_n^j < H, \quad \text{(binomial scheme)} \tag{8.28a}$$

or

$$V_n^j = e^{-r_n \, dt} \left[P_U^{(n)} V_{n+1}^{j+1} + P_M^{(n)} V_{n+1}^j + P_D^{(n)} V_{n+1}^{j-1} \right]$$

$$\text{if } S_n^j < H, \quad \text{(trinomial scheme)} . \tag{8.28b}$$

Here $P_U^{(n)}$, $P_M^{(n)}$, and $P_D^{(n)}$ are risk-neutral probabilities,

$$V_n^j = 0 \quad \text{if } S_n^j \geq H, \tag{8.28c}$$

and

$$V_N^j = Max \left[S_N^j - K, 0 \right] \text{ if } S_N^j \leq K. \tag{8.28d}$$

In the case of an up-and-in call, the boundary conditions (8.28c) and (8.28d) are replaced by

$$V_n^j = \tilde{V}_n^j \quad \text{if } S_n^j \geq H, \tag{8.29a}$$

where $\tilde{V}_n^j$ represents the value of a vanilla call at the node (n, j), and

$$V_N^j = 0 \quad \text{if } S_N^j \leq K. \tag{8.29b}$$

The validity of these equations follows from (i) the terms of the barrier options, which determine their value at the barrier and at maturity, and (ii) the absence of arbitrage, which implies (8.28a). The values of barrier-puts are determined by making obvious modifications.

---

[10]This option consists of a regular knock-out option with an attached American digital option.

Closed-Form Solutions

We derive closed-from solutions for barrier options assuming that the underlying asset follows a log-normal random walk with constant coefficients σ, q, and r. As in the case of the American binary option, we will need some auxiliary results on the properties of Brownian motion with drift.

LEMMA 8.2

(**"Reflection principle"**) *Let Z_t^μ, $t \geq 0$, represent a Brownian motion with drift μ. Then, if A and B are positive numbers with $B \leq A$,*

$$\mathbf{P} \left\{ \underset{0 \leq t \leq T}{Max} \ Z_t^\mu \geq A \ and \ Z_T^\mu \in (B, B + dB) \right\}$$

$$= \frac{1}{\sqrt{2\pi T}} e^{-\frac{(2A-B)^2}{2T}} e^{B\mu - \frac{1}{2}\mu^2 T} \, dB, \quad dB \ll 1. \tag{8.30}$$

We shall prove this lemma later.

To apply this result, let τ_H denote the first time that the lognormal walk (8.20) hits the level $S = H$. Then, the value of a down-and-out put with strike price K, knock-out H ($H < K$), and maturity T satisfies

$$P_{KO} (S, T; K, H) = e^{-rT} \mathbf{E} \{ Max [K - S_T, 0]; \tau_H > T \}$$

$$= e^{-rT} \mathbf{E} \left\{ Max [K - S_T, 0]; \underset{0 \leq t \leq T}{Min} \ S_t > H \right\} \tag{8.31}$$

This last expression can be rewritten as

$$e^{-rT} \mathbf{E} \left\{ K - S_T; H < S_T < K; \underset{0 \leq t \leq T}{Min} \ S_t > H \right\}$$

$$= e^{-rT} \mathbf{E} \{ K - S_T; H < S_T < K \}$$

$$- e^{-rT} \mathbf{E} \left\{ K - S_T; H < S_T < K; \underset{0 \leq t \leq T}{Min} \ S_t \leq H \right\}$$

$$= e^{-rT} \mathbf{E} \{ K - S_T; S_T < K \}$$

$$- e^{-rT} \mathbf{E} \{ K - S_T; S_T < H \}$$

$$- K e^{-rT} \mathbf{P} \left\{ H < S_T < K; \underset{0 \leq t \leq T}{Min} \ S_t \leq H \right\}$$

$$+ e^{-rT} \mathbf{E} \left\{ S_T; H < S_T < K; \underset{0 \leq t \leq T}{Min} \ S_t \leq H \right\}. \tag{8.32}$$

Notice that the first term corresponds to the value of a standard European put with strike K. The second term (recall $H < K$) can be calculated easily using the same reasoning as in the derivation of the Black–Scholes formula. To calculate the two remaining terms, we will use the result of Lemma 8.2. Introducing the parameters

$$A_H \equiv \frac{1}{\sigma} \ln \left(\frac{H}{S} \right) \, ,$$

$$A_K \equiv \frac{1}{\sigma} \ln \left(\frac{K}{S} \right) \, ,$$

and

$$\mu \equiv \frac{r - q}{\sigma} - \frac{1}{2} \sigma \, ,$$

we have, using Lemma 8.2,

$$\mathbf{P} \left\{ H < S_T < K \, ; \, \underset{0 \le t \le T}{Min} \, S_t \le H \right\}$$

$$= \mathbf{P} \left\{ A_H < Z_T^\mu < A_K \, ; \, \underset{0 \le t \le T}{Min} \, Z_t^\mu \le A_H \right\}$$

$$= \mathbf{P} \left\{ -A_K < Z_T^{-\mu} < -A_H \, ; \, \underset{0 \le t \le T}{Max} \, Z_t^{-\mu} \ge -A_H \right\}$$

$$= \int_{-A_K}^{-A_H} e^{- \frac{(-2A_H - B)^2}{2T}} e^{-B\mu - \frac{1}{2}\mu^2 T} \frac{dB}{\sqrt{2\pi T}} \, . \tag{8.33}$$

Here, to derive the third equation in (8.33) we use the fact that $-Z_t^\mu = -Z_t - \mu t$ and $Z^{-\mu} = Z_t - \mu t$ have the same probability distribution, which in turn follows easily from the symmetry of the Gaussian distribution. Similarly [using (8.22)], we have

$$\mathbf{E} \left\{ S_T \, ; \, H < S_T < K \, ; \, \underset{0 \le t \le T}{Min} \, S_t \le H \right\}$$

$$= S \cdot \mathbf{E} \left\{ e^{-\sigma Z_T^{-\mu}} \, ; \, -A_K < Z_T^{-\mu} < -A_H \, ; \, \underset{0 \le t \le T}{Max} \, Z_t^{-\mu} \le -A_H \right\}$$

$$= S \cdot \int_{-A_K}^{-A_H} e^{-\sigma B} \cdot e^{- \frac{(2A_H + B)^2}{2T}} e^{-B\mu - \frac{1}{2}\mu^2 T} \frac{dB}{\sqrt{2\pi T}} \, . \tag{8.34}$$

Calculating explicitly the two integrals in (8.33) and (8.34) and using the Black–Scholes formula to calculate the first two terms in (8.32), we arrive at the final result:

$$P_{KO}(S, T; K, H) = K e^{-rT} \cdot N(-d_2^K) - S e^{-qT} \cdot N(-d_1^K)$$

$$- K e^{-rT} \cdot N(-d_2^H) + S e^{-qT} \cdot N(-d_1^H)$$

$$- K e^{-rT} \left(\frac{H}{S} \right)^{\left(\frac{2(r-q)}{\sigma^2} - 1 \right)} \cdot \{ N(d_4) - N(d_5) \}$$

$$+ S e^{-qT} \left(\frac{H}{S} \right)^{\left(\frac{2(r-q)}{\sigma^2} \right) + 1} \cdot \{ N(d_6) - N(d_7) \} , \qquad (8.35)$$

where

$$d_1^K = \frac{1}{\sigma \sqrt{T}} \left\{ \ln \left(\frac{S}{K} \right) + \left(r - q + \frac{1}{2}\sigma^2 \right) T \right\} ,$$

$$d_2^K = \frac{1}{\sigma \sqrt{T}} \left\{ \ln \left(\frac{S}{K} \right) + \left(r - q - \frac{1}{2}\sigma^2 \right) T \right\} ,$$

$$d_1^H = \frac{1}{\sigma \sqrt{T}} \left\{ \ln \left(\frac{S}{H} \right) + \left(r - q + \frac{1}{2}\sigma^2 \right) T \right\} ,$$

$$d_2^H = \frac{1}{\sigma \sqrt{T}} \left\{ \ln \left(\frac{S}{H} \right) + \left(r - q - \frac{1}{2}\sigma^2 \right) T \right\} ,$$

$$d_4 = \frac{1}{\sigma \sqrt{T}} \left\{ \ln \left(\frac{H}{S} \right) + \left(r - q - \frac{1}{2}\sigma^2 \right) T \right\} ,$$

$$d_5 = \frac{1}{\sigma \sqrt{T}} \left\{ \ln \left(\frac{H^2}{SK} \right) + \left(r - q - \frac{1}{2}\sigma^2 \right) T \right\} ,$$

$$d_6 = \frac{1}{\sigma \sqrt{T}} \left\{ \ln \left(\frac{H}{S} \right) + \left(r - q + \frac{1}{2}\sigma^2 \right) T \right\} ,$$

and

$$d_7 = \frac{1}{\sigma \sqrt{T}} \left\{ \ln \left(\frac{H^2}{SK} \right) + \left(r - q + \frac{1}{2}\sigma^2 \right) T \right\} .$$

The formula for an up-and-out call is obtained immediately by a change of numeraire: an up-and-out call on the risky asset with strike K is nothing but a down-and-out put on cash, with the underlying asset viewed as the unit of account. (This conversion is best understood in the foreign-exchange world, where an option to buy dollars at 1.70 deutschemarks per dollar is (logically) equivalent.)

The pricing formulas for up-and-out puts/down-and-out calls are obtained using very similar techniques. We leave them as an exercise for the interested reader.

Finally, the fair values of knock-in options can be obtained using the parity relation (8.27). For instance, using (8.27) and (8.35) we find that the value of a down-and-in put is

$$P_{KI}(S, T; K, H) = + K e^{-rT} \cdot N(-d_2^H) - S e^{-qT} \cdot N(-d_1^H)$$

$$+ K e^{-rT} \left(\frac{H}{S} \right)^{\left(\frac{2(r-q)}{\sigma^2} - 1 \right)} \cdot \{ N(d_4) - N(d_5) \}$$

$$- S e^{-rT} \left(\frac{H}{S}\right)^{\left(\frac{2(r-q)}{\sigma^2} + 1\right)} \cdot \{N(d_6) - N(d_7)\} .$$

We end this section by presenting some numerical values for a particular barrier option (this example was mentioned earlier).

Example 8.3
A reverse-knock-out dollar put/DEM call.
USD interest rate: 5.85%
DEM interest rate: 4.00%
Volatility: 13.00%
Strike: 1.42 DEM/USD
Knock-out at: 1.27 DEM/USD

180 days to maturity

Spot	1.28	1.30	1.32	1.34	1.36	1.38	1.40	1.4225	1.44
Val.	.0015	.0043	.0068	.0090	.0105	.0115	.0116	.0118	.0117
Δ	.1893	.1823	.1623	.1326	.0968	.0588	.0219	−.0148	−.0111

90 days to maturity

Spot	1.28	1.30	1.32	1.34	1.36	1.38	1.40	1.4225	1.44
Val.	.0035	.0102	.0133	.0193	.0211	.0210	.0196	.0169	.0146
Δ	.4511	.4084	.3185	.2012	.0782	-.0315	−.1160	−.0174	−.0194

30 days to maturity

Spot	1.28	1.30	1.32	1.34	1.36	1.38	1.40	1.4225	1.44
Val.	.0107	.0298	.0411	.0438	.0398	.0320	.0232	.0145	.0135
Δ	1.3610	1.0327	.4989	−.0250	−.0390	−.5579	−.5668	−.4645	−.3544

7 days to maturity

Spot.	1.28	1.30	1.32	1.34	1.36	1.38	1.40	1.4225	1.44
Val.	.0308	.0713	.0737	.0600	.0445	.0411	.0169	.0067	.0025
Δ	3.6459	1.3226	−.4863	−.9666	−.9900	−.9417	−.7798	−.4676	−.2282

[]

We will return to this example later.

Hedging Barrier Options

The risk management of barrier options should take into consideration the mixed-Gamma exposure of these instruments (for reverse-knock-outs and knock-ins) as well as the pin risk at the barrier.

The risk exposure of a reverse knock-out put option can be understood intuitively as follows: from Eq. (8.27), the holder of this option is

- Long a standard put with strike K

- Short an American digital option with barrier at H and with payoff equal to one put with strike K upon hitting the barrier, in other words, a "knock-in put."

Near the expiration date ignoring the difference between a knock-in put and an American digital option with payoff $H - K$ at the barrier is not a bad approximation. We then see immediately that the option has mixed Gamma exposure: far away from the barrier, the standard put dominates and the holder of the knock-out is long Gamma/Vega, whereas near $S = H$ the "digital" dominates and the holder is short Gamma/Vega.

The seller who wishes to hedge faces the mirror-image position: short Vega and Gamma near the strike or in-the-money and long Gamma/Vega closer to the barrier. However, *at the barrier* the Gamma risk is complex: if the spot price is just below the barrier, the hedger must adjust his Delta in order to "earn time decay" (his liability is that of a standard put if the options fails to knock-out). The Delta increases without bounds near the barrier. On the other hand, if the option does not knock out, the large Delta position may be detrimental in case of a large market, because this would lead to a loss in the spot market.

Example 8.4
Consider the option described in the previous section, assuming that an agent sold the option with 180 days to expiration, when the spot price was 1.4225, at 0.0118 per dollar notional. If 7 days before expiration the spot trades at 1.28 DEM/USD and the agent Delta-hedges according to the above tables, his spot position in USD would be long $3.6459 per dollar notional. A drop of 0.03 in the exchange rate in 1 day will result in a loss of $0.1094 per dollar notional. To make this more concrete, assume that the notional amount is $100,000,000. The premium collected for the option was $1,180,000. The spot position, on the other hand is (a whopping) $364,559,000. The loss in the dollar position if the market moves suddenly down by 0.03 through the barrier would be $10,940,000! To have a better idea of the likelihood that this happens, note that this represents a 2.4% move of spot in 1 day. At an annual volatility of 13%, this would be a three-standard deviation move in 1 day. The event has low probability but is not impossible. Would you risk a loss of $9 million given the odds?[11]Moreover, let us mention the important point of **liquidity**. A selling order of nearly $400 million as the exchange-rate goes through the barrier may cause a further drop in the dollar, as there will be few buyers and many sellers. This will have dire consequences for the "hedger."[12]   ▯

8.3   Double Barrier Options

Options with two barriers are widely marketed by financial institutions to their customers. One of the main reasons for the use of double-barrier features is that the price of the option *vis-a-vis* standard options can be reduced substantially. Double-barrier options can be used to express the investor's view about the **trading range** of the underlying asset.

---

[11]One should also take into account that the annualized volatility may very well underestimate the daily move of the exchange rate.
[12]To find out more about the risk-management of exotic options, see Taleb [6], where the liquidity issue in the trading of barrier options is discussed in great depth.

The most common structures that contain two barriers are the following:

- **Range discount note.** The holder receives a cash-flow at a specified date if the underlying asset remains between two predefined levels until that date. If the underlying asset exits the range prior to this date, the holder receives either nothing or a cash rebate.

- **European range accrual note.** The holder receives a specified amount of money each day that the underlying asset remains inside a range, until the contract expires. No payments are made during the days that the asset trades outside the range.

- **American range accrual note.** The holder receives a specified amount of money each day that the underlying asset remains inside a range, until the contract expires or the price exits the range for the first time.

- **Double knock-out options.** The holder receives the payoff of a standard option if the underlying asset remains inside a specified range until the option expires. The holder receives nothing if the option lies outside the range.

These represent some of the most elementary structures. In practice, contracts may involve multiple ranges, complex cash-flows, and multiple underlying assets. In all cases presented above, the structures can be priced with a trinomial lattice with boundary conditions along the barriers. Closed-form solutions for lognormal underlying assets are also available in a limited number of cases.

Range Discount Note

This derivative pays a fixed amount of money if the underlying asset remains inside the range until maturity. It can sometimes be "embedded" or attached to the coupon of a standard bond as a means of enhancing the returns for investors (that bet correctly). Assume that the range is defined by

$$L < S_t < H , \quad t < T ,$$

where L and H are given levels and T is the maturity date.

The numerical scheme (trinomial lattice) for the range note is (with $N = T/dt$)

$$V_n^j = e^{-r_n \, dt} \left[P_U^{(n)} V_{n+1}^{j+1} + P_M^{(n)} V_{n+1}^j + P_D^{(n)} V_{n+1}^{j-1} \right]$$

$$\text{if } L < S_n^j < H , \tag{8.36a}$$

for $n \leq N$, where $P_U^{(n)}$, $P_M^{(n)}$, and $P_D^{(n)}$ are risk-neutral probabilities. The boundary conditions are

$$V_n^j = 0 \ \text{if } S_n^j \geq H \text{ or } S_n^j \leq L \tag{8.36b}$$

and

$$V_N^j = \text{final payoff} . \tag{8.36c}$$

For example, consider the case of a note that pays the holder USD 1 million if (and only if) the USD/JPY remains inside the range 115–130 for 90 days. The payoff of this security is represented in Figure 8.7 below, assuming than the yen and dollar interest rates are, respectively, 1% and 5% and that the volatility is 16%. The fair value of this security is approximately JPY 15.9 MM (USD 132,500) at the exchange rate of 120 yen per dollar. This security gives therefore a great return to the investor if the range is not exited in 90 days.

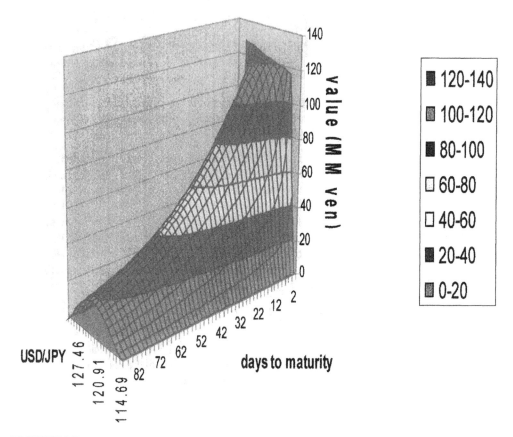

FIGURE 8.7
Fair value of a range discount note computed using a trinomial scheme. The parameters used were $r_{JPY} = 1\%$, $r_{US} = 5\%$, $\sigma = 16\%$.

Range Accruals

European range accrual notes pay a cash-flow for each day that the underlying security remains in a range, but do not "knock out" if the underlying asset exits the range. Therefore, they can be viewed as portfolios of European "digital" options with payoffs

$$f(S) = \begin{cases} 1 & \text{if } L < S < H \\ 0 & \text{otherwise} \end{cases}$$

which expire on each day between inception and maturity. Pricing can be done by closed-form solutions. In fact, the value of such an option in the log-normal setting is

$$V = e^{-rt} \mathbf{Pr}\{L < S_t < H\}$$

$$= e^{-rt} \, \mathbf{Pr}\{L < S_t\} - e^{-rt} \, \mathbf{Pr}\{H < S_t\}$$

$$= e^{-rt} \, [\, N(d_L) - N(d_H) \,] \tag{8.37}$$

where $N(\cdot)$ is the cumulative normal distribution,

$$d_L = \frac{1}{\sigma\sqrt{t}} \ln\left(\frac{S e^{-qt}}{L e^{-rt}}\right) - \frac{1}{2}\sigma\sqrt{t}$$

and

$$d_H = \frac{1}{\sigma\sqrt{t}} \ln\left(\frac{S e^{-qt}}{H e^{-rt}}\right) - \frac{1}{2}\sigma\sqrt{t}. \tag{8.38}$$

(We used Eq. (8.4) to derive this result.)

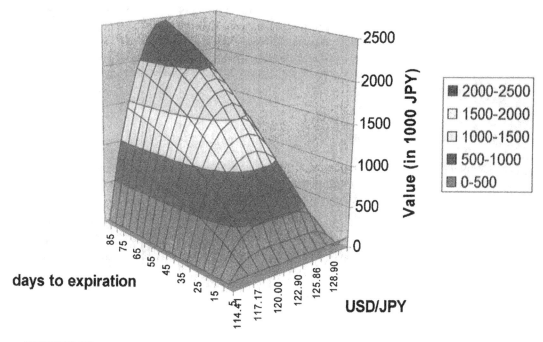

FIGURE 8.8
Fair value of a range-accrual option computed using a trinomial scheme. The parameters used were $r_{JPY} = 1\%$, $r_{US} = 5\%$, $\sigma = 16\%$.

The American range-accrual note "knocks out" if the price of the underlying asset crosses one of the barriers. Assuming a year of 360 days, for example, the recursive equations for pricing this derivative security are

$$V_n^j = e^{-r_n \, dt} \left[\, P_U^{(n)} \, V_{n+1}^{j+1} + P_M^{(n)} \, V_{n+1}^{j} + P_D^{(n)} \, V_{n+1}^{j-1} \,\right]$$

$$+ \text{daily payoff} \cdot dt \cdot 360 \quad \text{if } L < S_n^j < H, \tag{8.39a}$$

for $n \leq N$, with boundary conditions

$$V_n^j = 0 \text{ if } S_n^j \geq H \text{ or } S_n^j \leq L \tag{8.39b}$$

and

$$V_N^j = 0. \tag{8.39c}$$

Figure 8.8 displays the graph of the value of a range-accrual option that pays JPY 50,000 for each day that USD/JPY remains in the range 115–130 for a maximum period of 90 days. Notice that this option has negative convexity: the seller of the option becomes "long volatility," i.e., long Gamma and long Vega. If the exchange rate volatility increases, the value of the option decreases. Conversely, if volatility decreases, the value of the option increases. This type of product is often sold to investors who wish to increase their "income" and believe that the underlying security (the exchange rate, in this case) will trade in a narrow range. For this reason this corridor option is sometimes known as a **boost** option (as in boosting the yield or return on investment). The down side for the investor is that the market becomes volatile and the price gets out of range.

Double Knock-out Options

The numerical scheme for computing the fair value of these options is (8.36a) with boundary conditions (8.36b) and terminal payoff equal to that of the corresponding vanilla option. See Figure 8.9 for the graph of a call option with two barriers.

References and Further Reading

[1] Bowley, G. (November 16, 1996), New Breed of Exotic Thrives, *Financial Times,* supplement on derivatives.

[2] Billingsley, P. (1968), *Convergence of Probability Measures,* John Wiley & Sons, New York.

[3] Hull, J. (1997), *Introduction to Futures and Options Markets,* Prentice-Hall, Englewood Cliffs, NJ.

[4] Rubinstein, M. (1992), Exotic Options. HAAS School of Business. Working Paper.

[5] (1992), *From Black–Scholes to Black Holes,* Risk Magazine.

[6] Taleb, N. (1997), *Dynamic Hedging: Managing Vanilla and Exotic Options,* John Wiley & Sons, New York.

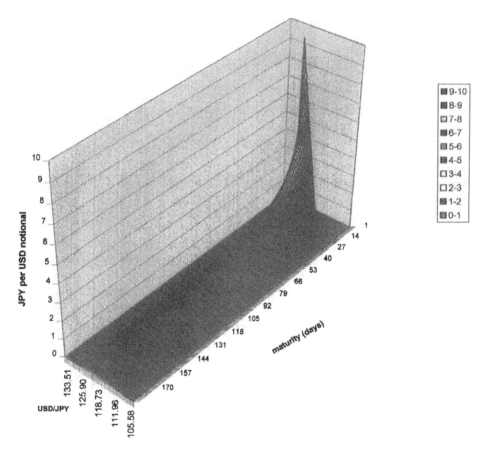

FIGURE 8.9
Fair value of a 180-day USD call/JPY put with strike 120 and KO barriers at 110 and 130. The parameters used were $r_{JPY} = 1\%$, $r_{US} = 5\%$, $\sigma = 16\%$.

Appendix A: Proofs of Lemmas 8.1 and 8.2

This section contains sketches of the proofs of the two lemmas used to derive closed-form solutions for single-barrier options and American digitals.

A.1 A Consequence of the Invariance of Brownian Motion Under Reflections

LEMMA A.3

Let Z_t denote standard Brownian motion on the interval $[0, T]$. Then, for all $A > 0$ and

$B < A$, we have

$$P\left\{\max_{0\le t\le T} Z_T > A \; ; \; Z_T \in (B, B+dB)\right\} = \frac{1}{\sqrt{2\pi T}}\, e^{-\frac{(2A-B)^2}{2T}}\, dB. \qquad (A.1)$$

PROOF   Consider a simple random walk defined by $X_0 = 0$ and

$$X_n = X_{n-1} \pm \sqrt{dt} \quad , n = 1, 2, \ldots, N$$

where dt represents a small positive number. The probabilities for $+\sqrt{dt}$ and $-\sqrt{dt}$ are assumed to be $1/2$. Set

$$A' \equiv \left[\frac{A}{\sqrt{dt}}\right] \sqrt{dt},$$

where $[X]$ represents the *integer part* of X. Therefore, A' represents the largest integer multiple of $\sqrt{dt}$ that is $\le A$.

Assume that a given path, or realization, of the random walk is such that $X_n = A'$ for some $n \le N$ and that $X_m < A'$ for $m < n$. We observe that the path that coincides with this realization of the random walk for $m \le n$ and is *reflected about the line* $X = A'$ for $m > n$ occurs with the same probability as the original one (namely $\left(\frac{1}{2}\right)^N$).

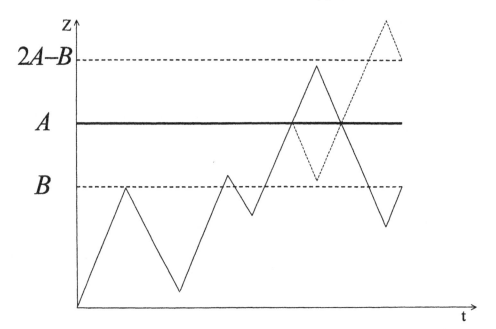

FIGURE A.10
The dotted line represents the reflection of the original path about the level $Z = A$. If the original path ends at $Z(T) = B$ the reflected path ends at $Z'(T) = A+(A-B) = 2A-B$.

Therefore, we conclude that for $B < A'$,

$$P\left\{\max_{1\le j\le N} X_j \ge A' \; ; \; X_N = B\right\} = P\left\{\max_{1\le j\le N} X_j \ge A' \; ; \; X_N = 2A' - B\right\}$$

$$= \mathbf{P} \left\{ X_N = 2A' - B \right\} . \tag{A.2}$$

(This last equality holds because $2A' - B > A'$.)

Let $T = N \, dt$. By the Central Limit theorem, the joint distribution of the random variables $X_{[nt]}$ approaches that of a Brownian motion as $dt \to 0, N \to +\infty$. Therefore, if we replace formally $\max_{1 \le j \le N} X_j$ by $\max_{0 \le t \le T} Z_t$, X_N by Z_T, and A' by A, we conclude from (A.2) that Eq. (A.1) holds. The lemma is proved.[13] ∎

Notice that we have established the analogue of Lemma 8.2 in the case $\mu = 0$.

A.2 The Case $\mu \neq 0$

To prove Lemma 8.2, we will need the following result about Brownian motion with drift:

THEOREM A.1
(**Cameron – Martin**) *Let* $F(z_1, z_2, \ldots, z_n)$ *be a continuous function. Then,*

$$\mathbf{E} \left\{ F \left(Z_{t_1}^{\mu}, Z_{t_2}^{\mu}, \ldots, Z_{t_n}^{\mu} \right) \right\}$$

$$= \mathbf{E} \left\{ F \left(Z_{t_1}, Z_{t_2}, \ldots, Z_{t_n} \right) \cdot e^{\mu Z_{t_n} - \frac{1}{2} \mu^2 t_n} \right\} . \tag{A.3}$$

This result states that the expectation of a function of Brownian motion with drift is equal to the expectation of the same function of regular Brownian multiplied by an exponential factor, namely

$$e^{\mu Z_T - \frac{1}{2} \mu^2 T} .$$

Proof of the Theorem   Define the increments of the Brownian path

$$Y_j^{\mu} = Z_{t_j}^{\mu} - Z_{t_{j-1}}^{\mu}$$

$$= \left(Z_{t_j} - Z_{t_{j-1}} \right) + \mu \left(t_j - t_{j-1} \right) .$$

Also, set

$$G(y_1, y_2, \ldots, y_n) \equiv F(y_1, y_1 + y_2, \ldots, y_1 + y_2 + \ldots y_n)$$

Using the explicit form of the Gaussian distribution and the fact that the increments Y_j are independent random variables with mean $\mu (t_j - t_{j-1})$ and variance $t_j - t_{j-1}$, we obtain

$$\mathbf{E} \left\{ F \left(Z_{t_1}^{\mu}, Z_{t_2}^{\mu}, \ldots, Z_{t_n}^{\mu} \right) \right\} = \mathbf{E} \left\{ G \left(Y_1^{\mu}, Y_2^{\mu}, \ldots, Y_n^{\mu} \right) \right\}$$

---

[13]For a rigorous proof of the formal passage to the limit (A.2) $\Longrightarrow$ (A.1), see for instance Billingsley, *Convergence of Probability Measures*, Wiley, 1968.

$$= \frac{1}{(2\pi)^{n/2}} \int_{\mathbf{R}^n} G(y_1, y_2, \ldots, y_n) \cdot \exp \left\{ -\sum_{j=1}^n \frac{(y_j - \mu(t_j - t_{j-1}))^2}{2(t_j - t_{j-1})} \right\}$$

$$\frac{dy_1 \, dy_2 \ldots dy_n}{\sqrt{(t_1 - t_0) \cdot \ldots (t_n - t_{n-1})}}$$

$$= \frac{1}{(2\pi)^{n/2}} \int_{\mathbf{R}^n} \tilde{G}(y_1, y_2, \ldots, y_n) \cdot \exp \left\{ -\sum_{j=1}^n \frac{y_j^2}{2(t_j - t_{j-1})} \right\}$$

$$\cdot \frac{dy_1 \, dy_2 \ldots dy_n}{\sqrt{(t_1 - t_0) \cdot \ldots (t_n - t_{n-1})}}, \tag{A.4}$$

where

$$\tilde{G}(y_1, y_2, \ldots, y_n) = G(y_1, y_2, \ldots, y_n) \cdot e^{\mu \sum_{j=1}^n y_j - \frac{1}{2} \mu^2 t_n}.$$

Making a change of variables, we find that the last integral in (A.4) is equal to

$$\mathbf{E} \left\{ F(Z_{t_1}, Z_{t_2}, \ldots, Z_{t_n}) \cdot e^{\mu Z_{t_n} - \frac{1}{2} \mu^2 t_n} \right\},$$

which is what we wanted to show. This concludes the proof of the theorem. ∎

Proof of Lemma 8.2   Using an approximation argument, which we omit, it can be shown that the above theorem can also be applied to the *functional* of the path

$$F(Z_\cdot^\mu) \equiv \underset{0 \le t \le T}{Max} Z_t^\mu.$$

(This is a continuous function of the path that can be approximated in a suitable sense by continuous functions of n variables, as in the previous theorem.[14])

Applying the theorem to this functional, we conclude that if $C < A$, then

$$\mathbf{P} \left\{ \underset{0 \le t \le T}{Max} Z_t^\mu \ge A; \; Z_T^\mu < C \right\}$$

$$= \mathbf{E} \left\{ \underset{0 \le t \le T}{Max} Z_t \ge A; \; Z_T < C; \; e^{\mu Z_T - \frac{1}{2} \mu^2 T} \right\}$$

$$= \int_0^C \mathbf{E} \left\{ \underset{0 \le t \le T}{Max} Z_t \ge A; \; Z_T = B \right\} \cdot e^{\mu B - \frac{1}{2} \mu^2 T} \, dB$$

$$= \int_0^C e^{-\frac{(2A-B)^2}{2T}} e^{\mu B - \frac{1}{2} \mu^2 T} \frac{dB}{\sqrt{2\pi T}}, \tag{A.5}$$

---

[14]See Billingsley [2].

where we used Lemma A.3 to derive the last equality. Lemma 8.2 follows immediately by differentiating both sides of (A.1). ∎

Finally, we prove Lemma 8.1 on the distribution of the first-passage time for Brownian motion with drift.

Proof of Lemma 8.1   Using the notation of Lemmas 8.1 and 8.2, we find that

$$\mathbf{P}\{\tau_A < T\}$$

$$= \mathbf{P}\left\{ \underset{0 \le t \le T}{Max} \; Z_t^\mu \ge A \right\}$$

$$= \int_0^A \mathbf{P}\left\{ \underset{0 \le t \le T}{Max} \; Z_t^\mu \ge A \; ; \; Z_T^\mu = B \right\} dB$$

$$= \int_0^A \mathbf{P}\{Z_T^\mu = B\} \, dB - \int_0^A \mathbf{P}\left\{ \underset{0 \le t \le T}{Max} \; Z_t^\mu < A \; ; \; Z_T^\mu = B \right\} dB$$

$$= \int_0^A e^{-\frac{B^2}{2T}} \frac{dB}{\sqrt{2\pi T}} - \int_0^A e^{-\frac{(2A-B)^2}{2T}} \frac{dB}{\sqrt{2\pi T}} . \tag{A.6}$$

The conclusion of Lemma 8.1 follows by evaluating this last expression in terms of the cumulative normal distribution. ∎

Appendix B: Closed-Form Solutions for Double-Barrier Options

B.1 Exit Probabilities of a Brownian Trajectory from a Strip $-B < Z < A$

We derive closed-form solutions for the values of double-barrier options assuming lognormal dynamics for the underlying asset price. For this purpose, we study the exit probabilities of a Brownian motion path from a strip.

We denote by τ_A the first time that a path crosses the level $Z = A$. To study the probability of exit from a strip, it is convenient to consider first the case $\mu = 0$. Let $\tau_{-B, A}$ denote the first time that the path exits the *strip* $\{-B < Z < A\}$, where $A, B > 0$. We set $\delta = A + B$.

Clearly, for $-B < X < A$, we have

$$\mathbf{P}\{ Z_T \in (X, X + dX) \; ; \; \tau_{-B,A} \ge T\}$$

$$= \mathbf{P}\{ Z_T \in (X, X + dX)\}$$

$$-\mathbf{P}\{ \ Z_T \in (X, X + dX) \ ; \ \tau_A < T \ ; \ \tau_A < \tau_{-B}\}$$

(exit first through A)

$$-\mathbf{P}\{ \ Z_T \in (x, x + dx) \ ; \ \tau_{-B} < T \ ; \ \tau_{-B} < \tau_A\}$$

(exit first through $- B$) .   (B.7)

The last two probabilities can be estimated using the Brownian Reflection Principle discussed in Appendix A. Consider the case $\tau_A < \tau_{-B}$, for example. We have

$$\mathbf{P}\{ \ Z_T \in (X, X + dX) \ ; \ \tau_A < T \ ; \ \tau_A < \tau_{-B}\}$$

$$= \mathbf{P}\{ \ Z_T \in (X, X + dX) \ ; \tau_A < T \ \}$$

$$-\mathbf{P}\{ \ Z_T \in (X, X + dX) \ ; \ \tau_{-B} < \tau_A < T\}$$

$$= \mathbf{P}\{ \ Z_T \in (2A - X, 2A - (X + dX))\} \ -$$

$$\mathbf{P}\{Z_T \in (2A - X, 2A - (X + dX)) \ ; \tau_{-B} < \tau_A < T\} .$$ (B.8)

To obtain the second equality, we used the Reflection Principle about the level $Z = A$. Since Z_T is normally distributed, the first probability in the last equation can be written explicitly. Let us then focus on the second term. We obtain

$$\mathbf{P}\{Z_T \in (2A - X, 2A - (X + dX)) \ ; \tau_{-B} < \tau_A < T\}$$

$$= \mathbf{P}\{Z_T \in (2A - X, 2A - (X + dX)) \ ; \tau_{-B} < T\}$$

$$-\mathbf{P}\{Z_T \in (2A - X, 2A - (X + dX)) \ ; \tau_A < \tau_{-B} < T\}$$

$$= \mathbf{P}\{Z_T \in (X - 2\delta, (X + dX) - 2\delta)\}$$

$$-\mathbf{P}\{Z_T \in (X + 2\delta, (X + dX) - 2\delta) \ ; \tau_A < \tau_{-B} < T\} .$$ (B.9)

Again, the first probability can be expressed in terms of the normal distribution. The second term in the last equation is similar to the expression we started with [top of (B.8)], except for the fact that this time it was obtained by reflecting about $Z = -B$ and that the endpoint is different. Applying this procedure iteratively, i.e., making successive reflections about A and $-B$, we obtain

$$\mathbf{P}\{ \ Z_T \in (X, X + dX); \ \tau_A < T \ ; \ \tau_A < \tau_{-B}\}$$

$$= \sum_{n=0}^{+\infty} (-1)^n \ \mathbf{P}\{Z_T \in I_n\}$$

$$= \sum_{n=0}^{+\infty} (-1)^n \int_{I_n} e^{-\frac{x^2}{2T}} \frac{dx}{\sqrt{2\pi T}} \tag{B.10}$$

where I_n are intervals of width dX given by

$$I_n = \begin{cases} (X - n\delta, \ X + dX \ - \ n\delta) & \text{for } n \text{ even} \\ \\ (2A + (n-1)\delta \ - \ X, \ 2A + (n-1)\delta \ - X - dX) & \text{for } n \text{ odd}. \end{cases}$$

The series in (B.10) converges very fast. In fact, the remainder of the nth partial sum decays like $e^{\frac{-\delta^2 n^2}{2T}}$ as $n \to \infty$. To see this, notice that the interval I_n lies at a distance of order $n\,\delta$ from the origin after n reflections. Therefore, the remainder of the nth partial sum (assuming that n is odd, for example) satisfies

$$\mathbf{P}\{ \ Z_T \in I_n \ ; \ \ \tau_A < T \ ; \ \tau_A < \tau_{-B}\} \leq \mathbf{P}\{ \ Z_T \in I_n\} \approx \frac{1}{\sqrt{2\pi T}} e^{-\frac{n^2 \delta^2}{2T}} dX \ . \tag{B.11}$$

Clearly, this estimate also holds if n is even (interchanging A and $-B$). The same argument can be made for the term in (B.8) corresponding to $\tau_{-B} < \tau_A$, which means that the probability of not exiting the strip and ending at position X can be expressed as a series involving the Gaussian distribution.

PROPOSITION B.1

For $A, B > 0$ and $Z(t)$ a standard Brownian motion, we have

$$\mathbf{P}\{ \ Z_T \in (X, X + dX) \ ; \ \tau_A < T \ ; \ \tau_A < \tau_{-B}\} \ = \ p_{-B, A} \ (X, T) \, dX \ ,$$

where

$$p_{-B, A} \ (X, \ T) \ = \ \sum_{n=-\infty}^{\infty} \ (G_T(X + 2n\delta) - G_T(2A - X + 2n\delta)) \tag{B.12}$$

with

$$G_T(X) = \frac{1}{\sqrt{2\pi T}} e^{-\frac{X^2}{2T}} \ .$$

PROOF   Since the series corresponding to the terms with $\tau_A < \tau_{-B}\}$ and $\tau_{-B} < \tau_A\}$ in Eq. (B.8) involve only the distribution of $Z(T)$, all terms are given explicitly in terms of the normal distribution. Formula (B.12) follows by adding the two series and rearranging terms.∎

This result is easily extended to the case of a Brownian motion with drift, by using the Cameron–Martin theorem.

PROPOSITION B.2

Let Z^μ be a Brownian motion with drift. Then

$$\mathbf{P}\{ \ Z_T^\mu \in (X, X + dX) \ ; \ \tau_A < T \ ; \ \tau_A < \tau_{-B}\} \ = \ p_{-B, A}^\mu(X, T) \, dX$$

where

$$p^{\mu}_{-B,A}(X,\,T) \;=\; p_{-B,A}(X,\,T) \cdot \exp\left(\mu X \,-\, \frac{1}{2}\mu^2 T\right).$$

PROOF   Apply the Cameron–Martin theorem.   ■

REMARK B.1   In numerical computations, a truncation of the series (B.12) must be made. Here, we can make use of the explicit error estimate in (B.10), in terms of δ, n, and T, which allows for very good control of the error. For standard values of the parameters A, B, σ, μ, and T, adding only a few terms is enough to give an extremely accurate result.   ■

B.2   Applications to Pricing Barrier Options

Closed-form solutions for the prices of double-barrier options can be obtained using the function $p^{\mu}_{-B,\,A}(X)$. Consider the price process

$$S(t) \;=\; S\,e^{\sigma\,(Z_t + \mu t)} \;=\; S\,e^{\sigma\,Z^{\mu}_t}\,, \quad \mu \;=\; \frac{r-q}{\sigma} \,-\, \frac{1}{2}\sigma\,.$$

We define the quantities

$$A_H \;=\; \frac{1}{\sigma}\ln\left(\frac{H}{S}\right)$$

and

$$A_L \;=\; \frac{1}{\sigma}\ln\left(\frac{K}{S}\right).$$

The following formulas for the prices of double-barrier options follow immediately.

a. Range Notes

$$V \;=\; N \cdot d e^{-rT}\int_{A_L}^{A_H} p^{\mu}_{A_L,\,A_H}(X,\,T)\,dX$$

where N is the payoff.

b. American Range Accrual Notes

A range accrual note paying p dollars every Δt units of time with maturity $T = N\,\Delta t$ can be regarded as a portfolio of N range discount notes. Accordingly, we have

$$V \;=\; p\sum_{n=1}^{N} e^{-r n\,\Delta t}\int_{A_L}^{A_H} p^{\mu}_{A_L,\,A_H}(X,\,n\Delta t)\,dX\,.$$

If payments are made frequently (for instance daily), we can approximate the sum by an integral, setting $p = Q\,\Delta T$. In this case, the sum is well approximated by the integral

$$V \;=\; Q\int_{0}^{T} e^{-rt}\int_{A_L}^{A_H} p^{\mu}_{A_L,\,A_H}(X,\,t)\,dX\,dt\,.$$

Here Q can be interpreted as the payout that the holder would receive in 1 year if the price did not exit the range. (In the example given in Section 8.3, we had a daily payoff of 50K yen. Assuming a 365-day year, this would give $Q = 360 \times 50 = 18,250,000$.

c. Double Knock-out Options

For a call with strike K, we have

$$V = e^{-rT} \int_{A_L}^{A_H} \max(e^{\sigma X} K, \ 0) \, p_{A_L, A_H}^{\mu}(X, \ T) \, dX \ .$$

Chapter 9

Ito Processes, Continuous-Time Martingales, and Girsanov's Theorem

This short chapter introduces useful concepts in Probability Theory that are used in subsequent chapters. In particular, Girsanov's theorem will play an important role in the derivation of no-arbitrage theorems in the next chapter.

9.1 Martingales and Doob–Meyer Decomposition

We have seen previously in Chapter 7 that an Ito process can be represented by the stochastic differential

$$dX(t) = \sigma(t) \, dZ(t) + \mu(t) \, dt \,, \tag{9.1}$$

where σ and μ are (nonanticipative) local parameters. Equivalently, $X(t)$ can be written in integral form:

$$X(t) = X(0) + \int_0^t \sigma(s) \, dZ(s) + \int_0^t \mu(s) \, ds$$

$$\equiv X(0) + M(t) + B(t) \,. \tag{9.2}$$

Let us analyze the processes $M(\cdot)$ and $B(\cdot)$ in this decomposition. For each t, $M(t)$ has mean zero since it is a stochastic integral. A more fundamental property of the process $M(\cdot)$ is that it is a **martingale**. This means that for all $T > t$, we have

$$\mathbf{E}_t \{ M(T) \} = M(t) \,, \tag{9.3}$$

where $\mathbf{E}_t$ represents the **conditional expectation operator given the history up to time** t. A remark on this last point: in the last chapter, we assumed, for simplicity, that the nonanticipative functions (σ, μ) were completely determined at time t by the path $Z(s), s \leq t$. The "history up to time t" thus meant the observed values of the Brownian path used to define the stochastic integral. It is useful to generalize this framework by assuming that the functions σ and μ may depend on other random processes in addition to the Brownian path $Z(\cdot)$ in (9.1). (For instance, we expect the volatility of an asset to be affected by changes in other economically correlated variables.) In these lectures, the expression "history up to time t" will be understood to mean *the observed values up to time t of all processes that determine the local parameters.*

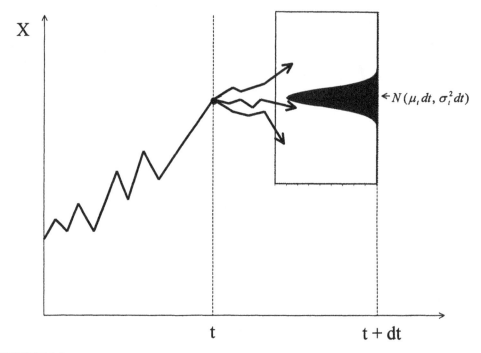

FIGURE 9.1
Intuitively, an Ito process is such that at time t the statistics for the increment $dX = X(t + dt) - X(t)$ is normally distributed with mean $\mu_t \, dt$ and variance $\sigma_t^2 \, dt$. The cumulative sum of the centered increments, $\sigma_t \, dZ_t$, given by $\int_0^t \sigma_s \, dZ_s$, is the martingale component in the Doob–Meyer decomposition.

The martingale property of $M(t)$ follows from Proposition B.2 in Chapter 8. In fact, note that

$$M(T) = M(t) + \int_t^T \sigma(s) \, dZ(s).$$

Since the stochastic integral has conditional expectation zero given the past up to time t,

$$\mathbf{E}_t \{ M(T) \} = M(t).$$

Moreover,

$$\mathbf{E}_t \{ \sigma(s) \, dZ(s) \} = \mathbf{E}_t \{ \mathbf{E}_s \{ \sigma(s) \, dZ(s) \} \}$$

$$= \mathbf{E}_t \{ \mathbf{E}_s \{ \sigma(s) \} \cdot \mathbf{E}_s \{ dZ(s) \} \}$$

$$= 0$$

due to the independence of $dZ(s)$ from the past up to time s.

The concept of martingale in Probability Theory formalizes the intuitive notion of *fair game.* For example, a "coin-tossing" game, in which a gambler bets on the outcome of a coin toss several times in a row, has an accumulated wealth process that is a martingale. In contrast, the accumulated wealth for the game of roulette (betting, say, on the color red each time) is not a martingale, because the house wins each time zero (green) occurs.[1]

Stochastic integrals are martingales with continuous paths. A well-known theorem in Probability Theory [3] shows that all such processes can be viewed as Brownian motions with "random time change."[2] In particular, we can study the quadratic variation of a martingale just like we considered the quadratic variation of Brownian paths. More precisely, the limit of the sums of squares of increments corresponding to a sequence of increasingly refined partitions satisfies

$$\lim_{\Delta t \to 0} \sum_j |\Delta M_j|^2 = \int_0^t \sigma^2(s)\,ds$$

in the mean-square sense.[3]

In contrast, $B(\cdot)$ has bounded first variation (and thus vanishing quadratic variation) since

$$\lim_{\Delta t \to 0} \sum_j |\Delta B_j| = \int_0^t |\mu(s)|\,ds$$

whenever the integral on the right-hand side exists.

The stochastic process $M(t)$ and $B(t)$ are known, respectively, as the **martingale component** and the **bounded variation component** of the Ito process $X(t)$. The last equation in (9.2) is often referred to as the **Doob–Meyer decomposition** of $X(\cdot)$.

9.2 Exponential Martingales

PROPOSITION 9.1
Let $\sigma(t)$ and $\mu(t)$ be bounded nonanticipative processes with respect to some Brownian motion $Z(\cdot)$. Then, the stochastic process $S(t)$ satisfies the stochastic differential equation

$$dS(t) = S(t)\,[\,\sigma(t)\,dZ(t) + \mu(t)\,dt\,] \tag{9.4}$$

with

$$S(0) = S_0$$

if and only if

$$S(t) = S_0\,e^{\int_0^t \sigma(s)\,dZ(s) - \frac{1}{2}\int_0^t \sigma^2(s)\,ds + \int_0^t \mu(s)\,ds}. \tag{9.5}$$

Setting $\mu = 0$ in (9.5), it follows from this proposition that the solution of the stochastic differential equation

$$dM(t) = M(t)\,\sigma(t)\,dZ(t) \tag{9.6}$$

---

[1]Theoretical results on gambling in games of chance have been sought since the days of Laplace and surely earlier. See Thorp [8], for interesting mathematical results on games of chance and speculative investing.
[2]More precisely, any continuous martingale $X(t)$ can be written in the form $X(t) = Z(\Theta(t))$ where $Z(\theta)$ is a Brownian motion and $\Theta(t)$ is nonanticipative and increasing.
[3]The difference between the successive sums and the integral has a second moment tending to zero as $\Delta t \to 0$.

with $S(0) = 1$, is given by

$$M(t) = e^{\int_0^t \sigma(s)\,dZ(s) - \frac{1}{2}\int_0^t \sigma^2(s)\,ds} \ . \tag{9.7}$$

The latter processes are called **exponential martingales**. Notice here again the difference between standard calculus and Ito calculus: in the "smooth world," the solution of the differential equation (9.6) is given by dropping the ds-integral from the exponent of (9.7).[4]

Proof of Proposition 9.1   We show that if $S(t)$ satisfies (9.4) then it must have the form (9.5). For this, let us apply the Generalized Ito Lemma to the function $\ln S(t)$. Accordingly, we find that

$$d\,(\ln S(t)) = \frac{1}{S(t)}\,dS(t) + \frac{1}{2}\frac{-1}{S(t)^2}\,(dS(t))^2$$

$$= \frac{1}{S(t)}\,dS(t) + \frac{1}{2}\frac{-1}{S(t)^2}\,\sigma^2(t)\,S(t)^2\,dt$$

$$= \sigma(t)\,dZ(t) + \mu(t)\,dt - \frac{1}{2}\sigma^2(t)\,dt \ ,$$

and thus

$$\ln S(t) = \int_0^t \sigma(s)\,dZ(s) - \frac{1}{2}\int_0^t \sigma^2(s)\,ds + \int_0^t \mu(s)\,ds + \text{const} \ .$$

This shows that (9.4) implies (9.5). The converse statement also follows from an application of the Generalized Ito Lemma; it is left to the reader as an exercise. ∎

REMARK 9.1

1. If $\sigma(t)$ and $\mu(t)$ are independent of the Brownian motion $Z(\cdot)$, the proposition gives an explicit solution of the stochastic differential equation (SDE) with linear coefficients (9.4). This is one of a few cases in which a closed-form solution for an SDE exists. If, on the other hand, $\sigma(t)$ or $\mu(t)$ depends on $S(t)$ (or $Z(t)$), then (9.5) should not be viewed as a "solution" of the differential equation, because the right-hand side of (9.5) may depend on $S(s)$, $s \le t$ (through σ and μ.) Nevertheless, this result will be very useful.

2. Stochastic differential equations such as (9.4) are often used to describe the accumulated wealth of investment strategies or the evolution of security prices: in fact, if we write Eq. (9.4) in the form

$$\frac{dS(t)}{S(t)} = \sigma(t)\,dZ(t) + \mu(t)\,dt \ ,$$

the parameters $\mu\,dt$ and $\sigma\sqrt{dt}$ can be interpreted as the infinitesimal mean and the infinitesimal volatility of returns.

---

[4]Notice also that if $\sigma \equiv 0$, the differential equation (9.4) has the "classical" solution $S(t) = S_0\,e^{\int_0^t \mu(s)\,ds}$.

3. Futures prices are often modeled as exponential martingales. The reason is that, in an ideal market, the expected return on an open futures position held for a small trading period should be exactly zero (after adjusting for the price of risk, as we shall see in the following chapter). We have already encountered this result in the context of the binomial pricing model.[5]

∎

9.3 Girsanov's Theorem

We present an important application of the concept of exponential martingale, which is related to the idea of *change of probability* in path-space.[6]

PROPOSITION 9.2

(Girsanov's theorem) *Consider a probability measure P on the space of paths $Z(t)$, $t \leq T$ such that $Z(\cdot)$ is a Brownian motion and assume that $b(\cdot)$ is a nonanticipative function. Set*

$$M(t) \equiv e^{\int_0^t b(s)\,dZ(s) - \frac{1}{2}\int_0^t (b(s))^2\,ds} \quad , \quad t \leq T ,$$

and define a new measure Q on the set of trajectories $\{Z(t),\ t \leq T\}$ by

$$Q\{S\} \equiv \mathbf{E}^P\{M(T); S\} = \int_S M(T)\,dP , \qquad (9.8)$$

where S represents an arbitrary set of paths and $\mathbf{E}^P$ is the expectation operator with respect to the probability P. Then, the random process

$$W(t) = Z(t) - \int_0^t b(s)\,ds \quad, t \leq T ,$$

is a Brownian motion under the measure Q.

REMARK 9.2   An important use of Girsanov's theorem is in transforming Ito processes with general drifts into martingales. Indeed, suppose $X(t)$ is an Ito process that satisfies

$$dX(t) = \sigma(t)dZ_t + \mu(t)\,dt$$

and define the process $W(t)$ as above, letting $b(t) = -\frac{\mu(t)}{\sigma(t)}$. Then,

$$W(t) = Z(t) + \int_0^t \frac{\mu(s)}{\sigma(s)}ds$$

---

[5]Recall that the cost of entering a futures contract is zero. Therefore, under the pricing measure the expectation of the cash-flows of the futures contract over a small period (during which interest rates are known) is zero.
[6]The Cameron–Martin theorem discussed in an earlier chapter can be viewed as a special case of Girsanov's theorem.

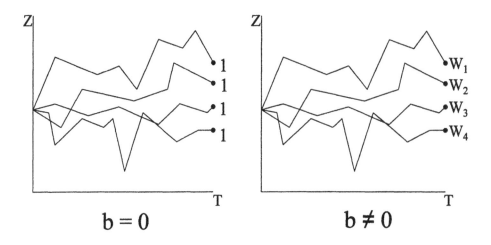

$$W = \exp\left[\int_0^T b_t \, dZ - \frac{1}{2}\int_0^T b_t^2 \, dt\right]$$

FIGURE 9.2
Illustration of Girsanov's theorem. Assume that the diagram on the left represents a random sample of Brownian paths without drift. Each path is equally probable under the Brownian motion statistics. On the right-hand side, we represent the same paths but we multiply the probabilities of the paths by the "weights" W_1, W_2, ..., computed using Girsanov's formula. The theorem says that the probability measure obtained in this way corresponds to a sample of Brownian motion with drift b over the interval $[0, T]$.

and we have

$$dX(t) = \sigma(t)\left(dW(t) - \frac{\mu(t)}{\sigma(t)}dt\right) + \mu(t)\,dt$$

$$= \sigma(t)\,dW(t)$$

so that $X(t)$ is a martingale with respect to the measure Q. ∎

Proof of Proposition 9.3   We will verify that, for all real numbers λ, we have

$$\mathbf{E}_t^Q\left\{e^{\lambda\,(W(T) - W(t))}\right\} = e^{\frac{\lambda^2}{2}(T-t)}. \tag{9.9}$$

This property actually characterizes $W(\cdot)$ as being Brownian motion. First of all, Eq. (9.9) implies that W is Gaussian, because the (unconditional) moment-generating function of $W(T) - W(t)$ is also given by $e^{\frac{\lambda^2 (T-t)}{2}}$.[7]

---

[7] The moment-generating function of a Gaussian random variable with mean zero and variance σ^2 is $\mathbf{E}\left\{e^{\lambda X}\right\} = e^{\frac{\lambda^2 \sigma^2}{2}}$. Therefore (9.9) implies that $W(T) - W(t)$ is Gaussian with mean zero and variance $T - t$.

Equation (9.9) also implies that two successive increments, say, $W(t + a) - W(t)$ and $W(t + a + a') - W(t + a)$, are statistically independent (if $W(\cdot)$ has probability distribution Q). In fact, for any a, $a' > 0$ and all λ_1, λ_2, we have

$$\mathbf{E}^Q \left\{ e^{\lambda_1 (W(t+a) - W(t))} \cdot e^{\lambda_2 (W(t+a+a') - W(t+a))} \right\}$$

$$= \mathbf{E}^Q \left\{ \mathbf{E}^Q_{t+a} \left\{ e^{\lambda_1 (W(t+a) - W(t))} e^{\lambda_2 (W(t+a+a') - W(t+a))} \right\} \right\}$$

$$= \mathbf{E}^Q \left\{ e^{\frac{\lambda_1^2}{2} a} \mathbf{E}^Q_{t+a} \left\{ e^{\lambda_2 (W(t+a+a') - W(t+a))} \right\} \right\}$$

$$= e^{\frac{\lambda_1^2}{2} a} \cdot e^{\frac{\lambda_2^2}{2} a'}$$

$$= \mathbf{E}^Q \left\{ e^{\lambda_1 (W(t+a) - W(t+a))} \right\} \cdot \mathbf{E}^Q \left\{ e^{\lambda_2 (W(t+a+a') - W(t+a))} \right\}.$$

These two properties—Gaussianity and independence of increments—characterize Brownian motion. We will therefore establish (9.9). Consider first the case $t = 0$. Then, from (9.8), we find that the moment-generating function of $W(\cdot)$ under Q is

$$\mathbf{E}^Q \left\{ e^{\lambda W(T)} \right\} = \mathbf{E}^P \left\{ e^{\lambda W(T)} M(T) \right\}$$

$$= \mathbf{E}^P \left\{ e^{\lambda \left(Z(T) - \int_0^T b(s) \, ds \right) + \int_0^T b(s) \, dZ(S) - \frac{1}{2} \int_0^T (b(s))^2 \, ds} \right\}$$

$$= \mathbf{E}^P \left\{ \exp \cdot \left[\int_0^T (\lambda + b(s)) \, dZ(S) - \int_0^T \lambda \, b(s) \, ds - \frac{1}{2} \int_0^T (b(s))^2 \, ds \right] \right\}$$

$$= \mathbf{E}^P \left\{ \exp \cdot \left[\int_0^T (\lambda + b(s)) \, dZ(S) - \frac{1}{2} \int_0^T (\lambda + b(s))^2 \, ds + \frac{1}{2} \lambda^2 T \right] \right\}$$

$$= \mathbf{E}^P \left\{ \exp \cdot \left[\int_0^T (\lambda + b(s)) \, dZ(S) - \frac{1}{2} \int_0^T (\lambda + b(s))^2 \, ds \right] \right\} \cdot e^{\frac{1}{2} \lambda^2 T}$$

$$= e^{\frac{1}{2} \lambda^2 T} . \tag{9.10}$$

Notice that in this calculation we used the fact that

$$\exp \cdot \left[\int_0^t (\lambda + b(s)) \, dZ(S) - \frac{1}{2} \int_0^t (\lambda + b(s))^2 \, ds \right]$$

is a martingale so, in particular, it has expectation one. We have established Eq. (9.9) in the case $t = 0$. The calculation of the conditional moment-generating function of the increments

$W(t+a) - W(t)$ is analogous to (9.10). The key fact that needs to be used is a "conditional" version of (9.8): the conditional probability Q_t is given explicitly by

$$Q_t \{S\} \equiv \mathbf{E}_t^P \left\{ S; \frac{M(T)}{M(t)} \right\} ,$$

$$= \mathbf{E}_t^P \left\{ S; e^{\int_t^T b(s)\,ds - \frac{1}{2} \int_t^T (b(s))^2\,ds} \right\} , \qquad (9.11)$$

where S represents a set of paths. (Equation (9.11) is a consequence of (9.8) and the fact that $M(t)$ is a martingale. This point will be explained further in the appendix to this chapter.) Using (9.11), the moment-generating function (9.11) can be computed as in (9.10). ∎

Girsanov's theorem shows that the probability distributions of standard Brownian motion and of Brownian motion with "drift" $-b(s)$ are related in a simple way; namely, the following equation holds:

$$Q = M(T) \cdot P .$$

In simple words, *one probability can be deduced from the other by multiplication by the exponential factor $M(T)$.* Therefore, the paths corresponding to Brownian motion with drift over a finite time-interval can be viewed as standard Brownian paths after a change of measure. This theorem has an interesting consequence: events with probability zero for standard Brownian motion have probability zero for Brownian motion with drift and vice-versa. The two probability measure are said to be **mutually absolutely continuous** or **equivalent**.[8]

References and Further Reading

[1] Bickel, P. and Doksum (1991), *Mathematical Statistics: Basic Ideas and Selected Topics*, Prentice-Hall, Englewood Cliffs, NJ.

[2] Breiman, L. (1968), *Probability*, Addison-Wesley, Reading, Massachusetts.

[3] Doob, J. (1953), *Stochastic Processes*, Wiley, New York.

[4] Girsanov, I. (1960), On Transforming a Certain Class of Stochastic Processes by Absolute Substitution of Measures, *Theory of Probability and Applications*, **5**, pp. 285–301.

[5] Ikeda, N. and Watanabe, S. (1981), *Stochastic Differential Equations and Diffusion Processes*, North Holland, Amsterdam.

[6] Karatzas, I. and Shreve, S.E. (1988), *Brownian Motion and Stochastic Calculus*, Springer Verlag, Berlin.

---

[8]This is true only for events that depend on values taken by paths up to a *finite* time-horizon $T < \infty$. For instance, Brownian motion with positive drift converges with probability 1 to $+\infty$ as $t \to +\infty$ and standard Brownian motion does not.

[7] Oksendaal, B. (1995), *Stochastic Differential Equations, An Introduction with Applications,* 4th ed., Springer Verlag, Berlin.

[8] Thorp, E. (1969), Optimal Gambling Systems for Favorable Games, *Rev. Intl. Statistical Institute,* **37**, pp. 3.

[9] Varadhan, S.R.S. (1980), *Diffusion Processes and Partial Differential Equations,* Tata Institute Lectures, Springer Verlag, New York.

Appendix: Proof of Equation (9.11)

We will use the elementary properties of conditional expectation operators with respect to a given sub-σ-algebra of events.[9] The conditional expectation operator with respect to the history of paths up to time t, $\mathbf{E}_t\{\cdot\}$, is characterized by the following property: let X be an arbitrary random variable. Then $\mathbf{E}_t\{X\}$ is the unique random variable that is measurable with respect to the past up to time t and satisfies

$$\mathbf{E}\{XY\} = \mathbf{E}\{\mathbf{E}_t\{X\}Y\}. \tag{A.1}$$

for all random variables Y that are measurable with respect to the past up to time t.

Let us use this characterization to compute $\mathbf{E}_t^Q\{X\}$, where Q is defined in (9.8). From the definition of Q, we have

$$\mathbf{E}^Q\{XY\} = \mathbf{E}^P\{XYM(T)\}$$

$$= \mathbf{E}^P\left\{X \cdot \left(\frac{M(T)}{M(t)}\right) \cdot M(t) \cdot Y\right\}$$

$$= \mathbf{E}^P\left[\mathbf{E}_t^P\left\{X\left(\frac{M(T)}{M(t)}\right)\right\} \cdot M(t)Y\right]$$

$$= \mathbf{E}^P\left[\mathbf{E}_t^P\left\{X\left(\frac{M(T)}{M(t)}\right)\right\} \cdot M(T)Y\right]$$

$$= \mathbf{E}^Q\left[\mathbf{E}_t^P\left\{X\left(\frac{M(T)}{M(t)}\right)\right\}Y\right]. \tag{A.2}$$

Here, we applied Eq. (A.1) and the martingale property of $M(t)$ to pass from the third to the fourth equation.

---

[9]The reader unfamiliar with these notions should consult, for instance, Breiman [2], or the first chapter of Bickel and Doksum [1].

From the characterization of the conditional probability operator, we conclude that

$$\mathbf{E}_t^Q \{ X \} = \mathbf{E}_t^P \left\{ X \left(\frac{M(T)}{M(t)} \right) \right\} .$$

This is precisely what we wanted to show.

Chapter 10

Continuous-Time Finance: An Introduction

10.1 The Basic Model

We present a model for a securities market with inter-temporal investment opportunities.[1]. We assume an idealized world in which the prices of securities (stocks or bonds) and the dividends that they provide are functions of an M-dimensional random process

$$\mathbf{X}(t) = (X_1(t),\ X_2(t),\ X_3(t),\ \ldots,\ X_M(t))\ ,\quad t > 0 . \tag{10.1}$$

The random variables $X_i(t)$ are **state variables** that describe the economy at any given time. We do not assume necessarily that these variables represent security prices: for instance $X_1(t)$ could represent the value of a consumer price index, the number of laptop computers sold by IBM in 1995, or the inches of rainfall in Iowa in 1994. State variables can represent, more abstractly, *investors' beliefs* about future states of the market. Some state variables $X_i(\cdot)$ may not be tradeable or even observable.

We assume that the state variables follow a system of stochastic differential equations

$$dX_i(t) = \sum_{0}^{\nu} \alpha_{i,k}\ (\mathbf{X}(t),\ t)\ dZ_k(t) + \beta_i\ (\mathbf{X}(t),\ t)\ dt\ , \tag{10.2}$$

for $i = 1, \ldots, M$. Here, $Z_k(\cdot), k = 1, \ldots, \nu$ are independent Brownian motions. Notice that $\alpha_{i,k}$ and β_i are functions of $(\mathbf{X}(t),\ t)$ at time t. Thus, in particular, the current state of the system completely determines the local parameters (means and volatilities) and hence the statistics of future moves. The latter property is called the Markov property.[2] The Markovian assumption on the evolution of economic factors is not unreasonable, provided that all relevant past information is reflected in the current state variables.

Another consequence of (10.2) is the continuity of the paths $X_i(\cdot)$ (recall that Ito processes have continuous paths). This assumption may not apply, in practice, if we expect sudden large-scale jumps in indices or prices (e.g., a sudden transition from centralized to free-market economy, devaluation of a currency, sudden bankruptcy, etc.).

We assume that the dividends of traded securities and their market prices are functions of the current state variables. Specifically, every security is characterized by its **dividend** process

$$D(t) = \widetilde{D}\ (X_1(t),\ X_2(t)\ \ldots\ X_M(t),\ t)$$

---

[1]This section is based partly on Cox et al. [4], and Duffie [7]. See also Chapter 12 in Hull [10].
[2]Ito processes that have the Markov property are called *diffusion processes*.

and by its **market price,**

$$P(t) = \tilde{P} \; (X_1(t), \; X_2(t) \; \ldots \; X_M(t), \; t)$$

$$= \tilde{P} \; (\mathbf{X}(t), \; t) \; .$$

By definition, the dividend process represents the cash-flow that an investor receives from holding the security. For mathematical simplicity, we think of this cash-flow as an annualized yield. Thus, the holder of one unit of this security over a period dt will be able to purchase

$$1 + D(t) \, dt$$

units of the security after this period.[3]

We will assume that there is short-term lending, i.e., that there exist short-term, default-free notes (e.g., Treasury bills) with maturity dt that pay the holder

$$1 + r(t) \, dt \; = \; 1 + \tilde{R} \; (\mathbf{X}(t), \; t) \; dt$$

times face value at maturity. Here, $r(t)$ is the annualized short-term interest rate, or **short rate.** "Rolling over" this short-term note and reinvesting the interest income produces a cumulative wealth of

$$B(t) \; = \; \exp . \left[\int_0^t r(s) \, ds \right] \tag{10.3}$$

times the original investment. (Thus, a money-market account can be viewed as a "security" with $P(t) = 1$ and $D(t) = r(t)$.)

A consequence of this model is that *security prices are Ito processes.* In fact, from the Generalized Ito's Lemma, we have

$$dP(t) \; = \; \sum_{i=1}^{\nu} \frac{\partial \tilde{P}}{\partial X_i} \cdot dX_i \; + \; \frac{\partial \tilde{P}}{\partial t} \cdot dt + \frac{1}{2} \sum_{i=1,\, j=1}^{\nu} \frac{\partial^2 \tilde{P}}{\partial X_i \, \partial X_j} \cdot \left(dX_i \, dX_j \right) \; . \tag{10.4}$$

Using Ito's multiplication rule,

$$dZ_k(t) \, dZ_l(t) \; = \; \begin{cases} dt & \text{if } k = l \\ \\ 0 & \text{if } k \neq l \, , \end{cases}$$

and (10.2), we find that

$$dX_i(t) \, dX_j(t) \; = \; \left(\sum_{k=1}^{\nu} \alpha_{i,k} \, \alpha_{j,k} \right) dt \; .$$

---

[3]This concept applies in a straightforward way to investment in foreign currencies or in coupon-bearing bonds. It can be generalized to stocks by using more complicated dividend functions.

Therefore, the price $P(t)$ satisfies the stochastic differential equation

$$dP = P \cdot \sigma_P(t)\, dZ_P + P \cdot \mu_P\, dt , \tag{10.5}$$

with

$$\sigma_P = \frac{1}{P} \sqrt{ \sum_{k=1}^{\nu} \left[\sum_{i=1}^{M} \alpha_{i,k} \frac{\partial \widetilde{P}}{\partial X_i} \right]^2 } , \tag{10.6}$$

and

$$\mu_P = \frac{1}{P} \sum_{i=1}^{M} \beta_i \frac{\partial \widetilde{P}}{\partial X_i} + \frac{1}{2P} \sum_{i=1}^{M} \sum_{j=1}^{\nu} \sum_{k=1}^{} \alpha_{i,k} \alpha_{j,k} \frac{\partial^2 \widetilde{P}}{\partial X_i \partial X_j} , \tag{10.7}$$

where Z_P is a Brownian motion defined by the stochastic differential equation

$$dZ_P = \frac{1}{\sigma_P} \sum_{k=1}^{\nu} \left[\sum_{I=1}^{M} \alpha_{i,k} \frac{\partial \widetilde{P}}{\partial X_i} \right] dZ_k$$

$$= \frac{1}{\sigma_P} \sum_{k=1}^{\nu} \sigma_{k,P}\, dZ_k .$$

Equations (10.6) and (10.7) show that the local parameters driving the price Equation (10.5), (σ_P, μ_P), can be complicated functions of the state-variables. A major challenge in implementing this basic model is to decide which variables determine security prices. Another problem, of econometric nature, is the estimation of the local parameters $\alpha_{i,k}$ and β_i appearing in the dynamical Equations (10.2). Even if it were possible to solve these problems (a big "if"), one must still find the prices of securities (the functions $\widetilde{P}$) from their cash-flows $\widetilde{D}$. The latter problem has been studied considerably in the Mathematical Economics literature. Economists have found that price functions $\widetilde{P}$ can be derived under the assumption that investors behave rationally and maximize some aggregate utility function.[4]

In this chapter, we take the point of view that prices and financial indices follow Ito processes, making different assumptions about the local parameters σ and μ according to the problem of interest.

10.2 Trading Strategies

To fix ideas, we shall assume that there are N traded securities, with prices $P_1(t)$, $P_2(t)$, ..., $P_N(t)$. A **trading strategy** consists of an N-tuple of nonanticipative processes (with respect to the economic factors X_i, the Brownian motions Z_k that drive the prices, etc.)

$$\Theta(t) = (\theta_1(t), \theta_2(t), \ldots \theta_N(t)) . \tag{10.8}$$

---

[4]See Cox, Ingersoll, and Ross [4] and/or Duffie [7].

The entry $\theta_i(t)$ represents the number of units of the ith security held in an investment portfolio at time t. The value of this portfolio at time t is

$$V(t) = \sum_{j=1}^{N} \theta_i(t) P_i(t) . \qquad (10.9)$$

We will assume that investors can purchase and short-sell arbitrary numbers of securities.[5] We also neglect transaction costs and bid-offer spreads in this first analysis.

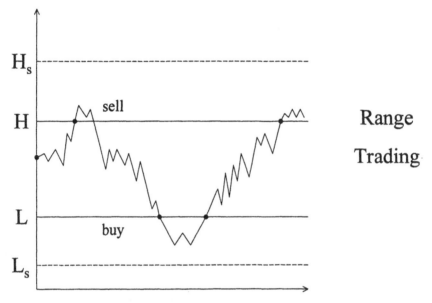

FIGURE 10.1
Schematic "range trading" strategy. The investor goes long stock when $S_t < L$, $(\theta_t = +\theta_0)$, and shorts the stock when $S_t > H$, $(\theta_t = -\theta_0)$. Profits are generated from successive crossings of the band (L, H). We also display levels H_s and L_s where the trader may "unwind" the position the position to limit losses, terminating the trading strategy.

For simplicity, it is convenient to restrict our attention to trading strategies that are **self-financed**. These are strategies in which the investor reallocates his wealth among different investments without adding or withdrawing capital. As a consequence of this, the change in value of the portfolio is entirely due to the change in the prices and the change in wealth due to the dividends received. After a small period of time dt, and before trading, the value of the portfolio (10.8) would be

$$V(t+dt) \approx V(t) + \sum_{j=1}^{N} \theta_i(t) \, dP_i(t) + \sum_{j=1}^{N} \theta_i(t) P_i(t) D_i(t) \, dt ,$$

where $P_i(t) D_i(t) dt$ represents the dividend flow from the ith security. If the strategy $\Theta(\cdot)$ is

---

[5]In particular, the variables θ_i can take arbitrary real values.

self-financed, then the value of the portfolio at time $t + dt$ after trading should also satisfy

$$V(t + dt) = \sum_{j=1}^{N} P_i(t + dt) \, \theta_i(t + dt) \, .$$

Thus, a self-financed trading strategy is a nonanticipative process that satisfies the stochastic differential equation

$$dV = d \left(\sum_{i=1}^{N} \theta_i \, P_i \right)$$

$$= \sum_{j=1}^{N} \theta_i \, dP_i + \sum_{j=1}^{N} \theta_i \, P_i \, D_i \, dt \, . \tag{10.10}$$

Often, it is convenient to analyze the profits/losses of trading strategies in constant dollars, i.e., discounting with respect to the short-term interest rate. The value of a portfolio worth $V(t)$ at time t in dollars-at-time-0 is

$$\hat{V}(t) \equiv \frac{V(t)}{B(t)} = e^{-\int_0^t r(s) \, ds} \, V(t) \, .$$

Using this equation and (10.10), we find that the self-financing equation in constant dollars is

$$d \left(\frac{V(t)}{B(t)} \right) = \frac{1}{B(t)} \sum_{j=1}^{N} \theta_i(t) \, dP_i(t) + \frac{1}{B(t)} \sum_{j=1}^{N} \theta_i(t) \, P_i(t) \, (D_i(t) - r(t)) \, dt$$

$$= \sum_{j=1}^{N} \theta_i(t) \, d \left(\frac{P_i(t)}{B(t)} \right) + \sum_{j=1}^{N} \theta_i(t) \left(\frac{P_i(t)}{B(t)} \right) D_i(t) \, dt \, , \tag{10.11}$$

or, with $\hat{P}_i = P_i / B$,

$$d\hat{V} = \sum_{j=1}^{N} \theta_i \, d\hat{P}_i + \sum_{j=1}^{N} \theta_i \, \hat{P}_i \, D_i \, dt \, . \tag{10.12}$$

This equation is analogous to (10.10); it just expresses the self-financing relation in a different unit of account.

10.3   Arbitrage Pricing Theory

DEFINITION 10.1   *A self-financed trading strategy* $\{\Theta(t), 0 \le t \le T\}$ *is said to be an arbitrage strategy if the profit/loss that it generates,*

$$\hat{V}(T) - V(0) = \sum_{j=1}^{N} \int_0^T \theta_i(t)\, d\hat{P}_i(t) + \sum_{j=1}^{N} \int_0^T \theta_i(t)\, \hat{P}_i(t)\, D_i(t)\, dt\ ,$$

is (i) nonnegative with probability 1, and (ii) positive with positive probability.

In previous chapters, we discussed arbitrage strategies from both the theoretical and practical points of view. The most common "arbitrages" are cash-and-carry trades, which try to exploit discrepancies between spot and forward prices. Other types of arbitrage opportunities may arise if, say, options can be replicated with market instruments (statically or dynamically) at a profit. Arbitrage opportunities may arise sporadically, but they cannot exist for long in efficient markets due to the forces of supply and demand.

The following result shows that the securities market model must satisfy certain constraints if there are no arbitrage opportunities. Roughly speaking, the constraints can be viewed as a relation between the volatility and the returns of traded securities.

PROPOSITION 10.1
Under the assumptions of the basic model, assume that security prices satisfy

$$dP_i(t) = P_i(t) \left\{ \sum_{k=1}^{\nu} \sigma_{k,i}(t)\, dZ_k(t) + \mu_i(t)\, dt \right\}\ , \quad i = 1, \ldots, N \qquad (10.13)$$

where $Z_k(\cdot)$ *are independent Brownian motions. Let* $D_i(t)$ *denote the corresponding dividend processes and let* $r(t)$ *be the short-term interest rate. Then, if there are no arbitrage opportunities, there exist nonanticipative processes* $\lambda_1(t), \lambda_2(t), \ldots, \lambda_\nu(t)$ *such that*

$$\mu_i(t) + D_i(t) - r(t) = \sum_{k=1}^{\nu} \sigma_{k,i}(t)\, \lambda_k(t)\ . \qquad (10.14)$$

This proposition, which we will prove shortly, has an important corollary (sometimes referred to as the Fundamental Theorem of Arbitrage Pricing Theory):

PROPOSITION 10.2
Under the assumptions of the basic model, let **P** *represent the probability measure on path-space associated with the security prices (10.13). A necessary and sufficient condition for the existence of no-arbitrage opportunities over the time-interval* $(0, T)$ *is that there exists an equivalent probability measure* **Q** *such that all security prices satisfy*

$$P_i(t) = \mathrm{E}_t^Q \left\{ e^{-\int_t^T (r(s) - D_i(s))\, ds}\, P_i(T) \right\}\ , \quad i = 1, 2, \ldots, N. \qquad (10.15)$$

Proof of Proposition 10.1   First, we will show that if there are no-arbitrage opportunities, Eq. (10.14) holds. For this, we consider the possibility of a specific "single-period" arbitrage at some point in time t. More precisely, assume that there exist quantities $\theta_i(t)$ $(i = 1, \ldots, N)$ such that the profit/loss over a small time-interval dt,

$$dV(t) \;=\; d\left(\sum_{i=1}^{N} \theta_i(t) \, dP_i(t) \right) ,$$

has *zero variance*, given the past up to time t. Consider a strategy that corresponds to holding the portfolio $\Theta(t) = (\theta_i(t))_{i=1}^{N}$, over the period $(t, t + dt)$ and subsequently closing the position after dividends are paid out. The return on this investment should equal the return of a riskless money-market account over the same period. Indeed, the existence of two different riskless rates of return in the market would give rise to an obvious arbitrage opportunity.[6] Therefore, if $dV(t)$ has variance zero we must have

$$dV(t) \;=\; r(t) \, V(t) \, dt . \tag{10.16}$$

Now, using Eq. (10.13), we find that

$$dV(t) = \sum_{i=1}^{N} \theta_i(t) \, P_i(t) \left(\sum_{k=1}^{v} \sigma_{k,i}(t) \, dZ_k(t) \right)$$

$$+ \sum_{i=1}^{N} \theta_i(t) \, P_i(t) \, (\mu_i(t) + D_i(t)) \, dt$$

$$= \sum_{k=1}^{v} \left(\sum_{i=1}^{N} \theta_i(t) \, P_i(t) \, \sigma_{k,i}(t) \right) dZ_k(t)$$

$$+ \sum_{i=1}^{N} \theta_i(t) \, P_i(t) \, (\mu_i(t) + D_i(t)) \, dt . \tag{10.17}$$

It follows that $dV(t)$ has zero variance if and only if

$$\sum_{i=1}^{N} \theta_i(t) \, P_i(t) \, \sigma_{k,i}(t) \;=\; 0 \quad , \quad k = 1, \ldots v , \tag{10.18}$$

and, moreover, that in the latter case Eq. (10.16) can be written as

$$\sum_{i=1}^{N} \theta_i(t) \, P_i(t) \, (\mu_i(t) + D_i(t)) \, dt \;=\; \sum_{i=1}^{N} \theta_i(t) \, P_i(t) \, r(t) \, dt ,$$

---

[6]If the return is higher than $r(t)$, borrow money and purchase the portfolio. If the return is lower, short-sell the portfolio and invest the proceeds in short-term funds.

or

$$\sum_{i=1}^{N} \theta_i(t) P_i(t) (\mu_i(t) + D_i(t) - r(t)) = 0. \qquad (10.19)$$

The conclusion is that if there are no-arbitrage opportunities, then *whenever the ν equations in (10.18) hold, Eq. (10.19) must hold as well.*

It is useful to reinterpret this in "geometric" terms. Define $\nu + 1$ vectors in N-dimensional Euclidean space by

$$s_k = (\sigma_{k,1}, \ldots \sigma_{k,N}) \qquad , k = 1, \ldots \nu$$

and

$$\mathbf{m} = (\mu_1(t) + D_1(t) - r(t), \ldots, \mu_N(t) + D_N(t) - r(t)).$$

The statement "(10.18) implies (10.19)" is equivalent to saying that "whenever a vector is orthogonal to $s_k, k = 1, \ldots, \nu$, then it is also orthogonal to $\mathbf{m}$." It follows from linear algebra that this condition holds if and only if $\mathbf{m}$ is contained in the linear subspace generated by the vectors s_k. We conclude that there must exist functions $\lambda_k(t), k = 1, \ldots, \nu$, such that

$$\mathbf{m} = \sum_{k=1}^{\nu} \lambda_k(t) s_k,$$

or, for all i,

$$\mu_i(t) + D_i(t) - r(t) = \sum_{k=1}^{\nu} \lambda_k(t) \sigma_{i,k}. \qquad (10.20)$$

We have thus shown that (10.20) must hold in the absence of arbitrage. (Note that the scalars λ_k are generally functions of t, because the argument leading to (10.20) is "local in time.")

Conversely, let us establish that (10.20) is a sufficient condition for the absence of arbitrage. In fact, substituting the values for μ_i derived from this condition into the equation for security prices (10.13), we have

$$dP_i(t) = P_i(t) \cdot \left\{ \sum_{k=1}^{\nu} \sigma_{i,k}(t) (dZ_k(t) + \lambda_k(t) dt) + (r(t) - D_i(t)) dt \right\}$$

$$= P_i(t) \cdot \left\{ \sum_{k=1}^{\nu} \sigma_{i,k}(t) dW_k(t) + (r(t) - D_i(t)) dt \right\}, \qquad (10.21)$$

where we set

$$W_k(t) \equiv Z_k(t) + \int_0^t \lambda_k(s) ds. \qquad (10.22)$$

By Girsanov's theorem, the processes $W_k(\cdot)$ are distributed like independent Brownian motions under the modified probability

$$Q\{S\} =$$

$$\mathbf{E} \left\{ S; \exp. \left[-\sum_{k=1}^{\nu} \int_0^T \lambda_k(s) dZ_k(s) - \frac{1}{2} \sum_{k=1}^{\nu} \int_0^T \lambda_k^2(s) ds \right] \right\}. \qquad (10.23)$$

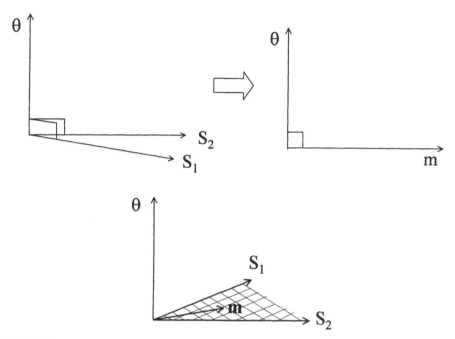

FIGURE 10.2
Schematic view of the argument leading to Eq. (10.20): If each vector θ that is orthogonal to s_1 and s_2 is also orthogonal to m, then m must lie in the plane generated by s_1 and s_2. This result generalizes to an arbitrary number of vectors and dimensionalities.

Using Proposition 9.1 of the previous chapter (generalized to multidimensional Brownian motion), we deduce from Eq. (10.21) that $P_i(\cdot)$ satisfies the equation

$$P_i(t) = P_i(0) \cdot M_i(t) \cdot e^{\int_0^t (r(s) - D_i(s)) \, ds} \tag{10.24}$$

where

$$M_i(t) \equiv \exp \cdot \left[\int_t^T \sum_{k=1}^{\nu} \sigma_{i,k}(s) \, dW_k(s) - \frac{1}{2} \int_t^T \sum_{k=1}^{\nu} \left(\sigma_{i,k}(s) \right)^2 \, ds \right] \tag{10.25}$$

is an exponential martingale under **Q**. Hence, the process

$$e^{- \int_0^t (r(s) - D_i(s)) \, ds} P_i(t) = e^{\int_0^t D_i(s) \, ds} \cdot \hat{P}_i(t)$$

is also a martingale under **Q**.

From the self-financing equation (10.12), we conclude that the discounted value of any self-financed trading strategy satisfies

$$d\hat{V}(t) = \sum_{i=1}^{N} \theta_i(t) \left[d\hat{P}_i(t) + D_i(t) \, \hat{P}_i(t) \, dt \right]$$

$$= \sum_{i=1}^{N} \theta_i(t)\, e^{-\int_0^t D_i(s)\,ds}\, d\left(e^{-\int_0^t (r(s)-D_i(s))\,ds}\, P_i(t)\right)$$

$$= \sum_{i=1}^{N} \theta_i(t)\, e^{-\int_0^t D_i(s)\,ds}\, P_i(0)\, dM_i(t)\,.$$

In particular, $\hat{V}(t)$ is a martingale under Q and hence

$$\mathbf{E}^Q\left\{\hat{V}(t) - \hat{V}(0)\right\} = 0. \qquad (10.26)$$

The fact that self-financed strategies are martingales under $\mathbf{Q}$ implies that arbitrage strategies cannot exist. In fact, since $\mathbf{P}$ and $\mathbf{Q}$ are equivalent, if the profit/loss generated by a strategy, $\hat{V}(T) - \hat{V}(0)$, is nonnegative with $\mathbf{P}$-probability 1, then it must be nonnegative with $\mathbf{Q}$-probability 1. But then, by (10.26), it must vanish with $\mathbf{Q}$-probability 1. Using the equivalence of the two measures, we conclude that $\hat{V}(T) - \hat{V}(0)$ must vanish with $\mathbf{P}$-probability 1.

The proof of Proposition 10.1 is complete. ∎

Proof of Proposition 10.2. In the proof of Proposition 10.1, we established that a necessary condition for the absence of arbitrage is that

$$e^{-\int_0^t (r(s)-D_i(s))\,ds}\, P_i(t) \quad , \quad 0 \le t \le T\,, \qquad (10.27)$$

is a martingale under the measure $\mathbf{Q}$ defined in (10.23). Therefore,

$$\mathbf{E}_t^Q\left\{e^{-\int_0^T (r(s)-D_i(s))\,ds}\, P_i(T)\right\} = e^{-\int_0^t (r(s)-D_i(s))\,ds}\, P_i(t)\,.$$

Multiplying both sides of this equation by $e^{\int_0^t (r(s)-D_i(s))\,ds}$ (which is measurable with respect to the past up to time t), we find that

$$\mathbf{E}_t^Q\left\{e^{-\int_t^T (r(s)-D_i(s))\,ds}\, P_i(T)\right\} = P_i(t) \quad , \quad t \le T\,. \qquad (10.28)$$

This establishes (10.28) as a necessary condition for no-arbitrage.

Conversely, if Eq. (10.28) holds for some probability $\mathbf{Q}$ that is equivalent to $\mathbf{P}$, then the process (10.27) is a martingale under $\mathbf{Q}$. But then, we can follow verbatim the argument presented in the proof of Proposition 10.1 to conclude that there are no-arbitrage strategies. ∎

Another useful result that we have established along the way is

PROPOSITION 10.3
Under the assumptions of the basic model, a necessary and sufficient condition for the absence of arbitrage over the time interval [0, T] *is that there exists an equivalent probability* $\mathbf{Q}$ *such*

that the value of any self-financed trading strategy in constant dollars $\hat{V}(t)$ is a martingale under **Q**. *In particular,*

$$V(t) = \mathbf{E}_t^Q \left\{ e^{-\int_t^T r(s)\, ds} \, V(T) \right\}. \tag{10.29}$$

In most applications of Arbitrage Pricing Theory to pricing derivative securities, we will focus our attention primarily on the measure **Q**. This is because Proposition 10.2 relates the values of traded securities with their expected future cash-flows **under Q**. Since derivatives are securities with cash-flows that depend on the values of other securities, Eq. (10.15) provides a framework for pricing them in terms of their "payoffs" at a future date. In a similar vein, the knowledgeable reader will interpret (10.29) as a pricing formula for contingent claims based upon the notion of *replicating portfolio:* if a derivative security delivers a (single) payoff of $V(T)$ dollars at a date T, and if $V(T)$ is also the value at time T of a self-financed strategy $\Theta(s), t \leq s \leq T$, then the value of the derivative security at time t represents the cost of entering into this dynamic strategy at time t, i.e., $V(t)$.

In conclusion, one may say that the risk-neutral measure that derives from no-arbitrage is first and foremost a **mathematical device**. It is a tool that allows us to express prices in terms of expected cash-flows, hence of relating the prices of different securities that trade in the same economy. As we observed in Chapter 1 of this book, the risk-neutral measure or, rather, *measures* (because such measure is not unique in practice), should not be confused with the econometric or subjective statistical measure used to forecast the market based on historical data. In this chapter, we saw how Arbitrage Pricing Theory relates the risk-neutral measure to the "prior" statistical measure driving the economy, by means of the vector of *market prices of risk* $\lambda_1(t), \ldots, \lambda_N(t)$. A detailed study of the relation between risk-neutral measures and objective measures has been made in very few cases [3]. This is an area of current research, which is beyond the scope of this book.

Some concrete applications of Arbitrage Pricing Theory are presented in the next chapter.

References and Further Reading

[1] Chatelain, G. and Stricker, C. (1994), On Componentwise and Vector Stochastic Integration, *Mathematical Finance,* **4**, pp. 57–66.

[2] Chen, R.R. and Scott, L. (December 1993), Maximum Likelihood Estimation for a Multifactor Equilibrium Model of the Term Structure of Interest Rates, *Journal of Fixed Income,* pp. 14–31.

[3] Chen, R.R. and Scott, L. (Winter 1995), Interest Rate Options in Multifactor Cox-Ingersoll-Ross Models of the Term Structure, *Journal of Fixed Income,* pp. 53–72.

[4] Cox, J., Ingersoll, J., and Ross, S. (1985), An Inter-Temporal General Equilibrium Model of Asset Prices, *Econometrica,* **53**.

[5] Delbaen, F. (1992), Representing Martingale Measures When the Asset Prices are Continuous and Bounded, *Mathematical Finance,* **2**, pp. 107–130.

[6] Delbaen, F. and Schachermayer, W. (1994), A General Version of the Fundamental Theorem of Asset Pricing, *Math. Annalen,* **300**, pp. 463–520.

[7] Duffie, D. (1992), *Dynamical Asset Pricing Theory,* Princeton University Press, Princeton, NJ.

[8] Harrison, J.M. and Kreps, D. (1979), Martingales and Arbitrage in Multi-Period Securities Markets, *Journal of Economic Theory,* 20, pp. 381–316.

[9] Harrison, J.M. and Pliska, S.R. (1981), Martingales and Stochastic Integrals in the Theory of Continuous Trading, *Stochastic Process. Appl.,* **11**, pp. 215–260.

[10] Hull, J. (1993), *Options, Futures and Other Derivative Securities,* 2nd ed., Prentice-Hall, Englewood Cliffs, NJ.

[11] Merton, R. (1991), *Continuous Time Finance,* Blackwell, Oxford.

[12] Muller, S. *Arbitrage Pricing of Contingent Claims,* Lecture Notes in Economics and Mathematical Systems 254, Springer Verlag, Berlin.

[13] Musiela, M. and Rutkowski, M. (1997), *Arbitrage Pricing Theory of Derivative Securities: Theory and Applications,* Springer Verlag, Berlin.

[14] Stricker, C. (1989), Arbitrage et Lois de Martingale, *Annales Inst. Poincaré,* **26**, pp. 451–460.

Chapter 11

Valuation of Derivative Securities

This chapter discusses derivative asset pricing from the point of view of Arbitrage Pricing Theory (APT). Under idealized market conditions—i.e., neglecting transaction costs, liquidity constraints, or trading restrictions—the absence of arbitrage implies the existence of a probability measure such that the value of any derivative security is equal to the expectation of its discounted cash-flows. We discuss the Black–Scholes model from the APT point of view and revisit the important notion of dynamic hedging, which was introduced previously in the discrete framework (Chapter 3). In particular, we distinguish between dynamically complete and dynamically incomplete pricing models, and discuss an example involving stochastic volatility. At the end of the chapter we present a general connection that exists between expected values of diffusion processes and partial differential equations.

11.1 The General Principle

If a market has no-arbitrage opportunities, there exists a probability on forward market scenarios such that the prices of all traded securities are martingales, after adjusting for the cost of money and dividends. This is the basic proposal of APT. In mathematical terms, there exists a probability Q, defined on the set of forward price paths in the time-interval $[0, T]$, such that, for all $0 < t < T$,

$$P(t) \;=\; \mathbf{E}_t^Q \left\{ e^{-\int_t^T (r(s) - d(s))\, ds} \; P(T) \right\}, \tag{11.1}$$

where $P(t)$ represents the price of any traded security at time t, $d(s)$ is the dividend yield, and $r(s)$ is the riskless rate.[1] The symbol $\mathbf{E}_t^Q$ denotes the conditional expectation under the probability measure Q given the past up to time t. A measure Q having this property is called a **martingale measure** or an **Arrow–Debreu measure** in Mathematical Finance. The reason is that (11.1) is equivalent to saying that the "adjusted" price processes

$$e^{-\int_0^t (r(s) - d(s))\, ds} \; P(t) \tag{11.2}$$

---

[1] Strictly speaking, Q is a probability defined on the collection of possible paths that can be described by the state-variables defining the economy.

are martingales. This last quantity represents the wealth accumulated by investing in one unit of a security at time $t = 0$, with dividend reinvestment, measured in constant dollars.

APT implies that a derivative security that entitles the holder to receive a series of cash-flows, say,

$$F(\tau_1), \ F(\tau_2), \ F(\tau_3), \ \ldots, \ F(\tau_N) \, ,$$

contingent on the value of other securities at a sequence of fixed or random times $\tau_1 < \tau_2, < \cdots < \tau_N$ must have fair value

$$V(t) = \mathbf{E}_t^Q \left\{ \sum_{i \, : \, \tau_i \geq t} e^{-\int_t^{\tau_i} r(s) \, ds} \, F(\tau_i) \right\} . \tag{11.3}$$

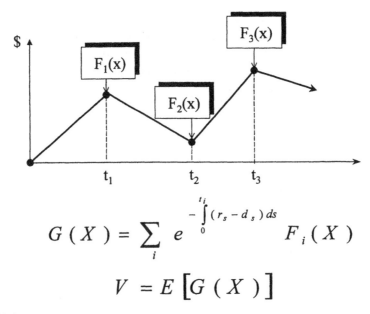

$$G(X) = \sum_i e^{-\int_0^{t_i} (r_s - d_s) \, ds} F_i(X)$$

$$V = E\left[G(X)\right]$$

FIGURE 11.1
Schematic representation of the valuation of uncertain cash-flows. The broken line represents a possible path X followed by the economy. A security delivers cash-flows at times t_1, t_2, t_3. The fair value of the security is the expected value of the sum of the discounted cash-flows along each path, where the expectation is taken with respect to a risk-neutral measure.

A key feature of APT is that (11.2) and (11.3) must hold with the *same* Q, independently of which security we consider. A martingale measure can be viewed as a *representation of the market's current opinion on the evolution of one or more indices and the prices of all derivatives contingent on them.* This includes not only traded derivatives but, more importantly, *any other derivative that may be traded in the future that matures before time T.* The knowledge of the martingale measure is all that is required, in principle, to value arbitrary derivative securities.

From a practical point of view, it is important to recognize that

(i) *The martingale measure is not unique.* The information available on a particular commodity and its derivatives consists of finitely many numbers (prices, interest rates, implied volatilities, volume traded, open interest, etc.). On the other hand, a probability measure defined on the space of Ito processes is determined by an infinite number of parameters. It is thus easy to understand nonuniqueness. Given a set of "underlying" indices or prices of assets that are correlated with the value of a derivative security, there is usually more than one martingale measure that is consistent with the data.[2]

(ii) *The martingale measure varies with time.* Derivatives markets are strongly affected by information shocks (elections, political announcements, takeover announcements, etc.), as well as by the price dynamics. Nonuniqueness of the martingale measure implies that the *measure itself* can change with time, as the market changes its expectations about the relative value of assets in the future.[3]

In view of these considerations, the art of pricing of derivative securities consists of

- Selection of a (simple) martingale measure consistent with (11.2) and (11.3)—i.e., including information embedded in the prices of traded derivatives;

- Evaluation of the expectation value in (11.3) for the derivative of interest;

- Estimation of the sensitivity of the resulting price with respect to the choice of probability.

This last point is very important for risk-management. For instance, bond options have different prices according to the model used for pricing them, and there is no clear consensus on which model should be used.[4]

11.2 Black–Scholes Model

The Black and Scholes model for pricing stock options has been applied to many different commodities and payoff structures.[5] We will assume that there is a single underlying asset, which is traded in the market, and that the price of this asset satisfies the stochastic differential equation

$$dS = S \sigma dZ + S \mu dt \qquad (11.4)$$

---

[2]This is known as **market incompleteness:** in general, there is not enough information available for the market to price every future state of the world. Recall that a **complete market** is an Arrow–Debreu market in which any state-contingent claim can be replicated, or synthesized, with traded securities. Market incompleteness is one of the forces that drives securities and commodities trading.

[3]This variation of the martingale measure in time is not incompatible with the basic securities market model described in the previous section. The reader should distinguish between the probability describing the *likelihood* of different economic states and the martingale measure that describes the *relative value* of future economic states, i.e., a probability that makes adjusted prices (11.2) into martingales. These are two different concepts. Uniqueness of the martingale measure is consistent with market equilibrium with respect to some (aggregate) utility, in which case security prices are equal to the marginal utilities for investment. (See Duffie [2].) In finance, one is typically dealing with a relatively short time-scale. It is therefore important to keep in mind that expectations and valuations change in time due to the arrival of new information.

[4]This issue is discussed in Wong [6].

[5]This includes not only financial options but also "real options," which are the options of building plants, drilling for oil, etc. (See Dixit and Pindyck [1].) Options pricing theory can be seen, in a broader sense, as a tool for decision-making.

where σ and μ are the local volatility and mean of short-term returns. We will assume that the interest rate r, the dividend yield d, and σ remain constant over the period. However, all the mathematical arguments extend without modifications to more complicated models in which r, d, and σ are functions of the spot price and of the current time (or time to maturity).

We will use Proposition 10.1 of Chapter 10. If there are no-arbitrage opportunities, we know that there exists an Ito process $\lambda(t)$ such that the local expected return μ_P, the dividend yield d_P, and the volatility σ_P of any derivative security satisfy the relation

$$\mu_P(t) + d_P(t) - r = \sigma_P(t)\,\lambda(t)\,. \tag{11.5}$$

This relation is true for *any* traded security that depends on $S(t)$. In particular, this includes the asset itself since it is a traded security. Thus,

$$\mu + d - r = \sigma\,\lambda(t)\,,$$

or,

$$\lambda = \frac{\mu + d - r}{\sigma}\,. \tag{11.6}$$

Since there is a single Brownian motion and there exists a traded security driven by it, the function $\lambda(t)$ can be computed explicitly. It is equal to the ratio of the interest-adjusted rate of return of the security and its volatility. The variable λ is called the **Sharpe ratio** of the underlying asset. From (11.5) and (11.6), we conclude that, for any derivative security contingent on S, we have

$$\frac{\mu_P + d_P - r}{\sigma_P} = \frac{\mu + d - r}{\sigma}\,. \tag{11.7}$$

Equation (11.7) states that the Sharpe ratio of any derivative is equal to the Sharpe ratio of the underlying asset.

Now, assume that the fair value of a derivative can be expressed as a deterministic function of the value of the underlying index and current time, i.e., that

$$P(t) = \tilde{P}(S(t), t)\,.$$

Let us show that this function can be characterized using (11.7). In fact, applying the Generalized Ito Lemma to $\tilde{P}(S(t), t)$, we find that

$$dP = P\left(\sigma_P\,dZ + \mu_p\,dt\right) \tag{11.8}$$

where

$$\sigma_P = \frac{1}{\tilde{P}}\frac{\partial \tilde{P}}{\partial S}\,S\sigma \tag{11.9}$$

and

$$\mu_P = \frac{1}{\tilde{P}} \cdot \left\{\frac{\partial \tilde{P}}{\partial t} + \mu S\frac{\partial \tilde{P}}{\partial S} + \frac{1}{2}\sigma^2 S^2\frac{\partial^2 \tilde{P}}{\partial S^2}\right\}\,. \tag{11.10}$$

Assume, for simplicity, that the derivative security pays no intermediate dividends to its holder, i.e., that $d_P = 0$. If we substitute the values from Eqs. (11.9) and (11.10) for σ_P and μ_P into

Eq. (11.7), we find that

$$\frac{1}{\tilde{P}} \cdot \left\{ \frac{\partial \tilde{P}}{\partial t} + \mu S \frac{\partial \tilde{P}}{\partial S} + \frac{1}{2} \sigma^2 S^2 \frac{\partial^2 \tilde{P}}{\partial S^2} \right\} - r = \frac{1}{\tilde{P}} \frac{\partial \tilde{P}}{\partial S} S (\mu + d - r) .$$

Rearranging terms, we conclude that $\tilde{P}(S, t)$ should satisfy

$$\frac{\partial \tilde{P}}{\partial t} + (r - d) S \frac{\partial \tilde{P}}{\partial S} + \frac{1}{2} \sigma^2 S^2 \frac{\partial^2 \tilde{P}}{\partial S^2} - r \tilde{P} = 0 . \tag{11.11}$$

This is just the Black–Scholes partial differential equation derived from the APT point of view. Notice that the parameters that enter into the equation are the riskless rate of return, the dividend rate, and the volatility. In particular, the local rate of return of the underlying asset, μ, does not affect the price of a derivative security.

We can draw a connection between the Black–Scholes equation and the theory of martingale measures. Let us imagine a fictitious market in which the price of the underlying security satisfies the modified SDE

$$dS = S (\sigma dZ + (r - d) dt) .$$

The solution of this stochastic differential equation is the well-known "geometric Brownian motion,"

$$S(t) = S(0) e^{\sigma Z(t) - \frac{\sigma^2}{2} t + (r - d) t} . \tag{11.12}$$

Applying Ito's Lemma to $P(t) = \tilde{P}(S(t), t)$, we find that (11.8) holds with σ_P given by (11.9) and with a modified "return"

$$\mu_P = \frac{1}{\tilde{P}} \cdot \left\{ \frac{\partial \tilde{P}}{\partial t} + (r - d) S \frac{\partial \tilde{P}}{\partial S} + \frac{1}{2} \sigma^2 S^2 \frac{\partial^2 \tilde{P}}{\partial S^2} \right\} . \tag{11.13}$$

Using (11.13) and the Black–Scholes equation (11.11), we find that $P(t)$ satisfies

$$dP = P \sigma_P dZ + P (r - d) dt . \tag{11.14}$$

This implies, in turn, that

$$d \left(e^{-(r-d)t} P(t) \right) = \left(e^{-(r-d)t} P(t) \right) dZ ,$$

i.e., $e^{-(r-d)t} P(t)$ is a martingale. In particular, we have

$$\tilde{P}(S, t) = \mathbf{E}^Q \left\{ e^{-r(T-t)} \tilde{P}(S(T), T) | S(t) = S \right\} , \tag{11.15}$$

where Q represents the probability associated with the risk-adjusted SDE in (11.14).

In conclusion, under the martingale measure, the underlying asset is a geometric Brownian motion with $\mu = r - d - \frac{\sigma^2}{2}$.

Example 11.1

Re-derivation of the Black–Scholes formula. The most important application of this theory is the Black–Scholes option pricing formula. Since the value of a European option is a known function of the price of the underlying asset at the expiration date, it can be found by solving a **final-value problem** for the Black–Scholes equation. Namely, the function $\tilde{P}(S, t)$ satisfies

$$
\begin{cases}
\frac{\partial \tilde{P}}{\partial t} + (r - d)\, S \frac{\partial \tilde{P}}{\partial S} + \frac{1}{2}\sigma^2 S^2 \frac{\partial^2 \tilde{P}}{\partial S^2} - r\,\tilde{P} = 0 \\[2mm]
\tilde{P}(S, T) = F(S),
\end{cases}
\tag{11.16}
$$

where T is the maturity date and $F(S) = \max(S - K, 0)$ or $\max(K - S, 0)$ according to whether the option is a call or a put. This problem has a closed-form solution when the parameters σ, r, and d are constant or dependent only on time. Otherwise, the equation must be solved numerically, as indicated in Section 11.4 below.

In the case of constant coefficients, the problem reduces to calculating the expected value of the random variable $F\left(e^Y\right)$, where Y is normal. For instance, the value of a call option with strike K expiring at time T is

$$
\tilde{P}(S(t), t) = \mathbf{E}\left\{ e^{-r\,(T-t)}\left(S(t)e^{Y(T)} - K \right) ; \ S(t)\,e^{Y(T)} > K \right\},
\tag{11.17}
$$

with

$$
Y(T) = \sigma\sqrt{T - t}\, X - \frac{1}{2}\sigma^2\,(T - t) + (r - q)\,(T - t),
$$

where X is a standard normal. The right-hand side can be evaluated explicitly in terms of the cumulant of the normal distribution

$$
N(x) = \frac{1}{2\pi}\int_{-\infty}^{x} e^{-\frac{y^2}{2}}\, dy.
$$

Accordingly, the value of the European call is

$$
C(S, t; K, T) = S\,e^{-d\,(T-t)}\, N(d_1) - K\,e^{-r\,(T-t)}\, N(d_2)
\tag{11.18}
$$

where

$$
d_1 = \frac{1}{\sigma\sqrt{T - t}}\, \ln\left(\frac{S\,e^{(r-q)\,(T-t)}}{K} \right) + \frac{\sigma\sqrt{T - t}}{2}
$$

and

$$
d_2 = d_1 - \sigma\sqrt{T - t}.
$$

We re-derive the Black–Scholes formula. The elegance and versatility of the Black–Scholes theory have resulted in its wide acceptance in securities markets as a pricing and hedging tool.
□

11.3 Dynamic Hedging and Dynamic Completeness

Let us pursue the subject of the previous section from a slightly different point of view. Suppose that a trader sells a European option (or, more generally, a European contingent claim) and wishes to implement a trading strategy using the underlying security to hedge the option exposure. Let $\Delta(t)$ and $B(t)$ represent, respectively, the number of units of the underlying asset held in the portfolio and the current balance in a money-market account. We assume that the strategy is self-financing.

The change in the value of the portfolio after a small period of time dt is

$$dV = \Delta\,dS + q\,\Delta\,S\,dt + r\,B(t)\,dt\,, \tag{11.19}$$

to leading order in dt. Let us assume momentarily that the value of the contingent claim is $P(t) = \tilde{P}(S(t),\, t)$, where $\tilde{P}$ is an unknown function. According to Ito's Lemma, the change in P from time t to time $t + dt$ is, to leading order,

$$dP = \frac{\partial \tilde{P}}{\partial S}\,dS + \left(\frac{\partial \tilde{P}}{\partial t} + \frac{1}{2}\sigma^2 S^2 \frac{\partial^2 \tilde{P}}{\partial S^2}\right) dt\,. \tag{11.20}$$

Suppose that the strategy followed by the trader is such that

- initially, $V(0) = \tilde{P}(S(0),\, 0)$ and $\Delta(t) = \frac{\partial \tilde{P}(S(0),\, 0)}{\partial S}$;

- subsequently, $\Delta(t) = \frac{\partial \tilde{P}(S(t),\, t)}{\partial S}$;

- dividends produced by the risky asset are reinvested in the asset.

Under these conditions, the value of the dynamic portfolio will remain equal to $\tilde{P}(S(t),\, t)$ (to leading order in dt) if and only if the dt-contributions from (11.19) and (11.20) are equal. That is, if

$$\frac{\partial \tilde{P}}{\partial t} + \frac{1}{2}\sigma^2 S^2 \frac{\partial^2 \tilde{P}}{\partial S^2} = d \cdot S \frac{\partial \tilde{P}}{\partial S} + r\,B$$

$$= d \cdot S \frac{\partial \tilde{P}}{\partial S} + r\left(\tilde{P} - S \frac{\partial \tilde{P}}{\partial S}\right)$$

$$= r\,\tilde{P} - (r - d)\,S \frac{\partial \tilde{P}}{\partial S}\,. \tag{11.21}$$

This condition is equivalent to requiring that $\tilde{P}$ satisfy the Black–Scholes equation

$$\frac{\partial \tilde{P}}{\partial t} + (r - d)\,S \frac{\partial \tilde{P}}{\partial S} + \frac{1}{2}\sigma^2 S^2 \frac{\partial^2 \tilde{P}}{\partial S^2} - r\,\tilde{P} = 0\,. \tag{11.22}$$

We have proved

PROPOSITION 11.1

Assume that an index or asset price follows the stochastic differential equation

$$dS = S(\sigma \, dZ + \mu \, dt) \tag{11.23}$$

and that investing in this asset produces a dividend yield d, where σ, r, and d are functions of time and the spot price. Suppose that $\tilde{P}$ satisfies

$$\begin{cases} \frac{\partial \tilde{P}}{\partial t} + (r - d) S \frac{\partial \tilde{P}}{\partial S} + \frac{1}{2}\sigma^2 S^2 \frac{\partial^2 \tilde{P}}{\partial S^2} - r\tilde{P} = 0, & t < T, \\[2mm] \tilde{P}(S, T) = F(S). \end{cases} \tag{11.24}$$

Then, the trading strategy

$$B(t) = \tilde{P}(S(t), t) - S(t) \frac{\partial \tilde{P}(S(t), t)}{\partial S} \quad, \quad \Delta(t) = \frac{\partial \tilde{P}(S(t), t)}{\partial S} \quad 0 \le t \le T, \tag{11.25}$$

is self-financed. Furthermore, the portfolio $(\Delta(T), B(T))$ has value $V(T) = F(S(T))$. In particular, the no-arbitrage value of a European-style contingent claim with payoff $F(S(T))$ is $\tilde{P}(S(t), t)$.

This proposition gives a concrete interpretation of APT in the context of the Black–Scholes model: the value of a derivative security is equal to its *replication cost*. If not, there is a clear arbitrage. For instance, if a derivative security trades below the value of its replicating portfolio, $(\Delta(t), B(t))$ at time t, a trader can purchase the derivative and implement a self-financing dynamic hedging strategy with $\Delta(t) = -\frac{\partial \tilde{P}(S(t).t)}{\partial S}$. The difference in prices implies that the trade produces an initial profit and is also riskless, since the position will have net value zero at time T (by Proposition 11.1). The reverse strategy can be implemented if the derivative is more expensive than the dynamic portfolio.

Is APT *equivalent* to pricing derivatives by dynamic replication? The answer is no, in general. The Black–Scholes model and more general parameterizations of the type

$$\sigma(t) = \tilde{\sigma}(S(t), t) \ , \quad r = \tilde{r}(S(t), t) \quad , d(t) = \tilde{d}(S(t), t) \ , \tag{11.26}$$

in which $\tilde{\sigma}, \tilde{r}$, and $\tilde{d}$ are deterministic functions of S and t, are examples of **dynamically complete models**. These are models in which every contingent claim can be replicated by dynamic trading with cash market instruments. A key property that distinguishes dynamically complete models from others is that the dimensionality of the space of "effective" factors is equal to the dimensionality of the space of cash instruments. We will formalize this statement below.

Notice that in the case of the Black–Scholes model with constant parameters or its generalization (11.26), the spot price is the only source of risk. In particular, for any other security contingent on $S(t)$, there exists a proportion, or hedge-ratio, of holdings among the two assets which results in a completely diversified portfolio (i.e., such that its local variance is zero). This property of the Black–Scholes model—the possibility of total diversification of local risk in the cash market—is the mechanism that allows for replication by dynamic hedging.

This discussion can be generalized to multifactor models. We state the following proposition for future use:

PROPOSITION 11.2

Consider a securities market model with M state variables, $X_1(t), X_2(t), \ldots, X_M(t)$, satisfying the dynamic equations

$$dX_i(t) = \sum_{0}^{\nu} \alpha_{i,k}(\mathbf{X}(t), t) \, dZ_k(t) + \beta_i(\mathbf{X}(t), t) \, dt , \quad for \; i = 1, \ldots, M , \quad (11.27)$$

where $Z_k(\cdot), k - 1, \ldots, \nu$ are independent Brownian motions. Suppose that the state-covariance matrix

$$A_{ij}(\mathbf{X}, t) \equiv \sum_{k=1}^{\nu} \alpha_{i,k}(\mathbf{X}, t) \alpha_{j,k}(\mathbf{X}, t) , \quad 0 \leq i, j \leq M \qquad (11.28)$$

has rank R ($R \leq M$). A necessary and sufficient condition for this market to be dynamically complete is that there exist N traded securities with prices $P_l(t) = \tilde{P}_l(\mathbf{X}(t), t)$ such that the local price-covariance matrix

$$\Sigma_{lm} \equiv \sum_{k=1}^{\nu} \sum_{i=1, j=1}^{M} \alpha_{i,k} \alpha_{j,k} \frac{\partial \tilde{P}_l}{\partial X_i} \frac{\partial \tilde{P}_m}{\partial X_j} \qquad (11.29)$$

has rank R. The condition can be expressed concisely in matrix notation as

$$\text{Rank} \left\{ \alpha \cdot \alpha^t \right\} = \text{Rank} \left\{ (\nabla \mathbf{P})^t \cdot \alpha \cdot \alpha^t \cdot (\nabla \mathbf{P}) \right\} , \qquad (11.30)$$

where the superscript $(\bullet)^t$ denotes matrix transposition.

The proof of this proposition follows essentially the same mathematical ideas as in the proof of Proposition 3.1 of Chapter 3. One way to visualize this theorem is to use the language of Differential Geometry. Accordingly, the rank of the matrix A can be viewed as the dimensionality of the tangent space to the "manifold" representing the economy, for which the state-variables are local coordinates. On the other hand, the rank of the price-covariance matrix corresponds to the dimensionality of the tangent subspace generated by portfolios of cash-market instruments. If the two dimensionalities were equal, a complete diversification of risk would be possible. In general, the dimensionality of the tangent space is higher, and one must introduce parameters $\lambda_i(t)$, which may not be directly computable from the dynamic equations (11.27) and the security prices $\tilde{P}_l$. The mathematical proof consists in checking that these statements are correct, using Ito calculus. The details are left to the interested reader.

Cox et al. call the parameters λ_i the **market prices of risk**. The number of unspecified market prices of risk corresponds therefore to the difference between the ranks or the matrices A and $(\nabla \mathbf{P})^t \cdot \alpha \cdot \alpha^t \cdot (\nabla \mathbf{P})$.[6]

Example 11.2
Stochastic volatility model. A simple model for a traded security with stochastic volatility was introduced a few years ago by Hull and White [4]. These authors considered a model for

---

[6]In the language of Differential Geometry, this number is the *codimension* of the tangent subspace generated by security prices in the tangent space.

the joint evolution of a stock price and its spot volatility, namely,

$$dS = \sigma \, S \, dZ + \mu_S \, S \, dt$$

$$d\sigma = \xi \, \sigma \, dW + \mu_\sigma \, \sigma \, dt \, , \tag{11.31}$$

where Z and W are Brownian motions with correlation coefficient $\rho = \mathbf{E} \{ dZ \, dW \} / dt \neq \pm 1$. This system can be regarded as a two-factor model in which the spot price and the volatility are the state-variables. Since spot volatility is not a traded asset, we are in a situation in which the volatility risk cannot be fully diversified. Proposition 11.2 does not apply, since we have Rank$(A) = 2$ and there is only one traded asset. "No-arbitrage" implies that the returns and "volatilities" of any traded asset must satisfy

$$\mu_P + d_P - r = \sigma_{P,1} \lambda_1 + \sigma_{P,2} \lambda_2 , \tag{11.32}$$

where the subscripts [1] and [2] refer, respectively, to the stock price and the stock volatility. Since the stock is a traded security, we have

$$\mu_S + d_S - r = \sigma \lambda_1 \quad , \quad \left(\text{since } \sigma_{S,2} = 0 \right). \tag{11.33}$$

We recover an old result: λ_1 is the Sharpe ratio of the stock. Substituting this into Eq. (11.32), we find that

$$\mu_P + d_P - r = \sigma_{P,1} \left(\frac{\mu_S + d_S - r}{\sigma} \right) + \sigma_{P,2} \lambda_2 . \tag{11.34}$$

If we assume that the price of a derivative security is given by $P(t) = \tilde{P}(S(t), \sigma(t), t)$, this last equation becomes a relation between the partial derivatives of $\tilde{P}$. Specifically, using Ito's Lemma, (11.31), and (11.34) we find, after some calculation, that $\tilde{P}$ satisfies the equation

$$\frac{\partial \tilde{P}}{\partial t} + (r - d_S) S \frac{\partial \tilde{P}}{\partial S} + (\mu_\sigma - \xi \lambda_2) \sigma \frac{\partial \tilde{P}}{\partial \sigma}$$

$$+ \frac{\sigma^2 S^2}{2} \frac{\partial^2 \tilde{P}}{\partial S^2} + \xi \rho \sigma^2 S \frac{\partial^2 \tilde{P}}{\partial S \partial \sigma} + \frac{\xi^2 \sigma^2}{2} \frac{\partial^2 \tilde{P}}{\partial \sigma^2} = 0. \tag{11.35}$$

Notice that this equation involves the parameter λ_2 in the coefficient of the first derivative of $\tilde{P}$ with respect to σ. Therefore, to obtain a pricing measure for this stochastic volatility model, we must make additional assumptions on the value of λ_2. The model is not dynamically complete.

Equation (11.35) shows that the martingale measure corresponding to (11.31) is such that the asset-volatility pair satisfies the SDE

$$dS = \sigma \, S \, dZ + (r - d) \, S \, dt$$

$$d\sigma = \xi \, \sigma \, dW + (\mu_\sigma - \xi \lambda_2) \, \sigma \, dt \, . \tag{11.36}$$

How can one specify the parameter λ_2 in order to generate a "closed" model? This is not an easy problem, for the following reason: to hedge volatility risk, one must obviously use options or some other volatility-sensitive instruments. But then we face the problem that in order to determine λ_2 we need to know the option value $\tilde{P}$ as a function of σ and S. But this is precisely what we are trying to find![7] ☐

We will return to the problem of "calibration" of the stochastic volatility model in other chapters, as we discuss further the relation between the volatilities implied from option prices and "spot" volatility.

11.4 Fokker–Planck Theory: Computing Expectations Using PDEs[8]

Most derivative pricing problems involve computing the expected value of a function of a diffusion path. In the previous sections, we derived the Black–Scholes PDE and the PDE for the stochastic volatility pricing model by applying Ito's formula to a function of the diffusion process. This gives a procedure for computing the expectations numerically or sometimes by closed-form solutions.

The following proposition establishes a general connection between expectations of diffusion processes and partial differential equations.

PROPOSITION 11.3

Assume that $X(t)$ is the diffusion process defined by

$$dX_i(t) \;=\; \sum_{0}^{\nu} \alpha_{i,k}\,(\mathbf{X}(t),\,t)\,dZ_k(t) \;+\; \beta_i\,(\mathbf{X}(t),t)\,dt\,, \quad \text{for } i = 1,\,...,\,M\,, \quad (11.37)$$

where $Z_k(\cdot), k = 1, \ldots, \nu$ are independent Brownian motions. Define

$$A_{ij}(\mathbf{X},\,t) \;\equiv\; \sum_{k=1}^{\nu} \alpha_{i,k}(\mathbf{X},\,t)\,\alpha_{j,k}(\mathbf{X},\,t)\,, \quad 0 \leq i,j \leq M$$

and the differential operator L_t by

$$\mathbf{L}_t\phi \;=\; \frac{1}{2}\sum_{i,j=1}^{M} A_{ij}(\mathbf{X},\,t)\,\frac{\partial^2\phi}{\partial X_i \partial X_j} \;+\; \sum_{i=1}^{M} \beta_i\,(\mathbf{X},t)\,\frac{\partial\phi}{\partial X_i}\,. \quad (11.38)$$

Let $\rho \;=\; \rho(\mathbf{X},\,t)$ and $F \;=\; F(\mathbf{X})$ be bounded functions. Given $T > 0$, suppose that there

---

[7]It can be shown that the option price cannot be determined using a "self-consistent" argument along these lines. See Scott [5]. Nevertheless, in Zhu and Avellaneda [7], a simple expression for the market price of risk of the volatility is derived. The authors assume that traders perform dynamic hedging of volatility risk with short-term options.

[8]This section is more technical than the rest of the chapter. It can be skipped in a first reading.

exists a smooth function $\phi(\mathbf{X}, t)$ satisfying

$$\begin{cases} \frac{\partial \phi}{\partial t} + \mathbf{L}_t \phi - \rho \phi = 0, & \text{for } t < T, \\ \phi(\mathbf{X}, T) = F(\mathbf{X}). \end{cases} \tag{11.39}$$

Then, we have

$$\mathbf{E}_t \left\{ e^{-\int_t^T \rho(\mathbf{X}(\mathbf{s}), s) \, ds} F(\mathbf{X}(T)) \right\} = \phi(\mathbf{X}(t), t). \tag{11.40}$$

This proposition gives a relation between certain functions of diffusion paths and partial differential equations. If we discretize the differential equation (11.39) we can generate a numerical scheme for pricing discounted cash-flows.

The partial differential equation in (11.39) is called the **backward Fokker–Plank equation** associated with the diffusion process (11.36) and the discounting function ρ (Figure 11.2).

Solving the backward Fokker-Planck equation

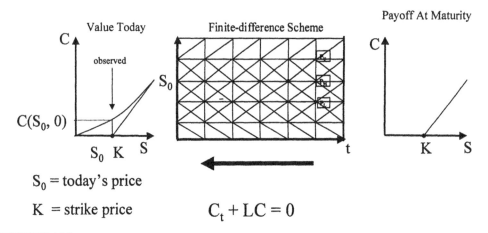

FIGURE 11.2
Computing expectations with the backward Fokker–Planck equation: numerical scheme for pricing a call option.

It is interesting to consider in more detail the case when $\rho = 0$ and

$$F(\mathbf{X}) = \delta(\mathbf{X} - \mathbf{Y}),$$

where $\mathbf{Y} \in \mathbf{R}^M$ is a given vector and δ is the **Dirac delta mass**.[9] Let $p(\mathbf{X}, t; \mathbf{Y}, T)$ be the

---

[9]The Dirac delta mass satisfies $\delta(\mathbf{X}) = 0$ for $\mathbf{X} \neq 0$ and

$$\int_{\mathbf{R}^M} \delta(\mathbf{X}) \, d\mathbf{X} = 1.$$

solution of the partial differential equation

$$\left(\frac{\partial}{\partial t} + \mathbf{L}_{t,\,X} \right) p(\mathbf{X},\, t;\, \mathbf{Y},\, T) = 0 \,, \text{ for } t < T \,,\; \mathbf{X} \in \mathbf{R}^M \qquad (11.41)$$

(the subscript $[X]$ in the operator $\mathbf{L}_{t,\,X}$ indicates that partial derivatives are taken with respect to $\mathbf{X}$) with final condition

$$p(\mathbf{X},\, T;\, \mathbf{Y},\, T) = \delta(\mathbf{X} - \mathbf{Y}) \,.$$

Applying formally Proposition 11.3 to the case $F(\cdot) = \delta(\cdot - \mathbf{Y})$, we conclude that

$$p(\mathbf{X},\, t;\, \mathbf{Y},\, T) = \mathbf{E}\{\delta(\mathbf{X}(T) - \mathbf{Y}) \mid \mathbf{X}(t) = \mathbf{X}\}$$

$$= \mathbf{P}\{\mathbf{X}(T) = \mathbf{Y} \mid \mathbf{X}(t) = \mathbf{X}\} \,.$$

Therefore, $p(\mathbf{X},\, t;\, \cdot,\, T)$ is identified as the probability density of $\mathbf{X}(T)$ conditional on the event $\mathbf{X}(t) = \mathbf{X}$. It follows that

PROPOSITION 11.4

The transition probability density associated with the diffusion process $\{\mathbf{X}(t),\, t > 0\}$ satisfies the backward Fokker–Planck equation (11.41) in $(\mathbf{X},\, t)$ for $t < T$.

Another useful result has to do with the behavior of the function $p(\mathbf{X},\, t;\, \mathbf{Y},\, T)$ with respect to the *forward variables* $(\mathbf{Y},\, T)$. Indeed, we have

PROPOSITION 11.5

The probability density function $p(\mathbf{X},\, t;\, \mathbf{Y},\, T)$ satisfies the equation

$$\left(\frac{\partial}{\partial T} - \mathbf{L}^*_{T,\,Y} \right) p(\mathbf{X},\, t;\, \mathbf{Y},\, T) = 0 \,, \text{ for } T > t \,, \qquad (11.42)$$

where

$$\mathbf{L}^*_{T,\,Y}\phi = \frac{1}{2} \sum_{i,j=1}^{M} \frac{\partial^2}{\partial Y_i \partial Y_j} \left(A_{ij}(\mathbf{Y},\, T)\, \phi \right) - \sum_{i=1}^{M} \frac{\partial}{\partial Y_i} \left(\beta_i\,(\mathbf{Y},\, t)\, \phi \right) \qquad (11.43)$$

is the formal adjoint of $\mathbf{L}_{t,\,X}$.

Equation (11.42) is called the **forward Fokker–Planck equation** (see Figure 11.3).

This last result is useful to compute the probability density $p(\mathbf{X},\, t;\, \mathbf{Y},\, T)$ *as a function of* $\mathbf{Y}$. An important application of the forward Fokker–Planck equation is that expected values of

---

Notice that δ is not, strictly speaking, a function. Nevertheless, the formal arguments that follow are true and can be made precise by assuming that $F(\mathbf{X})$ is a positive function with integral 1 localized in the vicinity of $\mathbf{Y}$.

Solving the forward Fokker-Planck equation

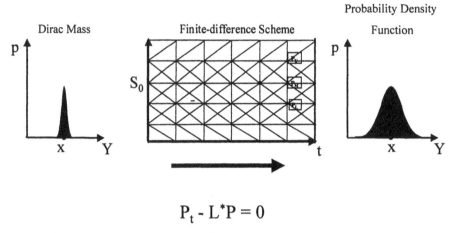

$$P_t - L^*P = 0$$

FIGURE 11.3
The transition probability function $p(X, t; Y, T)$ is obtained by "propagating forward"
a Dirac mass $\delta(Y - X)$ from date t to date T using the forward Fokker–Planck equation.

functions can be computed by numerical quadrature using the formula

$$E_t \{F(\mathbf{X}(\mathbf{T}))\} = \int_{\mathbf{R}^M} p(\mathbf{X}, t; \mathbf{Y}, T) F(\mathbf{Y}) d\mathbf{Y}. \tag{11.44}$$

(Notice that, in contrast, the backward FP equation gives the p as a function of the backward variable $\mathbf{X}$ and hence cannot be used to compute the probability density—only the expectation value.)

A sketch of the proof of Proposition 11.5 is given as an appendix to this chapter.

References and Further Reading

[1] Dixit, A. and Pindyck, R. (1994), *Investment under Uncertainty*, Princeton University Press, Princeton, NJ.

[2] Duffie, D. (1992), *Dynamical Asset Pricing Theory*, Princeton University Press, Princeton, NJ.

[3] Hull, J. (1998), *Introduction to Futures and Options Markets*, Prentice-Hall, Upper Saddle River, NJ.

[4] Hull, J. and White, A. (1987), The Pricing of Options on Assets with Stochastic Volatilities, *Journal of Finance*, **42**, pp. 281–300.

[5] Scott, L.O. (1986), *Option Pricing When the Variance Changes Randomly: Theory and Application.*

[6] Wong, M.A. (1991), *Trading and Investing in Bond Options: Risk Management, Arbitrage, and Value Investing,* Wiley, New York.

[7] Zhu, Y. and Avellaneda, M. (August 1997), A Risk-Neutral Stochastic Volatility Model, *Applied Mathematical Finance.*

Appendix: Proof of Proposition 11.5

We wish to show that $p(\mathbf{X}, t; \mathbf{Y}, T)$ satisfies the forward Fokker–Plank equation in the variables $(\mathbf{Y}, T)$. For this, consider an intermediate time θ such that $t < \theta < T$. It is clear that the density function satisfies the following integral equation:

$$p(\mathbf{X}, t; \mathbf{Y}, T) = \int_{\mathbf{R}^M} p(\mathbf{X}, t; \mathbf{Z}, \theta)\, p(\mathbf{Z}, \theta; \mathbf{Y}, T)\, d\mathbf{Z} \,. \tag{A.1}$$

In fact, this equation just expresses the fact that the diffusion process that starts at $\mathbf{X}$ at time t and ends at $\mathbf{Y}$ at time T must be at *some* point $\mathbf{Z}$ at time θ. The integral on the right-hand side just sums over all possible positions at time θ.

Notice also that the left-hand side of (A.1) is independent of θ. Differentiating both sides of the equation with respect to θ, we conclude that

$$0 = \int_{\mathbf{R}^M} \frac{\partial p(\mathbf{X}, t; \mathbf{Z}, \theta)}{\partial \theta} p(\mathbf{Z}, \theta; \mathbf{Y}, T)\, d\mathbf{Z}$$

$$+ \int_{\mathbf{R}^M} p(\mathbf{X}, t; \mathbf{Z}, \theta) \frac{\partial p(\mathbf{Z}, \theta; \mathbf{Y}, T)}{\partial \theta}\, d\mathbf{Z} \,. \tag{A.2}$$

Using the fact that in the second integral p satisfies the backward FP equation in the variables $(\mathbf{Z}, \theta)$, we have

$$\int_{\mathbf{R}^M} \frac{\partial p(\mathbf{X}, t; \mathbf{Z}, \theta)}{\partial \theta} p(\mathbf{Z}, \theta; \mathbf{Y}, T)\, d\mathbf{Z} = \int_{\mathbf{R}^M} p(\mathbf{X}, t; \mathbf{Z}, \theta) \mathbf{L}_{\theta, \mathbf{Z}}\, p(\mathbf{Z}, \theta; \mathbf{Y}, T)\, d\mathbf{Z}$$

$$= \int_{\mathbf{R}^M} \mathbf{L}^*_{\theta, \mathbf{Z}}\, p(\mathbf{X}, t; \mathbf{Z}, \theta)\, p(\mathbf{Z}, \theta; \mathbf{Y}, T)\, d\mathbf{Z} \,,$$

where we used integration by parts in passing from the first line to the second. (Of course, we are assuming here that the probabilities decay fast enough for a large $|\mathbf{Z}|$ to preclude boundary terms.) If we now let θ approach T, the function $p(\mathbf{Z}, \theta; \mathbf{Y}, T)$ approaches the Dirac mass $\delta(\mathbf{Z} - \mathbf{Y})$. Therefore, as $\theta \to T$, the last equation becomes

$$\int_{\mathbf{R}^M} \frac{\partial p(\mathbf{X}, t; \mathbf{Z}, T)}{\partial T} \delta(\mathbf{Z} - \mathbf{Y})\, d\mathbf{Z} = \int_{\mathbf{R}^M} \mathbf{L}^*_{T, \mathbf{Z}}\, p(\mathbf{X}, t; \mathbf{Z}, T)\, \delta(\mathbf{Z} - \mathbf{Y})\, d\mathbf{Z} \,,$$

or simply

$$\frac{\partial p(\mathbf{X}, t; \mathbf{Y}, T)}{\partial T} = \mathbf{L}^*_{T, Y} p(\mathbf{X}, t; \mathbf{Y}, T).$$

This shows that $p(\mathbf{X}, t; \mathbf{Y}, T)$ satisfies the forward Fokker–Planck equation in $(\mathbf{Y}, T)$, as desired.

Chapter 12

Fixed-Income Securities and the Term-Structure of Interest Rates

This chapter is an introduction to the modeling of fixed-income securities. We introduce standard terminology and conventions for debt securities and concepts from bond mathematics, such as yield and duration. We also define and give elementary properties of forward rates agreements, swaps, interest-rate futures, and options on interest-rate-sensitive securities. The value of fixed-income securities is determined by the *term-structure of interest rates*. The term-structure is usually represented as a curve that displays the value of an index (bond yield, swap rate, forward rate) as a function of the time-to-maturity.

Another objective of this chapter is to relate the valuation of fixed-income securities to APT and the concept of a "risk-neutral measure" on the term-structure of interest rates. The value of "straight debt" (bonds, swaps) is determined from the current (presently observed) term-structure. On the other hand, options and instruments with embedded options, such as callable bonds, derive their value from the volatility of interest rates, i.e., the evolution of the term-structure under a risk-neutral probability measure.

In the last section, we present several algorithms for constructing an *instantaneous forward-rate curve* from prices of bonds and/or swaps observed in the market. Procedures for curve construction are very important in the practical implementation of term-structure models.

12.1 Bonds

Bonds are perhaps the simplest fixed-income securities. When a bond is issued, the investor (bond buyer) is essentially lending money to the issuer in exchange for interest payments and the promise of repayment of the principal at a future date. Most bonds also trade in the secondary market, like stocks and other securities. Understanding the pricing of a bond and, in particular, the relative valuation of different bond issues is very important for investors and traders. The main considerations that enter the pricing of a bond are

- Principal (notional amount of the loan)
- Maturity
- Interest payments
- Call provisions and other features such as conversion to shares, etc.
- Credit quality of the issuer.

The first three points define the structure of cash flows that the investor expects to receive if the bond was held to maturity.

Call provisions exist in many bond issues: this means that the issuer has the option to "call" (retire) the bond after a certain time, stipulated in the bond agreement. For the investor, this means that there is uncertainty as to the true maturity of the loan.

Putable bonds are bonds that can be "put back" to the issuer at a fixed notional value at a given date or dates. In this case, the bondholder essentially owns an option.

Convertible bonds (usually issued by corporations) can be converted into stock at a given price (the "conversion ratio"). Conversion events complicate the pricing, because the investor is uncertain about the cash-flows that will be received.

Credit quality is a very important variable, because it represents our beliefs about the issuer's capacity to repay the principal and interest. U.S. government bonds are considered to be the most credit-worthy, since they are backed by the "full faith and power" of the government.[1]

Bonds issued by corporations have a certain probability of defaulting in the event that the corporation cannot meets its obligations. Thus, corporate bonds have a lower credit quality. Low-credit bonds have higher coupons than high-grade bonds trading at the same price. Investors demand a premium for the default risk.

In this chapter, we will ignore credit quality considerations altogether, because we wish to focus primarily on the interest rates and interest rate risk. We will therefore assume issuers will not default on bond payments or, alternatively, that issuers have the highest credit quality.

Also, for simplicity, we will not discuss call provisions or convertible bonds right away, since call and conversion features are "embedded options" and require option valuation methods. The question of pricing embedded options is postponed until we develop the mathematical machinery for modeling interest-rate volatility.

A **zero-coupon bond** (also called a "zero" or a "pure discount bond") is a bond that has a single payment of principal at maturity, without intermediate interest payments (Figure 12.1). Let us denote the principal amount by $Pr.$ and the bond maturity date by T. Intuitively, the fair value of the bond at date $t < T$ should be less than $Pr.$ dollars. The difference is sometimes called the **liquidity premium** in finance theory. It comes from the fact that the investor would rather have cash now, for investment or consumption purposes, as opposed to having the same amount in the future. In other words, interest rates are positive.

Arbitrage Pricing Theory gives a way of expressing the value of a zero-coupon bond in terms of a **risk-neutral measure** on the paths of short-term interest rates.[2] We define the short-term interest rate r_t as the rate associated with a money-market account, in the framework of APT. For mathematical simplicity, we assume that r_t is a continuously compounded interest rate. Since there is a single cash-flow of $Pr.$ dollars at time T, according to APT, the value of this zero coupon bond at time t is

$$Z_t^T = Pr. \times \mathbf{E}_t^P \left\{ e^{-\int_t^T r_s \, ds} \right\}. \tag{12.1}$$

Here, $\mathbf{E}_t^P$ represents the expected value operator with respect to the risk-neutral measure; the subscript t reminds us that this expectation is taken conditionally on the state of the market at

---

[1]More generally, the debt of sovereign issuers (countries) in the local currency usually has the highest rating among debt securities issued in the local currency.

[2]We are not making the claim that the risk-neutral measure is unique, i.e., that we are considering a complete market. Arbitrage Pricing Theory postulates the existence of *at least* one probability measure on the interest rates paths that is such that interest-rate-contingent claims are priced consistently with the expected values of discounted cash-flows. In the sequel, we will use the terminology "the risk-neutral measure" as referring to such a measure, without implying that it is unique.

FIGURE 12.1
**Cash-flow diagram for a zero-coupon bond. The holder receives a single payment of the
principal amount at maturity and no intermediate payments.**

date t. The expression

$$P_t^T = \mathbf{E}_t^P \left\{ e^{-\int_t^T r_s \, ds} \right\}$$ (12.2)

is called the **discount factor** or the present value of one dollar (**PV01**) corresponding to date
T. The function

$$T \longrightarrow P_t^T \ , \quad T > t$$

is the **discount curve** at time t. Notice that T represents "absolute time," i.e., a fixed date in
the future of t. It is often customary to use "relative time" or "time-to-maturity" $\tau = T - t$
when dealing with the discount curve. In the latter case, the function

$$\tau \longrightarrow P_t^{t+\tau} \ , \quad \tau > 0$$

is said to represent the discount curve at time t. We will use whichever representation of the
discount curve is more convenient according to the context.

Clearly, if the short-term interest rate is nonnegative, we have $P_t^T \leq 1$. The **yield** of the
zero-coupon bond is, by definition, the constant interest rate that would make the bond price
equal to the discounted value of the final cash-flow. The continuously compounded yield of
the zero is therefore

$$Y_t^T = -\frac{\ln\left(P_t^T\right)}{T - t} \ ,$$ (12.3)

so that

$$P_t^T = e^{-(T-t)Y_t^T} \ .$$

Equivalently, the yield is the continuously compounded (constant) rate of return that the investor
would receive if the zero-coupon bond was bought and held to maturity.[3]

---

[3]In Chapter 5, we discussed the notion of forward rates in the context of option pricing. There, we converted
term rates into forward rates. In practice, the term rates are often taken to be the yields on short-term debt
securities such as Treasury bills.

There is an important practical consideration regarding the calculation of yields. In fact, expressing the time-to-maturity, $T - t$, as a decimal number requires using a *day-count* (DCC), or convention to convert days and months into fractions of a year. A DCC determines unequivocally the decimal fraction of year that corresponds to the period between two calendar dates.[4] Notice that the price and the yield vary inversely to each other: an increase in price corresponds to a decrease in yield and vice-versa. Moreover, the price is a convex function of yield.

Most bonds have interest payments (coupons) as well as payment of the principal. The "generic" bond will therefore specify a maturity date, in which the principal payment is made, as well as a schedule of interest payments.

- Maturity date

- Principal ("face value")

- Coupon (interest)

- Frequency and payment dates

The coupon is the annualized interest rate of the bond. The frequency represents how many payments are made per year ($\omega = 1, 2, 4,$ or 12). Most bonds have annual or semiannual interest payments ($\omega = 1$ or 2). Thus, a 10-year bond with face value of $1,000 and a semiannual coupon of 6.25% will pay the investor an interest of $0.5 \times 0.0625 \times 1,000 = \$31.25$ every 6 months (20 payments) and the principal will be paid at the 20th payment date (Figure 12.2).

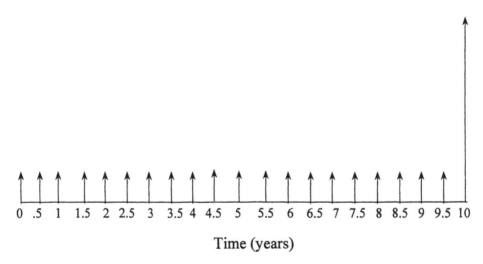

0   .5   1   1.5   2   2.5   3   3.5   4   4.5   5   5.5   6   6.5   7   7.5   8   8.5   9   9.5   10

Time (years)

FIGURE 12.2
The generic fixed-coupon bond has a stream of cash-flows that consist in coupon payments, represented by the small arrows. The last payment consists of the principal plus the last coupon.

To derive a mathematical formula for the value of a coupon-bearing bond in terms of a risk-neutral probability measure on future interest rate scenarios, we set $Pr. =$ principal, $C =$

---

[4]Day-count conventions are important when yields are used to quote bond prices. This is somewhat analogous to quoting option prices in terms of implied volatilities rather than in a currency. A day-count convention in fixed-income markets typically specifies (i) the number of days in a month and (ii) the number of days in a year. The reader interested in DCCs should consult, for example, Fabozzi [1] or ISDA (International Swap Dealers Association) publications.

coupon, ω = frequency.[5] Let t_n, $n \le N$, represent the cash-flow dates, assuming that $t = 0$ now. Then, the value of the bond is given by

$$B_t = \sum_{n:t_n \ge t}^{N} \frac{C \cdot Pr.}{\omega} E_t^P \left\{ e^{-\int_t^{t_n} r_s ds} \right\} + Pr. \cdot E_t^P \left\{ e^{-\int_t^{t_N} r_s ds} \right\} , \qquad (12.4a)$$

or, simply,

$$B_t = \sum_{n:t_n \ge t}^{N} \frac{C \cdot Pr.}{\omega} \cdot P_t^{t_n} + Pr. \, P_t^{t_N} , \qquad (12.4b)$$

where $P_t^{t_n}$ are the discount factors. Notice that this formula is an immediate consequence of the fact that a bond is equivalent to a series of zero-coupon bonds maturing at the coupon-payment dates and at maturity. In particular, *the value of a bond on a particular date is completely determined by the discount curve at that date.*

Although Eqs. (12.4a) and (12.4b) give the true *market value* of the security, notice that the function B has a jump of $\frac{C \cdot Pr.}{\omega}$ at each date $t = t_n$. Thus, the value of the bond changes discontinuously.

The bond "price" can be made continuous, however, if we subtract from expression (12.4b) the interest accrued to the bondholder between the last coupon date and the present date. The market convention is that the interest on the next coupon accrues *linearly*. Thus, if the present date is t and the last coupon date was t_n, the **accrued interest** "earned" by the bondholder up to time t is

$$AI(t, t_n) = \frac{C \cdot Pr.}{\omega} \frac{t - t_n}{t_{n+1} - t_n} = \frac{\cdot Pr.}{\omega} f ,$$

where f represents the fraction of the period elapsed since the last coupon date $(0 < f < 1)$ (Figure 12.3).

By definition, the **clean price** of a bond corresponds to the price at which the transaction takes place minus the accrued interest. Hence, we have

$$\text{transaction price} = \text{clean price} + AI(t, t_n) .$$

In an arbitrage-free economy, the transaction price should be equal to the theoretical value (12.4a)–(12.4b). In particular, the clean price can be expressed in terms of the term-structure of interest rates with the formula

$$\text{Clean price} = \sum_{n:t_n \ge t}^{N} \frac{C \cdot Pr.}{\omega} \cdot P_t^{t_n} + Pr. \, P_t^{t_N} - AI(t; t_n) .$$

The clean and transaction prices coincide on the coupon date after the coupon is paid (since $AI(t_n; t_n) = 0$).

Bond quotes in the U.S. Treasury, international, and corporate markets are usually in terms of clean prices.[6]

---

[5]It is customary to use $Pr. = 100$ for quoting bond prices.

[6]Since the clean price varies continuously, it is not uncommon for practitioners to model the clean bond price as a diffusion process for the purposes of pricing options.

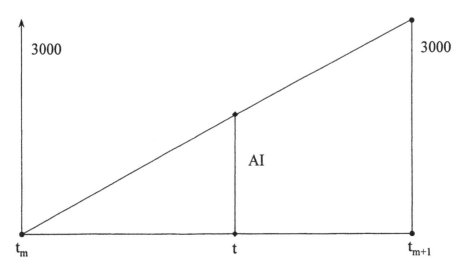

FIGURE 12.3
Schematic diagram of accrued-interest calculation. The market convention is that interest accrues *linearly*, so the coupon payment is "prorated" using the number of days elapsed between the last coupon payment and the next.

The yield of a bond (or yield-to-maturity) is the effective constant interest rate that makes the bond price equal to the future cash-flows discounted at this rate. The yield-to-maturity is usually computed using the same frequency as the bond's interest payments (e.g., semiannually), rather than the continuously compounded yield used for zero-coupon bonds.

For simplicity, let us assume that the current date coincides with a coupon payment date, so that $t = t_m$. In this case, we define the yield-to-maturity of the bond (after the coupon was paid) to be the value of Y such that[7]

$$B = \sum_{n=m+1}^{N} \frac{C \cdot Pr.}{\omega} \left(\frac{1}{1 + Y/\omega} \right)^{n-m} + Pr. \left(\frac{1}{1 + Y/\omega} \right)^{N-m}. \qquad (12.5a)$$

Notice that this formula just states that the market price of the bond is equal to its cash-flows discounted at the interest rate Y. Here the market price is equal to both the clean or the transaction price, because we assume that the current date is a coupon date.

If the current date does not coincide with a coupon date, we must take into account the interest accrued between coupon dates.

Accordingly, assume that $t_m < t < t_{m+1}$ and that f represents the ratio of the number of days remaining until the next coupon date to the number of days in the coupon period, using the appropriate DCC. The bond yield Y is defined by the relation

$$B = \sum_{n=m+1}^{N} \frac{C \cdot Pr.}{\omega} \left(\frac{1}{1 + Y/\omega} \right)^{f+n-m-1} + Pr. \left(\frac{1}{1 + Y/\omega} \right)^{f+N-m-1} \qquad (12.5b)$$

where B is the transaction price.

---

[7]Bond yields and bond prices vary with time over the life of the bond and therefore we should write B_t, Y_t, etc. However, we will sometimes omit the dependence on t to simplify notation.

Equations (12.5a) and (12.5b) define Y implicitly in terms of the bond value. It is easy to see that B is a decreasing function of Y. Moreover, B is convex in Y. To obtain the yield from the bond value, equations (12.5a),(12.5b) must be solved numerically. Nevertheless [recalling the price relation (12.4b)], the yield of a bond is a well-defined function of its theoretical value B and thus of the discount factors $P_t^{t_1}, \ldots, P_t^{t_N}$.

Example 12.1

Consider a 9% U.S. Treasury bond maturing on November 15, 2018. On the date of May 24, 1989, the bond traded at 103 28/32. (This means that the clean price per \$100 face value is 103.875.) To compute the transaction price, we must calculate the accrued interest. The number of days between coupon payments is 184. Taking into account that T-bonds settle 1 day after the transaction takes place, the bond is 9 days into its current coupon period and has 175 days remaining until the next coupon. Thus $f = 9/184 = 0.04891$. The accrued interest is AI $= 0.04891 \times 4.5 = 0.220$ and the transaction price is $B = 103.875 + 0.220 = 104.095$. We can apply formula (12.5b) to obtain a yield $Y = 8.63\%$. $\Box$

Notice that if $t = t_m$ we can use the summation formula for a geometric series to obtain

$$B = \frac{C \cdot Pr.}{Y} \cdot \left(1 - \left(\frac{1}{1 + Y/\omega}\right)^{N-m}\right) + Pr. \left(\frac{1}{1 + Y/\omega}\right)^{N-m}. \qquad (12.5c)$$

This formula shows that if the yield is equal to the coupon rate, the value of the bond is equal to its face value. From this fact and the monotonicity of the price/yield relationship, we can derive some elementary relationships between the price, yield, and coupon of a bond.

If, on a coupon date a bond trades at 100% of the principal, we say that the bond trades **at par**. In this case, its yield is exactly equal to the coupon rate. If the bond price is less than 100% of face value, we say that the bond trades **at a discount**. In this case, its yield is higher than the coupon rate. If the bond trades above 100% of face value, we say that bond trades **at a premium.** In this case, the yield is lower than the coupon rate.

In an arbitrage-free market, two bonds with the same price and same cash-flow dates cannot have different coupons (otherwise, we can short the one with the smaller coupon and buy the one with the larger one). Similarly, two bonds with the same price and payment dates cannot have different yields. The notion of **par yield**—the yield of a par bond—is sometimes used to represent the term structure of interest rates implied by the bond market. In this case, one speaks of the **par yield curve.**

The yield-to-maturity of a bond is often equated with its **internal rate of return**. This means that buying the bond is equivalent to an investment that "earns" an interest rate of Y on money. The price-yield relationship is clear from this point of view: all other things equal, a smaller price corresponds to a higher yield. Another relationship that emerges from this picture is that if the prices of two bonds are equal, the one with the larger coupon has the highest yield.

12.2 Duration

The price-yield relation gives rise to several quantities that are commonly used in bond risk-management. The most important notion is that of **duration** (or **average duration**, or **McCauley duration**). Suppose that $t_m < t < t_{m+1}$ and set $f = (t - t_m)/(t_{m+1} - t_m)$. Duration is defined as

$$D = \frac{1}{B} \cdot \left[\sum_{n=m+1}^{N} \frac{C \cdot Pr.}{\omega} \frac{n - m + f - 1}{\omega} \left(\frac{1}{1 + Y/\omega} \right)^{n-m+f-1} \right.$$

$$\left. + \frac{N - m + f - 1}{\omega} Pr. \left(\frac{1}{1 + Y/\omega} \right)^{N-m+f-1} \right] \tag{12.6}$$

$$= \frac{\sum_{n=m+1}^{N} \frac{C \cdot Pr.}{\omega} \cdot \frac{n-m+f-1}{\omega} \left(\frac{1}{1+Y/\omega} \right)^{n-m+f-1} + Pr.\frac{N-m+f-1}{\omega} \left(\frac{1}{1+Y/\omega} \right)^{n-m+f-1}}{\sum_{n=m+1}^{N} \frac{C \cdot Pr.}{\omega} \cdot \left(\frac{1}{1+Y/\omega} \right)^{n-m+f-1} + Pr. \left(\frac{1}{1+Y/\omega} \right)^{N-m+f-1}}.$$

Notice that $(n - m + f - 1)/\omega$ represents the time between t and t_n measured in years. Therefore, the duration of a bond is a *weighted average of the future cash-flow dates,* weighted by the cash-flows measured in constant dollars (Figure 12.4). (Mathematically, it is the "center of mass" of the future cash-flow dates.)

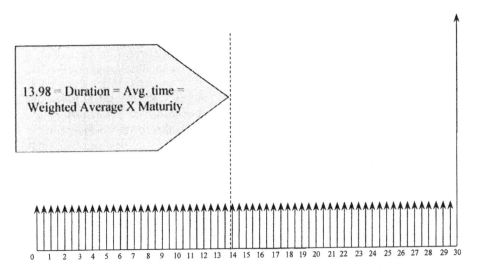

FIGURE 12.4
Cash-flow representation of 30-year bond with 6% coupon paid semiannually. The duration of the bond is approximately 13.8 years.

A closely related quantity is obtained by differentiating the bond price with respect to the yield-to-maturity. Going back to the price/yield relation (12.5b) and differentiating both sides of the equation with respect to Y, we obtain

$$\frac{\partial B}{\partial Y} = - \sum_{n=m+1}^{N} \frac{C \cdot Pr.}{\omega} \left(\frac{n - m + f - 1}{\omega} \right) \left(\frac{1}{1 + Y/\omega} \right)^{n - m + f}$$

$$- Pr. \left(\frac{N - m + f - 1}{\omega} \right) \left(\frac{1}{1 + Y/\omega} \right)^{N - m + f}$$

$$= - \frac{D \, B}{1 + Y/\omega} \, . \tag{12.7}$$

We conclude by rewriting (12.7) in form

$$\frac{1}{B} \, dB \; = \; - \frac{D}{1 + Y/\omega} dY,$$

that the percent change in the value of the bond due to a change in the bond yield is of opposite sign and proportional to the average duration. The quantity

$$D_{mod} \; = \; \frac{D}{1 + Y/\omega}$$

is known as the **modified duration**. The following equations express the fact that the longer the duration, the greater the sensitivity of a bond to a change in yield, in percentage terms

$$dB \; = \; - \frac{B \times D}{1 + Y/\omega} \cdot dY \; = \; -B \, D_{mod} \, dY \, . \tag{12.8}$$

Clearly, the duration of a zero-coupon bond is equal the time-to-maturity. Immediately after a coupon date ($t = t_m$), the duration of a coupon-bearing bond trading at par and having $v(= N - m)$ coupons remaining is given by

$$D = \frac{1}{\omega} \sum_{n=0}^{v-1} \frac{1}{(1 + Y/\omega)^n}$$

$$= \frac{1}{Y} \cdot (1 + Y/\omega) \cdot \left(1 - \frac{1}{(1 + Y/\omega)^v} \right) \, . \tag{12.9}$$

(The derivation of this formula is left as an exercise to the reader.)

It shows that, all things equal, duration decreases with frequency. In fact, if the bond matures in T years and makes only a single payment, we have $v = 1$, $\omega = 1/T$. Substituting these values into (12.9) we find $D = T$, the result for zeros. In the limit $\omega \gg 1$, setting $v = \omega T$, we have $D = \left(1 - e^{-YT} \right)/Y$.

The duration of a coupon-bearing bond is always smaller than the time-to-maturity, because far-away cash-flow dates are "discounted" more than nearby dates. From formula (12.9) the modified duration of a par bond is

$$D_{mod} = \frac{1}{Y} \cdot \left(1 - \frac{1}{(1 + Y/\omega)^\nu} \right) .$$ (12.10)

These formulas are useful for estimating the price-yield sensitivity of bonds.

For example, if $\nu \gg 1$ we can make the approximation $D_{mod} \approx 1/Y$. This approximation is exact for **perpetual** or **consol** bonds. These are fixed-income securities that pay a fixed coupon and have no redemption date (and thus no repayment of principal). Due to the fact that these are bonds with infinite maturity, the duration after a coupon date of a perpetual bond that trades at par is $1/Y$ [set $\nu = \infty$ in (12.10)]. Moreover, because there is no repayment of principal, the above formulas apply even if the consol bond is not trading at par, by simply scaling the coupon. The modified duration of a consol after the last coupon payment is exactly equal to $1/Y$. Moreover, it is easy to see that $Y = C \times (Pr./B)$, where C is the coupon, B is the price, and $Pr.$ is the face value. For example, a 9.25% consol bond trading at $102 immediately after the coupon date has a yield of $0.0925/1.02 = 9.07\%$. The modified duration is $\frac{1}{0.0907} = 11.02$ years. Therefore, according to (12.8), if the yield varies by 1 basis point (1bp = .01%), the price will vary by $102 \times 11.02 \times 10^{-4}$ dollars, or 11.24 cents.[8]

Treasury bond prices are usually quoted in yield and bonds usually trade close to par (this is true for recently issued bonds). For example, using formulas (12.8) and (12.10), we find that a 30-year U.S. Treasury bond yielding 5.90% and trading at par has a modified duration of 13.98 years. The $1/Y$ approximation would give instead 16.94 years. This means that a 1 basis point variation in yield gives rise to a variation in price of 13.98 cents per 100-dollar face value.

Historically, duration was introduced as a measure of the risk-exposure of a bond portfolio and hence as a hedging tool. The rationale for this is simple: suppose momentarily that all yields vary by the same amount when the bond market moves, i.e., that the yield curve shifts **in parallel**.[9] Under this assumption, we can use Eq. (12.8) to quantify the exposure of a portfolio to a shift in the yield curve. In fact, a portfolio consisting of M bonds with n_1 dollars invested in bond 1, n_2 dollars invested in bond 2, etc., has, under the parallel shift assumption, a first-order variation with respect to yield of

$$\sum_j n_j \frac{dB_j}{B_j} = - \left(\sum_j n_j D_{mod\,j} \right) dY .$$

Thus, the sensitivity to a parallel shift in yields is equal to the *dollar-weighted modified duration* of the portfolio. A portfolio with vanishing dollar-weighted modified duration has no exposure to parallel shifts in the yield curve.

It has been recognized now for quite some time that duration-based hedging (under the tacit or explicit assumption of parallel shifts of the yield curve) is often not sufficiently accurate to

---

[8]Corporations sometimes issue very long term bonds, with maturities as long as 100 years, but this is not a common occurrence. In such cases, the $1/Y$ approximation and the computation of the yield as coupon/price apply. Notice that such "century bonds" are closer in spirit to company preferred stock than to debt, since bond-holders receive income but will not redeem the principal from the issuer in the near future.

[9]By yield curve, we mean the par yield curve. The terminology "parallel move of the yield curve" means a change in the term structure of interest rates in which the yields of bonds of different maturities change by the same amount. This assumption, although not correct, is often used by bond investors. The reason is that bond yields are positively correlated, so that a move in yields in the same direction is more likely than a move in the opposite direction.

immunize a fixed income portfolio against interest-rate risk. The reason is that yields of bonds of different maturities generally do not move exactly by the same amount and occasionally exhibit independent variations. A more sophisticated modeling of yield correlations is needed to produce efficient portfolio hedges and to correctly price fixed-income derivatives that are contingent on more than one yield. This subject will be considered in depth in Chapter 13.

12.3 Term Rates, Forward Rates, and Futures-Implied Rates

This section discusses loans and forward-rate agreements between default-free counterparties. **Term rates** are interest rates corresponding to loans of different maturities starting at the present date. These rates are calculated using simple compounding. Hence, a loan of \$1 for a period of time ΔT years starting today can be expressed in terms of a simple interest rate $R(\Delta T)$ for that period. The relation between this rate and the discount factor of the same maturity is (assuming that today corresponds to $t = 0$),

$$\frac{1}{1 + R(\Delta T) \cdot \Delta T} = E_0^P \left\{ e^{-\int_0^{\Delta T} r_s \, ds} \right\} = P_0^{\Delta T} .$$

Solving for $R(\Delta T)$, we obtain

$$R(\Delta T) = \frac{1}{\Delta T} \left(\frac{1}{P_0^{\Delta T}} - 1 \right) . \tag{12.11}$$

The same computation applies to term rates at *future dates*, which are random variables. Let us denote by $R(t, t + \Delta T)$ the term rate at a future date t for a loan of \$1 over the period $(t, t + \Delta T)$. This rate is not known today, since, by definition, it depends on the cost of money at the future t. Nevertheless, generalizing (12.11), we can express the stochastic rate in terms of conditional expectations and discount factors, viz.,

$$\frac{1}{1 + R(t, t + \Delta T) \cdot \Delta T} = E_t^P \left\{ e^{-\int_t^{t + \Delta T} r_s \, ds} \right\} = P_t^{t + \Delta T} ,$$

where E_t^P represents the conditional expectation operator given the market information up to time t. Solving for $R(t, t + \Delta T)$, we obtain, as before,

$$R(t, t + \Delta T) = \frac{1}{\Delta T} \left(\frac{1}{P_t^{t + \Delta T}} - 1 \right) . \tag{12.12}$$

Notice that this equation gives a relation between two random variables, the unknown future term rate and the conditional expectation at time t of a functional of the short-term rate over the loan period $(t, t + \Delta T)$.[10]

---

[10]Note: it is easy to check from this equation that $\lim_{\Delta T \to 0} R(t, t + \Delta T) = r(t)$, a relation linking *term* (i.e., discretely compounded) rates with the mathematical, continuously compounded, short-term rate r_t.

A **forward rate agreement** (FRA) is a contract between two counterparties to enter into a loan at a future date for a specified period of time. The rate at which the counterparties agree to this loan in the future is called a **forward rate**. Forward rate agreements are used by corporations and financial institutions to hedge interest-rate risk in their portfolios.

A forward rate depends on two variables: the starting date of the loan and the ending date of loan. What is the arbitrage-free value of a forward rate? We consider the situation from the lender's perspective. Assuming a notional amount of \$1, and denoting the forward rate by $F(t, t + \Delta T)$, the lender agrees to pay 1 dollar to the borrower at time t and will receive

$$1 + \Delta T \ F(t, t + \Delta T)$$

dollars from the borrower at date $t + \Delta T$. The net present value for the lender is therefore

$$- P_0^{t+\Delta T} + (1 + \Delta T \ F(t, t + \Delta T)) \cdot P_0^{t+\Delta T} . \tag{12.13}$$

Since, by definition, a forward-rate agreement is entered at zero cost, the "fair" value for the contracted interest rate for the period $(t, t + \Delta T)$ should be

$$F(t, t + \Delta T) = \frac{1}{\Delta T} \left(\frac{P_0^t}{P_0^{t+\Delta T}} - 1 \right) . \tag{12.14}$$

Another way of phrasing this arbitrage relationship is that *in an FRA, the lender is long a zero with maturity $t + \Delta T$ and short a zero with maturity t.*

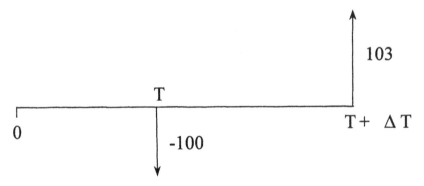

FIGURE 12.5
A forward-rate agreement is a contract for lending a specified amount in the future at a stipulated, fixed interest rate. The lender's position is equivalent to being short one zero and long another one with longer maturity.

Equation (12.14) can be rewritten in a different form, which is useful to compare FRAs with futures contracts. In fact, we have, from (12.14),

$$F(t, t + \Delta T) = \frac{1}{\Delta T} \frac{1}{P_0^{t+\Delta T}} \cdot \left(P_0^t - P_0^{t+\Delta T} \right)$$

$$= \frac{1}{\Delta T} \frac{1}{P_0^{t+\Delta T}} \cdot \mathbf{E} \left\{ e^{- \int_0^t r_s \, ds} \left(1 - P_t^{t+\Delta T} \right) \right\}$$

$$= \frac{1}{P_0^{t+\Delta T}} \cdot \mathbf{E}\left\{ e^{-\int_0^t r_s \, ds} \; P_t^{t+\Delta T} \; R(t, \, t+\Delta T) \right\},$$

where we used Eq. (12.12). Therefore, since $R(t, \, t+\Delta T)$ is known at time t, we have

$$F(t, \, t+\Delta T) = \frac{1}{P_0^{t+\Delta T}} \cdot \mathbf{E}\left\{ e^{-\int_0^{t+\Delta T} r_s \, ds} \; R(t, \, t+\Delta T) \right\}$$

$$= \frac{\mathbf{E}^P\left\{ R(t, \, t+\Delta T) \, e^{-\int_0^{t+\Delta T} r_s ds} \right\}}{\mathbf{E}^P\left\{ e^{-\int_0^{t+\Delta T} r_s ds} \right\}}. \qquad (12.15)$$

Thus, the forward rate can be viewed as the expected value of the corresponding term rate in the future, with respect to a new probability measure given by expression (12.15). This probability is not the risk-neutral measure, due to the presence of the discount term $e^{-\int^{t+\Delta t} r_s \, ds}$. This fact proves useful in comparing interest rate forwards with interest rate futures.

Interest rate futures are exchange-traded contracts that are closely related to FRAs. The most popular contract is the CME 3-month Eurodollar futures contract. In this case, the underlying rate at the settlement date, $R(t, \, t+0.25)$, represents the average 3-month LIBOR rate on that date.[11] The contract has a settlement price of $100 * (1 - R(t, \, t+0.25))$.

With this example in mind, consider a stylized futures contract with settlement date t, settling into the rate $R(t, \, t+\Delta T)$. The futures price today $(t = 0)$ implies therefore a rate for the loan over the period $(t, \, t+\Delta T)$ equal to

$$F_{fut.}(t, \, t+\Delta T) = \mathbf{E}^P\left\{ R(t, \, t+\Delta T) \right\}. \qquad (12.16)$$

This follows from the fact that, in a risk-neutral world, futures prices and the corresponding implied rates should be martingales.[12]

In particular the expressions for the interest rates implied by futures contracts and the forward rates for the same period are not equal [compare (12.15) and (12.16)]. From these equations, it follows that

$$F(t, \, t+\Delta T) - F_{fut.}(t, \, t+\Delta T) = \frac{\mathbf{E}(R\,D) - \mathbf{E}(R)\,\mathbf{E}(D)}{\mathbf{E}(D)}$$

where $R = R(t, \, t+\Delta T)$, $D = e^{-\int^{t+\Delta T} r_s \, ds}$, or

$$F(t, \, t+\Delta T) - F_{fut.}(t, \, t+\Delta T) = \frac{\mathbf{Cov}\left(R(t, \, t+\Delta T), \; e^{-\int_0^{t+\Delta T} r_s \, ds} \right)}{\mathbf{E}(D)}. \qquad (12.17)$$

---

[11] See Chapter 5 and also Siegel and Siegel [5].

[12] For example, in the 3-month Eurodollar contract, the implied interest rate is equal to (100-(futures price))%. The martingale property for the risk-neutral measure is due to the fact that futures are entered at *zero cost* and "marked-to-market" daily. Thus, since the cost of entering the contract is zero, the (risk-neutral) expected futures price tomorrow must be today's futures price.

If we assume a positive correlation between the short rate r_s, $s < t + \Delta t$, and the term rate $R(t, t + \Delta T)$, which is reasonable, we conclude that R and D are negatively correlated. (For example, this assumption is consistent with the "parallel movements" approximation alluded to earlier.) Negative correlation implies that

$$F(t, t + \Delta T) < F_{fut.}(t, t + \Delta T). \tag{12.18}$$

This result is consistent with empirical evidence of the spread between ED futures and forwards in the LIBOR market. The difference between futures and forwards is caused by the correlation between short-term financing and the rate under consideration. Table 12.1 shows the magnitude of the difference between futures-implied rates and forward rates.

Table 12.1   The column on the right gives the amount in (%) (that must be subtracted from the futures-implied rate to obtain the forward rate for the corresponding period). The period is 3 months starting from the maturity date. The futures-implied rate is 100-(futures price). The table was generated using market rates from late 1997.

CONVEXITY	ADJUSTMENT:	EURODOLLAR FUTURES
Maturity (months)	Futures price	Convexity Adjustment (%)
4	94.19	0.0012
10	94.14	0.0023
13	94.08	0.003
16	93.98	0.0044
19	93.98	0.0092
22	93.94	0.0131
25	93.91	0.0176
28	93.85	0.0234
31	93.87	0.0292
34	93.85	0.0371
37	93.83	0.0447
40	93.74	0.0522
43	93.79	0.0637
46	93.77	0.073
49	93.75	0.083

12.4   Interest-Rate Swaps

A swap is a contractual agreement between two parties in which they agree to make periodic payments to each other according to two different indices.[13].

In a "plain vanilla" interest-rate swap, one party (Counterparty A) makes *fixed interest rate* payments on a stipulated notional to the other party (Counterparty B). Counterparty B makes

[13]For an in-depth study of swaps, see Kapner and Marshall [4].

floating rate payments to Counterparty A based on the same notional (Figure 12.6). The swap contract specifies the **notional** amount, or face value of the swap, the payment frequency (quarterly, semiannual, etc), the **tenor** or maturity, the **coupon**, or fixed rate, and the floating rate that will be used. Most swaps are arranged so that their value is zero at the starting date.

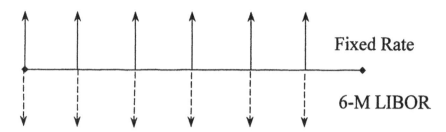

FIGURE 12.6
"Plain vanilla" interest-rate swap. The straight lines represent the fixed rate "leg" and the dashed lines represent the floating (LIBOR) leg.

For U.S. dollar swaps, floating rates are typically the 3-month or 6-month LIBOR rates prevailing over the period before the interest payment is made. (The interest rates are determined *in advance,* or equivalently, the payments are made *in arrears.*)

In practice, there are many variants of the basic swap structure. For instance, swaps can be such that the notionals are different for the two counterparties, or the notional(s) amortize, or where the floating leg is LIBOR plus or minus a fixed coupon, etc.[14] For simplicity, we will focus primarily on the standard, "plain vanilla" interest rate swaps.

Currently, swaps are the most popular fixed-income derivatives. They are used for a variety of purposes and by many different market participants. One of the most common applications used by financial institutions, corporations, and large institutional investors is for hedging interest rate risk. An often-cited motivation for using swaps is the theory of *comparative advantages:* institutions often have different abilities for borrowing in capital markets. This could be due to differences in credit rating or tax treatment, or for accounting reasons (Figure 12.7). Swaps can serve, for instance, to change the fixed-coupon debt into floating-rate debt and vice-versa. Due to the liquidity of the swap market, swaps can be used to hedge the interest rate risk of fixed-income portfolios at low cost.[15]

In this section, we derive the basic pricing formulas for swaps. We assume that there are N cash-flow dates in the swap, $t_1, t_2, \ldots, t_N$. Payments are made **in arrears**. At each date, the floating rate for the following period is set equal to the simple interest rate for that period. Accordingly, the floating rate for the period (t_{n-1}, t_n) is

$$R(t_{n-1}, t_n), \quad n = 1, 2, \ldots, N.$$

If we denote the fixed rate by F and the notional *Not.*, the stream of cash flows for a swap at inception ($t = t_0 = 0$) is as follows:

Fixed rate payment at date t_n:

$$\Delta_n \cdot F \cdot Not. \text{ at dates } t = t_1, \ldots, t_n \text{ where } \Delta_n = t_n - t_{n-1},$$

---

[14]Amortization means that the notional on which the payments are made can decrease as time progresses, according to some schedule or index.
[15]see Kapner and Marshall [4].

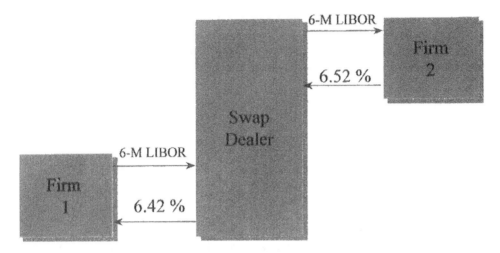

FIGURE 12.7
In general, nonfinancial counterparties enter into swaps with a dealer that makes two-way markets in swaps and earns a commission (spread). Notice that the counterparties are not exposed to each other's credit risk, which is assumed instead by the dealer separately with each counterparty.

Floating rate payment at date t_n:

$$\Delta_n \cdot R(t_{n-1}, t_n) \cdot Not.$$

Needless to say, the fixed rate payer is "long floating rate" and the floating rates payer is "short floating rate." The value at time $t_0 = 0$ of a swap from the point of view of the fixed rate payer is

$$Swap = \sum_{n=1}^{N} \Delta_n \, Not. \, \mathbf{E}^P \left\{ (\, R(t_{n-1}, t_n) \, - \, F \,) \, e^{-\int_0^{t_n} r_s \, ds} \right\}. \qquad (12.19)$$

This expression can be simplified considerably. Indeed, let us consider the series of floating-rate payments alone. If we add to this series a fictitious payment of principal (*Not.*) at time t_N, we obtain the same cash-flows as in a **floating-rate bond**. Therefore, the cash-flows of a swap are equivalent to those generated by being *long a coupon-bearing bond* and *short a floating-rate bond* with the same principal (Figure 12.8).

As it turns out, the value of the floating rate bond is equal to par. To see this, notice that the value of such a floating-rate bond is

$$B = \sum_{n=1}^{N} \Delta_n \, Not. \, \mathbf{E}^P \left\{ R(t_{n-1}, t_n) \, e^{-\int_0^{t_n} r_s \, ds} \right\}$$

$$+ \, Not. \, \mathbf{E}^P \left\{ R(t_{N-1}, t_N) \, e^{-\int_0^{t_N} r_s \, ds} \right\}. \qquad (12.20)$$

Each term in the series of interest payments can be rewritten using (12.12) and properties of

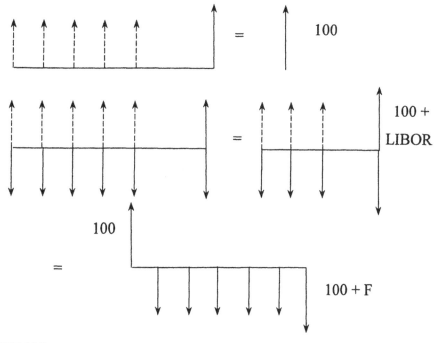

FIGURE 12.8
**Equivalence between *swaps* and *par bonds*. The first diagram shows that a floating rate
bond is worth par ($100) immediately after a floating rate payment. Adding and sub-
tracting a "fictitious" principal payment on both legs of the swap, we conclude that the
fixed-rate payer in a swap is effectively long $100 in cash and short a par bond with
principal $100.**

the conditional expectation, as

$$\Delta_n \, Not. \, \mathbf{E}^P \left\{ R(t_{n-1}, \, t_n) \; e^{-\int_0^{t_n} r_s \, ds} \right\}$$

$$= Not. \, \mathbf{E}^P \left\{ \left(\frac{1}{P^{t_n}_{t_{n-1}}} - 1 \right) e^{-\int_0^{t_n} r_s \, ds} \right\}$$

$$= Not. \, \mathbf{E}^P \left\{ \left(\frac{1}{P^{t_n}_{t_{n-1}}} - 1 \right) P^{t_n}_{t_{n-1}} \, e^{-\int_0^{t_{n-1}} r_s \, ds} \right\}$$

$$= Not. \left(P^{t_{n-1}}_0 - P^{t_n}_0 \right).$$

Therefore, summing over all payment dates, we have

$$B = \sum_{n=1}^{N} Not. \left(P^{t_{n-1}}_0 - P^{t_n}_0 \right) + Not. \, P^{t_N}_0 = Not.$$

Returning to the valuation of the swap, we have

$$Swap = \text{Floating rate bond} \quad - \quad \text{Fixed coupon bond} \quad (\text{Coupon} = F)$$

$$= Not. - \left(\Delta_n \cdot F \cdot Not. \sum_{n=1}^{N} P_0^{t_n} + Not. \, P_0^{t_N} \right). \tag{12.21}$$

From the point of view of the fixed rate payer, a swap is equivalent to *a synthetic position in which the investor issues coupon-bearing bond and invests the proceeds at the floating rate.* The crucial difference between the swap and the above transaction is that in a swap there is no exchange of principal; only interest payments are made. This means that

- There is less credit risk—a default will affect only future interest payments, but not the principal.

- The investor does not actually issue a bond—he does not need to establish the same degree of credit-worthiness of a bond issuer, since there is no risk of default on the principal.

By definition, the **swap rate** (corresponding to a particular maturity and payment schedule) is the fixed rate F that makes the swap value equal to zero. In other words, it is the rate at which counterparties would agree to enter the swap without paying or receiving any premium.[16] The swap rate is computed by setting Swap $= 0$ in Eq. (12.21) and solving for F. The result is

$$F_{swap} = \frac{1 - P_0^{t_N}}{\sum_{n=1}^{N} \Delta_n \cdot P_0^{t_n}}. \tag{12.22}$$

If this equation would not hold and a par bond with the same maturity and frequency were traded in the market, this would give rise to an arbitrage opportunity. For example, if the swap rate F prevailing in the market is *less* than the theoretical rate F_{swap}, the arbitrage strategy would consist in entering into the swap paying fixed. Such a position is equivalent to being long a floating-rate bond and short a coupon-bearing bond with coupon rate F. It can be offset by buying the fixed-rate par bond and financing the purchase at the floating rate. This trade has zero cost and produces a profit, because the coupon of the par bond (namely, F_{swap}) is greater than F. A mirror argument applies if $F > F_{swap}$.

This equivalence between swaps and long–short combinations of a fixed and a floating rate bonds implies that we can compute the duration, or sensitivity, of a swap to yield changes. In fact, notice that a floating rate bond has very little interest rate risk, since the interest rate resets after each period. We conclude that the sensitivity of a swap to yield curve movements is roughly equivalent to the modified duration of the fictitious coupon bond.

Another useful way to view a swap is as *a series of forward rate agreements.* This can be used to derive no-arbitrage relations that must hold between swaps and FRAs. In practice, traders "arbitrage away" any mispricing between swaps of different maturities using swaps, futures, and FRAs.

---

[16] We assume no-default here and neglect transaction fees.

12.5   Caps and Floors

Caps and floors are a basic type of option associated with swaps and FRAs. The parameters defining a cap are

- Notional

- Cash-flow dates

- Floating rate

- Strike rate (fixed rate)

Denoting the cash-flow dates by $t_1, t_2, \ldots, t_N$, the floating rate by $R(t_{n-1}, t_n)$, and the strike rate by F, the cash-flow at the date t_n is

$$Not. \times (\ t_n - t_{n-1}\ )\ \max\ (R(t_{n-1},\ t_n) - F,\ 0\ )$$

Notice that payments are made in arrears, to mimic the usual payment of interest rates (Figure 12.9). Therefore, a cap is analogous to a series of calls on the floating rate. Similarly, a floor corresponds to a series of puts on the floating rate and has cash-flows

$$Not. \times (\ t_n - t_{n-1}\ )\ \max\ (F - R(t_{n-1},\ t_n),\ 0\ )\ .$$

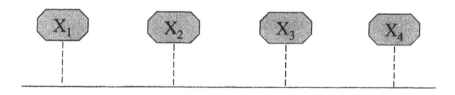

where $x_i = \max (F - R_i, 0)$

FIGURE 12.9
The holder of a cap receives the difference between the floating rate (LIBOR) and the stipulated fixed rate—if this difference is positive—at each cash-flow date. A cap is therefore equivalent to a series of calls on the floating rate. Payments are usually made in arrears.

Caps and floors are options and hence trade at a premium. The premia of cap and floors with the same strike rate are related by a put–call parity, analogous to the one of vanilla options. In fact, we have

$$(Fixed - payer)\,swap\ =\ Cap\ -\ Floor\ ,$$

where the fixed leg of the swap corresponds to the strike rate for the cap/floor, F. Given that the strike rate is arbitrary, and not necessarily equal to the corresponding swap rate, the swap may have non-zero value. More precisely, in the above formula, we have

$$(Fixed - payer)\, swap = Not. \sum_{n=1}^{N} \Delta_n \,(F_{swap} - F)\, P_{t_0}^{t_n} \,,$$

$$Cap = Not.\, (F_{swap} - F) \sum_{n=1}^{N} \Delta_n \, P_{t_0}^{t_n} \,+\, Floor \,.$$

The ingredients needed to price caps and floors are (i) the current level of interest rates, (ii) the strike level, and (iii) the volatility of future interest rates. The theoretical value of a cap in terms of a risk-neutral measure P is

$$Cap \,=\, Not. \sum_{n=1}^{N} (\,t_n - t_{n-1}\,)\, \mathbf{E}_t^P \left\{ e^{-\int_t^{t_n} r_s\, ds} \,\max\, (R(t_{n-1},\, t_n) \,-\, F\,,\, 0\,) \right\}\,.$$

The expectation values in this equation are determined by the *probability distribution* of interest rates under the risk-neutral measure. Models for interest-rate dynamics will be discussed in the next three chapters.

12.6   Swaptions and Bond Options

Swaptions are options to enter into swaps. In a **payer swaption**, the investor has the option to enter into a swap with given tenor (e.g., a 10-year swap) at the option's maturity date (e.g., 2 years from today) paying the fixed rate (strike) and receiving floating. A **receiver swaption** gives the right to enter into a swap receiving fixed and paying floating (Figure 12.10).

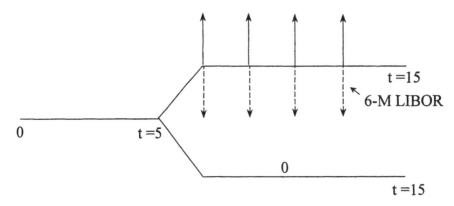

FIGURE 12.10
A 5 × 10 European receiver swaption. The holder has an option in 5 years to enter into a 10-year swap, in which she/he receives a fixed payment against LIBOR.

Swaptions can be structured as European or American options. Since swaptions are traded over the counter and not in organized exchanges, the terms of these contracts vary considerably. In general, American-style swaptions gives the holder the right to enter into a swap at a *discrete* set of dates, as opposed to the "continuous" exercise feature of stock options discussed earlier in the book. The discrete exercise dates correspond typically to the reset dates of an underlying swap. These options are sometimes called "mid-Atlantic" or "Bermudan."

The main parameters that define a swaption are therefore

- Notional
- Maturity of the option
- Payer or receiver
- Type: American or European
- Maturity of the swap
- Cash-flow dates of the swap
- Floating rate
- Fixed rate

For instance, a European **in-5-for-10** LIBOR payer swaption with strike 6.50% is an option to enter into a 10-year swap paying a fixed rate of 6.50% 5 years from now.

Another example of a contract that can be viewed as an American swaption is called a **cancellable swap**. The contract specifies an underlying swap (starting today, say) with given tenor (e.g., 15 years) and a **lockout period** (e.g., 5-years). The terms are the following: Counterparty A pays Counterparty B 6-month LIBOR minus 25 basis points (bp) semiannually over the duration of the contract. Counterparty B pays Counterparty A the 15-year swap rate semiannually. Moreover—and this is the part of the contract that involves optionality— Counterparty B has the right to terminate this agreement after the 5-year lockout period, at any reset date of the underlying swap. In other words, the counterparties in a cancellable swap have the following position between them:

- A 15-year swap (in-the-money for A because he pays less than LIBOR)
- B is long a forward-starting Bermudan option that gives him the right to enter into the opposing swap every 6 months after the 5-year lockout period.

If this cancellable swap is issued "at market," i.e., for zero cost to the counterparties, then the 25 bp rebate on the floating side that Counterparty A has can be viewed as the price that Counterparty B pays for the embedded Bermudan option.[17]

We consider pricing European swaptions. Let us denote the maturity date of the option by $T = t_0$ and the present date by $t = 0$. The payoff of a payer swaption can be expressed in the

---

[17]As the reader expects, there are infinitely many ways in which a swap can be modified by extending, inter-rupting, or modifying the cash-flows contingent to some event or decision of the counterparties. Most of these variants can be viewed as embedded cap/floors or embedded swaptions. The cancellable swap is one of the most common structures of this kind.

form

$$
Max \, (Swap, \, 0 \,) \; = \; Not. \times Max \left[1 - \left(\sum_{n=1}^{N} \Delta_n \, F \, P_{t_0}^{t_n} + P_{t_0}^{t_N} \right), \; 0 \right],
$$

where we used the fact that the underlying swap is equivalent to a long fixed-coupon bond–short floating-rate bond position.

Therefore, the theoretical value of a European payer swaption is, if we use (12.21) to value the swap,

$$
Payer \; Swaption =
$$

$$
Not. \, \mathbf{E}_0^P \left\{ e^{-\int_0^{t_0} r_s ds} \max \left[1 - \left(\sum_{n=1}^{N} \Delta_n F P_{t_0}^{t_n} + P_{t_0}^{t_N} \right), \; 0 \right] \right\}. \qquad (12.23)
$$

An equivalent expression, which follows from Eq. (12.15), is

$$
Payer \; Swaption =
$$

$$
Not. \, \mathbf{E}_0^P \left\{ e^{-\int_0^{t_0} r_s ds} \max \left[\sum_{n=1}^{N} \Delta_n \, \mathbf{E}_{t_0}^P \left\{ e^{-\int_{t_0}^{t_n} r_s \, ds} \, (R(t_{n-1}, \, t_n) - F \,) \right\}, \, 0 \right] \right\}.
$$

The latter formula can be used to establish a comparison between a European swaption and a cap that have the same expiration dates. Since $\max(X, \, 0)$ is an increasing function of X, we note that the right-hand side of the last equation can increase only if we replace $(R(t_{n-1}, \, t_n) - F)$ by its positive part $\max \, (R(t_{n-1}, \, t_n) - F, \, 0)$. This change increases the value of the conditional expectations in the sum and hence of the sum. However, the value of the sum after the substitution is nonnegative. We conclude that

$$
Payer \; Swaption \leq
$$

$$
Not. \times \mathbf{E}_0^P \left\{ e^{-\int_0^{t_0} r_s ds} \sum_{n=1}^{N} \Delta_n \, \mathbf{E}_{t_0}^P \left\{ e^{-\int_{t_0}^{t_n} r_s \, ds} \max \, (R(t_{n-1}, \, t_n) - F, \; 0) \right\} \right\}
$$

$$
= Not. \sum_{n=1}^{N} \Delta_n \, \mathbf{E}_0^P \left\{ e^{-\int_0^{t_n} r_s \, ds} \max \, (R(t_{n-1}, \, t_n) - F, \; 0) \right\} \; = \; Cap.
$$

We have just derived something that is financially obvious: the theoretical value of a European swaption cannot exceed the value of a cap with the same maturity and strike rate. In fact, the cap is essentially a series of calls on interest rates, whereas a payer swaption is a single option on a "basket" of interest rates. The cap guarantees protection *at each period* against the rise of interest rates above the level F; the swaption provides protection *on average* and therefore should be cheaper.

Finally, we consider bond options. In the "ideal world" that we are considering here, credit risk is neglected. Furthermore, recall that APT assumes traders can borrow and lend at the same rates in arbitrary amounts. The question of liquidity (i.e., availability of a particular security in the market) is not directly addressed by APT. Because of this, our model cannot address two important aspects of bond and bond option pricing: credit quality, on the one hand, and the existence of bonds that are "on special," i.e., that are in shorter supply than other bonds. Credit quality was discussed previously. The second consideration is also very important. Bonds that are in high demand trade at a premium over other bonds of similar maturities. The rates at which these bonds can be financed via repurchase agreements are lower than the standard bond financing rate (repo rate). Once these issues are separated from the modeling of interest rate risk, we are left with a simplified "world" where bond options can be priced using the same risk-neutral measure as swaptions.

Interestingly, many practitioners use the standard Black–Scholes theory or certain modifications such as Black '76 to price bond options. The latter approach is sometimes deemed to be more intuitive than option models based on term-structure modeling. Term-structure models are better fitted to a situation where there is a "homogeneity" of credit quality and liquidity across the term structure and where the modeler is interested in capturing the effect of *correlations* between different interest rates on the prices of interest rate derivatives.[18] On the other hand, applying the Black–Scholes formula to price a bond option puts the emphasis on the volatility and price of that particular bond and not so much on its correlation with other instruments. This is well suited to situations where credit and liquidity are the dominant risk factor and the correlation between different yields is not.[19]

Neglecting therefore credit and liquidity considerations, we can use APT to derive the price of a bond option, using the notation of Section 12.1. The intrinsic value of a European-style call option with expiration date T and strike K on a coupon-bearing bond with coupon rate C, payment frequency ω, cash-flow dates t_n, and principal $Pr.$ is (with the notation of Section 12.1)

$$Bond\ option\ =$$

$$\mathbf{E}^P \left\{ e^{-\int_0^T r_s\, ds} \max \left(\sum_{n:\, t_n \geq T}^{N} \frac{C \cdot Pr.}{\omega} P_T^{t_n} + Pr.\ P_T^{t_N} - K,\ 0 \right) \right\},$$

where P is the risk-neutral measure [compare this with Eq. (12.21)]. Notice that payer and receiver swaptions are equivalent, respectively, to calls and puts on a bond with coupon rate F, with strike equal to the principal of the bond (the par amount).

12.7 Instantaneous Forward Rates: Definition

The examples of the previous paragraphs show that, in general, the cash-flows of an interest-rate-sensitive security at time t can be expressed in terms of the collection of discount factors

$$P_t^T = \mathbf{E}_t^P \left\{ e^{-\int_t^T r_s\, ds} \right\}, \quad T \geq t.$$

---

[18]The primary example of this market is the U.S. dollar interest rate derivatives market.

[19]An example of this situation is the market in dollar-denominated debt issued by foreign governments.

The variable t is the **calendar date**—it corresponds to the date at which interest rates are observed. The variable T is often called the **forward date**. It labels the maturity date of a loan. The discount factors at time t represent the totality of term rates and forward rates prevailing at time t. In other words, all rates corresponding to swaps and FRAs as well as bond yields are functions of the discount factors. Because of this fact, any model suitable for pricing simultaneously *all* fixed-income securities must describe the joint evolution of the discount factors $\{\, P_t^T, \ T > t \,\}$. Such models are called **term-structure models** because they describe the complete evolution of the term structure of interest rates across time (Figure 12.12). The reader can think of a term structure model as describing the dynamics of a *curve* (a function of T) across time.

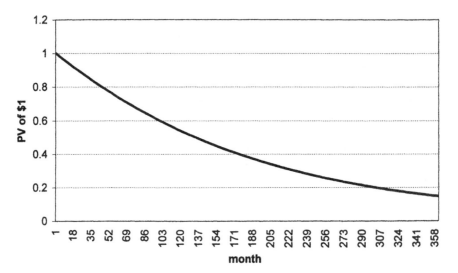

Spot Curve of Discount Factors

FIGURE 12.11
Theoretical spot curve of discount factors P_0^T, $0 < T < 30$, generated by smoothing USD LIBOR market data from November 1997. The data used to generate the curve consisted of 15 FRA rates and 7 swap rates.

An alternative parameterization of the term-structure of interest rates often favored by theoreticians is the so-called **term-structure of instantaneous forward rates**, or "term-structure of forward rates" for short. An instantaneous forward rate is, by definition, the forward rate that would prevail for an infinitesimal time-interval.

Consider first, for simplicity, the current ($t = 0$) term-structure of interest rates. We have seen in Section 12.2 that the forward rate for a loan over the period $(T, T + \Delta T)$ is given by

$$F(T, T + \Delta T) = \frac{1}{\Delta T} \left(\frac{P_0^T}{P_0^{T+\Delta T}} - 1 \right).$$

Hence, if we consider a very short time-interval in the future of T, the corresponding *instantaneous* forward rate should be defined by the limit

$$f(0; T) = \lim_{T \to 0} \frac{1}{\Delta T} \left(\frac{P_0^T}{P_0^{T+\Delta T}} - 1 \right) = -\frac{\partial}{\partial T} \log\left(P_0^T \right).$$

Here, the first variable in $f(0; T)$ is written to emphasize that these are *current* forward rates, corresponding to $t = 0$ (Figure 12.12). There is a one-to-one correspondence between discount factors and instantaneous forward rates. Indeed, we can recover the discount factors from instantaneous forward rates by the formula

$$P_0^T = e^{-\int_0^T f(0;s)\,ds} \ .$$

Spot Forward Rate curve f(0,T)

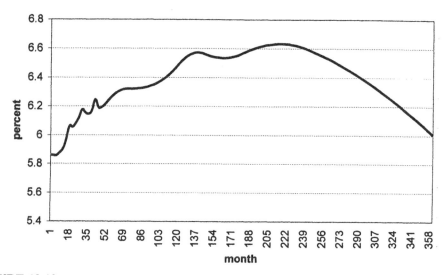

FIGURE 12.12
Spot 1-month forward-rate curve generated by setting $f(0, t) = \frac{1}{\Delta T} \log \left(P_0^T / P^{T+\Delta T} \right)$ **where** P_0^T **is the discount curve of Figure 12.11. The "wiggles" occurring before 50 months correspond to the fitting of 15 Eurodollar futures rates settling every 3 months.**

This equation applies to instantaneous forward rates at a *future* date t (which are now random variables). We define the instantaneous forward rate $f(t; T)$ at time t for a loan over an infinitesimal period after date T by

$$f(t; T) = -\frac{\partial}{\partial T} \log \left(P_t^T \right) ,$$

which is equivalent to

$$P_t^T = e^{-\int_t^T f(t;s)\,ds} \ .$$

The curves P_t^T and $f(t; T)$ are mathematically equivalent parameterizations of the term-structure of interest rates. As we will see in Chapter 13, the dynamics of the term-structure of interest rates is often modeled by assuming that, for each T, $f(t; T)$ is an Ito process in the variable t.

In practice, there are significant subtleties involved in modeling the term-structure of interest rates as a continuous curve, even at the level of P_0^T and $f(0; T)$, (i.e., without dynamics). In fact, the universe of fixed-income instruments used to infer the curve of instantaneous forward rates is always *finite* and sometimes rather small. It consists typically of bonds, bills, FRAs, futures, and swaps traded liquidly in the market. This means that we need to determine a function $f(0; \bullet)$ based on the knowledge of a few integrals of this function, which is an ill-posed problem. There are many possible algorithms for generating the forward-rate curve from data. In the next section, we discuss and analyze some of the most common approaches.

12.8   Building an Instantaneous Forward-Rate Curve

An important first step for modeling interest rate derivatives is the construction of a curve of instantaneous forward rates $f(0; T)$ from the prices of a given set of reference securities. From this curve, we can infer the theoretical values of discount factors via the formula

$$P_0^T = e^{-\int_0^T f(0; t)\, dt}\,, \quad T > 0\,,$$

and hence the values of arbitrary bonds and swap portfolios.

In Chapter 5, we described the construction of forward rates from a given set of term rates. That procedure can be applied only for calibrating to term deposits or to zero-coupon bonds. These instruments typically have maturities of less than 2 years. Fixed-income securities of longer maturities usually have coupons. Therefore, to extend the forward rate curve to longer maturities, we need to take into account the prices of coupon bonds and swaps.

To fix ideas, we shall assume that the benchmark instruments used to construct the curve are coupon-bearing bonds (modeled after U.S. bonds or plain-vanilla swaps). Let $T_1, T_2, \ldots, T_N$ be the maturities of the bonds, ordered chronologically. For simplicity, we assume that all bonds have the same payment frequency (e.g., semi-annual), but this is not necessary.

The simplest prescription of instantaneous forward rates consistent with bond prices is one in which the forward rates are *constant between the different bond maturities*. Accordingly, we seek a model in which the forward rate curve satisfies

$$f(0; T) = f(0; T_n) \equiv f_n \quad \text{for } T_{n-1} < T \le T_n\,, \ n = 1, \ldots N. \tag{12.24}$$

The algorithm for determining the sequence $\{f_n\}$ is inductive. Let us denote the coupon and the price of the nth bond by C_n and B_n, respectively. For simplicity, we assume a \$1 face amount. Suppose that the instantaneous forward rates $f_1, f_2, \ldots, f_n$ have been determined. To compute f_{n+1}, we write

$$B_{n+1} = B_{n+1}^{(0)} + B_{n+1}^{(1)}\,, \tag{12.25}$$

where the first term represents the present value of the coupons due before date T_n. Since the forward rates curve are known until date T_n, we can compute $B_{n+1}^{(0)}$ explicitly:

$$B_{n+1}^{(0)} = \sum_{t_j \le T_n} P_0^{t_j} C_n / \omega \tag{12.26}$$

where $P_0^{t_j}$ is a discount factor defined as follows: if $T_k \le t < T_{k+1}$,

$$P_0^t = e^{-\sum_{j=1}^{k} (T_j - T_{j-1})\, f_j - (t - T_k)\, f_{k+1}}\,.$$

[Notice that this is consistent with (12.24)]. We can also write the remainder $B_{n+1}^{(1)}$ in the form

$$B_{n+1}^{(1)} = P_0^{T_n} \left(\sum_{T_n < t_j \leq T_{n+1}} e^{-(t_j - T_n) f_{n+1}} C_n/\omega + e^{-(T_{n+1} - T_n) f_{n+1}} \right).$$

By virtue of (12.25) and (12.26), we obtain the equation

$$\frac{B_{n+1} - \sum_{t_j \leq T_n} P_0^{t_j} C_n/\omega}{P_0^{T_n}}$$

$$= \sum_{T_n < t_j \leq T_{n+1}} e^{-(t_j - T_n) f_{n+1}} C_n/\omega + e^{-(T_{n+1} - T_n) f_{n+1}}. \tag{12.27}$$

This equation can be used to determine f_{n+1} in terms of the other parameters.[20] The right-hand side is monotone in f_{n+1} and hence there is at most one solution. The solution can easily be determined using a root-finding algorithm such as Newton–Raphson. In this way, the rates $f_1, f_1, \ldots, f_N$ are determined successively. This approach is known as the **bootstrapping method** or sometimes as the Fama–Bliss method (see Bliss [1996]).

Figure 12.13 represents the forward-rate curve obtained by the bootstrapping method that corresponds to the data in Table 12.1. Using this curve, we can generate a curve of discount factors P_0^T, $T > 0$. In turn, the discount factors can be used to price portfolios of fixed-income instruments (without option features).

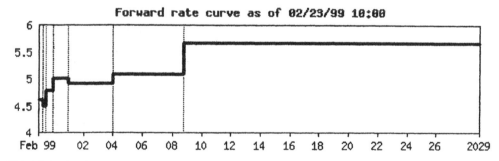

FIGURE 12.13
Forward rate curve obtained by the bootstrapping using seven on-the-run U.S. Treasury notes and bonds.[21]

Notice that the curve generated by the bootstrapping method exhibits jump discontinuities at the input maturities. These discontinuities are problematic, for several reasons. For example, if we assume the diffusion framework of APT, then jump discontinuities of the initial forward rate curve imply that, with probability 1, the short-term rate will jump by the same amount on these dates.[22] Hence, the volatility of short-term rates around these dates is extremely high. In particular, the implied volatility of forward-starting, short-dated options on $r(T_n)$ may be

---

[20]Notice that it is mathematically similar to the relation between bond yield and bond price.

[21]"On-the-runs" is the name given to the most recent Treasury security with a given "benchmark" maturity. They are the most liquidly traded and quoted securities.

[22]A rigorous proof of this statement follows from the results of Chapter 13.

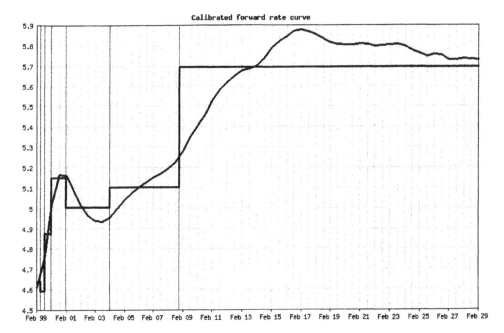

FIGURE 12.14
Smooth forward curve corresponding to the same U.S. Treasury data as Figure 12.13.
The bootstrapped curve is also shown for comparison purposes.

significantly overestimated. Another, perhaps more serious, criticism of the bootstrapping method is that jumps occur precisely at the maturities of the instruments used to calibrate the forward-rate curve. In the universe of dollar swaps, for example, the swaps quoted in the market have fixed tenors (e.g., 3 years, 5 years, 7 years, ... from the *current* date). Consequently, the swap maturities move with time. This implies that an instantaneous forward rate corresponding to any *fixed* date will jump (when the date corresponds to the expiration date of an input instrument). Given that a portfolio of fixed-income instruments has cash-flows that fall on well-defined (fixed) dates, the evolution of the jumps in time will generate an artificial variation in the value of the portfolio. This "noise" is produced by the method used to build the curve, not by a change in the true value of the portfolio.

One way to diminish the magnitudes of the jumps is to interpolate linearly the swap rates quoted in the market, introducing more swaps. For example, if the input swap tenors are 5, 7, 10, 12, 15, 20, and 30 years, with rates R_5, R_7, R_{10}, R_{12}, R_{15}, R_{20}, and R_{30}, we consider dates $t = 5 + n \times 0.5$, for $n = 0, \ldots, 48$, and the interpolated swap rates

$$R_t = \frac{t_2 - t}{t_2 - t_1} R_{t_1} + \frac{t - t_1}{t_2 - t_1} R_{t_2} \quad \text{for} \quad t_1 < t < t_2,$$

where t_1 and t_2 are the nearest dates corresponding to input swaps. The resulting curve will still exhibit discontinuities, but the jumps will be smaller.

Another class of methods for building forward-rate curves from market data consists in specifying a parametric family of curves, i.e., $f(T) = f(\theta; T)$, where $\theta = (\theta_0, \ldots, \theta_N)$ is a vector of parameters. For example, a possible parameterization could be

$$f(\theta; T) = \sum_{j=0}^{N} \theta_j \left(\frac{T}{1 + \theta_{N+1} T} \right)^j$$

where $f(\theta; 0) = \theta_0$ represents the short-term interest rate and

$$f(\theta; \infty) = \sum_{j=0}^{N} \theta_j \left(\frac{1}{\theta_{N+1}} \right)^j$$

represents the asymptotic value of the forward rate for $T \gg 1$. The coefficients θ_i are determined by fitting this curve to swap/FRA data. The precise method that is used for fitting the parameters, i.e., for solving the resulting system of equations for θ, will not be discussed. We point out, nevertheless, that parametric methods may give rise to multiple solutions for the fitting problem. This is a disadvantage of the parametric methods over the bootstrapping methods.

Finally, another approach consists in finding curves that minimize "penalization functions" of the type

$$\int_0^{T_{max}} \left| \frac{df(T)}{dT} \right|^2 dT \quad , \quad \int_0^{T_{max}} \left| \frac{d^2 f(T)}{dT^2} \right|^2 dT , \quad \text{etc.,}$$

subject to the price constraints. Figure 12.14 corresponds to such a construction, using a particular penalization algorithm.[23] A detailed description of penalization algorithms is beyond the scope of this book.

In summary, the construction of a forward-rate curve allows us to generate a "continuum" of discount factors and thus to price all "non-optionable" debt securities. Another important application of forward-curve modeling is for hedging a portfolio in terms of the securities used to build the curve. Hedge-ratios are derived by computing the sensitivities of the value of the portfolio to small changes in the bond prices (or the swap rates).

Which "cooking" method should we use in practice? The answer to this question ultimately lies in the user's preferences in terms of the prices and hedges that result from each type of construction. At the time of this writing, one of the most popular methods in use by swap traders is the bootstrapping approach with interpolated swap rates.

References and Further Reading

[1] Fabozzi, F. (1997), *Fixed Income Mathematics: Analytical and Statistical Techniques*, Irwin, Chicago.

---

[23] See, for example, the WWW Courant Finance Server (http://marco-pc.cims.nyu.edu) for an on-line yield curve calibrator. See also the references at the end of the chapter.

[2] Jarrow, R. (1995), *Modelling Fixed-Income Securities and Interest-Rate Options,* McGraw-Hill, New York.

[3] Jarrow, R. and Heath, D. (1998), *Fixed-Income Securities and Interest Rates,* McGraw-Hill, New York.

[4] Kapner, K.R. and Marshall, J.F. (1994), *Understanding Swaps,* Wiley, New York.

[5] Siegel, D.R. and Siegel D.F. (1990), *The Futures Markets: Arbitrage, Risk Management and Portfolio Strategies,* Probus, Chicago.

[6] Wong, M.A. (1991), *Trading and Investing in Bond Options,* John Wiley & Sons, New York.

Chapter 13

The Heath–Jarrow–Morton Theorem and Multidimensional Term-Structure Models

This chapter describes the fundamentals of term-structure modeling from the point of view of Arbitrage Pricing Theory. In the previous chapter we introduced the concept of *spot* term-structure on interest rates, which was illustrated in the construction of forward-rate curves and discount curves using market data. Here, we discuss probability measures that describe the *evolution* of these curves in an arbitrage-free economy. These probabilistic models are used for pricing interest-rate options, and other fixed-income derivatives with option features.

Term-structure (TS) models are often called "relative value" tools. The idea is that they can be used to price and hedge a variety of fixed-income derivatives with the *same* risk-neutral measure. One of their main uses is for pricing ("marking-to-market") and hedging large portfolio of derivatives. TS models imply theoretical relations between complex fixed-income securities such as European and American swaptions or bonds with callable/putable features. Traders often compare the prices of illiquid derivatives with the values derived from TS models and attempt to construct profitable trades.

At any time t, the term structure of interest rates can be represented by the collection of discount factors

$$ P_t^T \; = \; \mathbf{E}_t^P \left\{ e^{- \int_t^T r_s \, ds} \right\} , \quad T > t , \tag{13.1} $$

or, equivalently, by the instantaneous forward-rate curve

$$ f(t; T) \; = \; \lim_{\Delta t \to 0} \frac{1}{\Delta t} \left(\frac{ \mathbf{E}_t^P \left\{ e^{- \int_t^T r_s \, ds} \right\} }{ \mathbf{E}_t^P \left\{ e^{- \int_t^{T+\Delta t} r_s \, ds} \right\} } - 1 \right) \; = \; - \frac{\partial \log(P_t^T)}{\partial T} . \tag{13.2} $$

These representations were introduced in Chapter 12.

The main theoretical result of this chapter is the **Heath–Jarrow–Morton** (HJM) characterization of risk-neutral measures on forward rates. After discussing this theorem, we present examples of HJM-type models in which the statistical distribution of discount factors is log-normal. We then discuss **factor analysis** and the modeling of correlations between different sectors of the yield curve using Principal Component Analysis and the implementation of the HJM equations using Monte Carlo simulation. In the last section, we discuss certain pathologies associated with volatility specifications in the HJM framework.

13.1   The Heath–Jarrow–Morton Theorem

In addition to the standard assumptions of Arbitrage Pricing Theory, we make the following hypotheses:

(i)   Agents have the highest credit quality (credit "spreads" are neglected)

(ii)   At any time t, the market trades zero-coupon bonds of all maturities $T > t$ at prices that are known to all agents at this time.

The last assumption, which could be called "yield-curve completeness," is tantamount to assuming that there exist enough debt instruments traded in the market at any time so that investors are able to lend and borrow money for arbitrary maturities $T, < T_{max}$ at interest rates that are known. In more mathematical terms, agents agree on the curves P_t^T and $f(t; T$ at time t).

REMARK 13.1   Assumption (i) focuses the modeling on the problem of *interest-rate risk,* as opposed to credit risk, for example. Assumption (ii) goes further and declares that all spot rates (for arbitrary maturities) are observed at each date. This assumption is well suited for mature debt markets, such as the Eurodollar/LIBOR swaps market, the U.S. government bond market, the European Community, and Japan, all of which have liquid instruments with maturities spanning from a few months to 30 years. The existence of such a large class of debt instruments allows traders to determine with high precision the value of term rates along the yield curve.[1] It is important to note, however, that the hypothesis of yield-curve completeness does not apply to all markets. If there are only a few traded instruments and price information is not readily available, it may be difficult to reason in terms of a continuum of maturities or to generate a reliable forward-rate curve.   ∎

Our first goal will be to characterize the evolution of interest rates in a no-arbitrage economy under assumptions (i) and (ii). It is convenient for this purpose to focus on the discount factors $P_t^T, t < T < T_{max}$. Let us assume that there are ν factors of risk in the economy, modeled by ν uncorrelated Brownian motions $W_i(t), t > 0, i = 1, \ldots \nu$. We assume that P_t^T satisfy, under the risk-neutral measure, the equations

$$\frac{dP_t^T}{P_t^T} = \sum_{i=1}^{\nu} \sigma_i(t; T) \, dW_i(t) + \mu(t; T) \, dt \quad t \leq T. \tag{13.3}$$

Here $\mu(t; T)$ and $\sigma_i(t; T)$ are, respectively, the instantaneous means and variances, labeled by the maturity date (T).

REMARK 13.2   At this point, in Eq. (13.3), we make no *a priori* assumptions regarding the dependence of these parameters on the underlying risk factors, rates, etc. In other words,

---

[1] In practice, this is done by constructing, from no-arbitrage considerations, a forward-rate curve that is consistent with the prices of liquidly traded instruments. As we saw in the last chapter, there may be several forward-rate curves (actually, infinitely many) that are consistent with the prices of a finite number of securities. The point is that the uncertainty about the "fair value" of forward-rates diminishes as we increase the number of securities used for constructing the curve.

the instantaneous mean and variance are allowed to depend on state-variables, interest rates, forward rates, and so on. The only restriction imposed is that these functions be adapted processes, i.e., that $\sigma_i(t;\ T)$ and $\mu_i(t;\ T)$ are known given the information available at time t.[2] ∎

We postpone the discussion of how to determine ν and the specific forms of the parameters $\sigma_i(t;\ T)$. The question of interest is to determine what relations must exist between the parameters $\sigma_i(t;\ T)$ and $\mu(t;\ T)$ in the absence of arbitrage.

A simple observation, which leads to the characterization of risk-neutral (arbitrage-free) measures, is the fact that discount factors correspond to the prices of traded securities. From our assumptions, all zeros are traded securities (or, more precisely, synthethizable using tradeable securities, which has the same consequences). Since zeros do not pay dividends, the drift of a risk-neutral measure defined by (13.3) must satisfy

$$\mu(t;\ T) = r_t \ ,$$

where r_t is the short-term rate. In other words, risk-neutral dynamics for discount factors must be SDEs of the form

$$\frac{dP_t^T}{P_t^T} = \sum_{i=1}^{\nu} \sigma_i(t;\ T)\,dW_i(t) + r_t\,dt \quad t \leq T . \tag{13.4}$$

Notice that the short rate r_t is itself determined by the discount curve since, by definition,

$$r_t = \lim_{\Delta t \to 0} \frac{1}{\Delta t}\left(\frac{1}{P_t^{t+\Delta t}} - 1\right) = -\left[\frac{\partial \log(P_t^T)}{\partial T}\right]_{T=t} \tag{13.5}$$

To express the risk-neutral measure in more transparent form, we shall work with instantaneous forward rates instead of discount factors. Let us derive a stochastic differential equation for $\log(P_t^T)$ from Eq. (13.4). Applying Ito's Lemma, we have

$$d\,\log(P_t^T) = \sum_{i=1}^{\nu} \sigma_i(t;\ T)\,dW_i(t) + r_t\,dt - \frac{1}{2}\sum_{i=1}^{\nu} \sigma_i(t;\ T)\sigma_i(t;\ T)\,dt\,, \quad t \leq T .$$

Differentiating this equation with respect to T, and taking into account Eq. (13.2), we obtain

$$- d\,f(t;\ T) = \sum_{i=1}^{\nu} \dot{\sigma}_i(t;\ T)\,dW_i(t) - \left(\sum_{i=1}^{\nu} \dot{\sigma}_i(t;\ T)\sigma_i(t;\ T)\right)dt\,, \quad t \leq T , \tag{13.6}$$

where $\dot{\sigma}_i(t;\ T)$ represents the derivative of $\sigma_i(t;\ T)$ with respect to T. Using the fact that $-W_i(\cdot)$ is also distributed like a Brownian motion, we obtain in this way a stochastic differential

---

[2]Essentially, we assume here the general setting of continuous-time finance, in which prices are Ito processes. We do not discuss the case in which there are discontinuities (jumps) in asset prices. Another restriction imposed on the dynamics of prices, implicit in Eq. (13.3), is the fact that there are finitely many (ν) risk factors (a natural modeling assumption).

equation that describes the evolution of the forward-rate curve under the risk-neutral measure, namely,

$$d f(t; T) = \sum_{i=1}^{v} \dot{\sigma}_i(t; T) dW_i(t) + \left(\sum_{i=1}^{v} \dot{\sigma}_i(t; T) \int_t^T \dot{\sigma}_i(t; s) ds \right) dt \quad t \leq T . \quad (13.7)$$

This is known as the **Heath–Jarrow–Morton equation** (cf. Heath, Jarrow and Morton, [6, 7]).

The coefficient $\dot{\sigma}_i(t; T)$ represents the "portion" of the standard deviation of the forward rate $f(t; T)$ contributed by the ith factor. Notice that the coefficients $\dot{\sigma}_i(t; T)$ determine the *instantaneous covariance of forward rates*, since Eq. (13.7) implies that

$$\mathbf{Cov} (d f(t; T_1), d f(t; T_2)) = \sum_{i=1}^{v} \dot{\sigma}_i(t; T_1) \dot{\sigma}_i(t; T_2) dt . \quad (13.8)$$

The HJM equation shows that the drifts of forward rates can be expressed as functions of the variance–covariance matrix. It can be viewed as the analogue, in the context of fixed income, of the characterization of the risk-neutral probability for stocks and commodities (drift = riskless rate minus dividend rate) given by Black and Scholes [1973] and Merton [15].

Using the HJM Equation (13.7), we can derive a differential equation that describes the evolution of the short rate r_t under the risk-neutral measure. For this purpose, let us rewrite the equation in integral form, *viz.,*

$$f(t; T) = f(0; T) + \int_0^t \left(\sum_{i=1}^{v} \dot{\sigma}_i(s; T) \int_s^T \dot{\sigma}(s; u) du \right) ds$$

$$+ \int_0^t \sum_{i=1}^{v} \dot{\sigma}_i(s; T) dW_i(s) . \quad (13.9)$$

Recalling that $r_t = f(t; t)$, we have

$$r_t = f(0; t) + \int_0^t \left(\sum_{i=1}^{v} \dot{\sigma}_i(s; t) \int_s^t \dot{\sigma}(s; u) du \right) ds$$

$$+ \int_0^t \sum_{i=1}^{v} \dot{\sigma}_i(s; t) dW_i(s) . \quad (13.10)$$

This equation shows that r_t is an Ito process under the risk-neutral measure, satisfying the stochastic differential equation

$$dr_t = \sigma_r(t) dZ(t) + \mu_r(t) dt , \quad (13.11)$$

where $Z(t)$ is a Brownian motion,[3]

$$\mu_r(t) = \dot{f}(0; t) + \sum_{i=1}^{\nu} \int_0^t ds \left[(\dot{\sigma}_i(s; t))^2 + \sigma_i^{\cdot\cdot}(s; t) \int_s^t \dot{\sigma}_i(s; u) \, du \right]$$

$$+ \sum_{i=1}^{\nu} \int_0^t \sigma_i^{\cdot\cdot}(s; t) \, dW_i(s) \tag{13.12}$$

(here $(\bullet)^{\cdot\cdot} = \frac{\partial^2 \bullet}{\partial T^2}$), and

$$\sigma_r(t) = \sqrt{\sum_{i=1}^{\nu} (\dot{\sigma}_i(t; t))^2} \ .$$

We observe that if the interest-rate volatility were zero, we should have $r_t = f(0; t)$ and thus $\mu_r(t) = \dot{r}_t = \dot{f}(0; t)$, as predicted by Eq. (13.12). In general, we see from Eq. (13.12) that *the drift of the short-term rate is the sum of the slope of the forward-rate curve $\dot{f}(0; t)$ and a term depending on the volatility of interest rates.* Notice that the latter depends on the past history of interest-rate volatility.[4]

The HJM theorem is interesting for the following reasons:

- It focuses on the forward-rate curve and on the instantaneous covariance of forward rates, as opposed to more abstract state variables;

- It shows that there exists a unique risk-neutral probability measure on forward rates consistent with the initial curve $f(0; T)$ and the instantaneous covariance matrix for forward rates $\mathbf{Cov}\{df(t; T) \, df(t; T')\}$.[5]

REMARK 13.3 The HJM framework models directly the evolution of forward rates. Previous models inspired by the general APT framework (Cox, Ingersoll and Ross [4], Longstaff and Schwartz [13]; Vasicek [17]) are based on the evolution of *state-variables* $(X_1(t), \ldots, X_N(t))$ from which the dynamics of forward rates are derived. In this sense, the HJM formalism is "more direct." However, it requires a description of the evolution of the entire curve of rates, i.e., of an infinite number of variables. ∎

Unfortunately, there are few volatility specifications for the HJM SDEs that give rise to closed-form solutions for forward rates. In practice, the computation of expected cash-flows is usually done by Monte Carlo simulation. We discuss some of the few analytically tractable versions of HJM in the following sections.

---

[3]More precisely, $dZ = C \sum_{k=1}^{\nu} \dot{\sigma}(t; t) \, dW_k$, with $C = \left(\sum_{k=1}^{\nu} \dot{\sigma}(t; t)^2 \right)^{-1/2}$.

[4]This path-dependence leads to difficulties in the implementation of these equations that will be addressed later.
[5]Recall that, in general, there may exist several risk-neutral measures that are consistent with the prices of observed securities. The strong assumption of the HJM theorem, namely that we know the *entire* initial forward-rate curve and the *entire* covariance structure of forward rates, implies a unique risk-neutral measure, described by the SDE (13.7) [or Eq. (13.9) or (13.10)] consistent with a specified covariance matrix.

13.2 The Ho–Lee Model

The simplest illustration of the HJM theorem arises in the case of a single risk-factor ($\nu = 1$) and a constant variance of forward rates, i.e.,

$$\dot{\sigma}(t; T) = \sigma = \text{constant}.$$

In this case, the HJM equation becomes

$$df(t; T) = \sigma\, dW(t) + \left(\sigma \int_t^T \sigma\, ds\right) dt = \sigma\, dW(t) + \sigma^2\,(T - t)\, dt.$$

This equation admits an explicit solution, obtained by integrating with respect to t, namely

$$f(t; T) = f(0; T) + \sigma\, W(t) + \sigma^2 \left(T t - \frac{1}{2} t^2\right). \tag{13.13}$$

This model is known as the **Ho–Lee model.**

Let us examine the dynamics of the forward rate curve implied by Eq. (13.13). Clearly, the deformations of the curve associated with the Ho–Lee model correspond to a "shift" of the initial forward rates by the "linear shock" (Figure 13.7)

$$L(T) = \sigma\, W(t) + \sigma^2 \left(T t - \frac{1}{2} t^2\right)$$

$$= \sigma\, W(t) + \sigma^2 \frac{t^2}{2} + \sigma^2 t\,(T - t). \tag{13.14}$$

The *slope* of the line $L(T)$ is, from (13.14), equal to $\sigma^2 t$. In particular, the amplitudes of the shocks increase with T. The *intercept,* or value of the line at the shortest maturity $T = t$, represents the shock to the short rate. The latter is given by

$$r_t = f(0; t) + \sigma\, W(t) + \frac{\sigma^2 t^2}{2}. \tag{13.15}$$

Another way to express the dynamics of the forward-rate curve under the Ho–Lee model is through the equation

$$f(t; T) = r_t + \sigma^2 t \cdot (T - t) + f(0; T) - f(0; t).$$

In particular, forward rates are perfectly correlated with the short-term rate (and thus with each other). This is a common feature of all one-factor models ($\nu = 1$) since only one source of randomness drives all rates (Figure 13.2).

Perfect correlation is a limitation of one-factor models, which makes them of little use for pricing complex derivatives such as options on the slope of the forward rate curve. In the

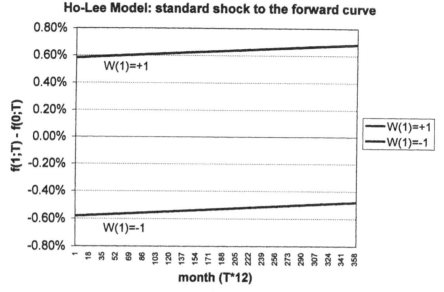

FIGURE 13.1
**Changes in instantaneous forward rates in 1 year according to the Ho–Lee model. The
higher line represents the displacements for a $+\sigma$ move; the lower line the displacements
corresponding to a $-\sigma$ move. The curves are not exactly symmetric about zero due to
the term $0.5\sigma^2 t^2$ in Eq. (13.14). The parameters used were $\sigma = 0.0058$ and $t = 1$. This
value of σ corresponds approximately to a relative volatility of 10% per year for a spot
rate of 5.8%.**

extreme case of the Ho–Lee model, the difference between two forward rates observed at time
t is, in fact, a *deterministic* function, since

$$f(t; T_1) - f(t; T_2) = f(0; T_1) - f(0; T_2) + \sigma^2 t (T_1 - T_2).$$

Market participants generally believe that forward rates are not perfectly correlated: an option
on the difference between two forward rates (or on the yields of two bond with different
maturities) should have a positive premium.[6] Another limitation of Ho–Lee, which we discuss
in the next section, is the fact that shocks to the forward curve increase with maturity.

One of the reasons for the interest in the Ho–Lee model is that it is very tractable analytically.
Recalling the relationship between zero-coupon bond volatility and forward rate volatility, [cf.
(13.4), (13.7)] we find that the volatility of a zero-coupon bond with maturity T is given by

$$\sigma(t; T) = \int_t^T \sigma \, ds = (T - t)\sigma .$$

Since the risk-neutral drift of a zero-coupon bond is equal to the short-term interest rate, we

---

[6]Econometric analysis shows that there is an imperfect correlation between forward and spot rates (see Chen
and Scott [3]). For instance, the correlation between spot 6-month LIBOR and 20-year forward 6-month LIBOR
is believed to be approximately 40%. This last statement concerns the "real world" probability, as opposed to
the risk-neutral measure. This fact notwithstanding, the prices of interest-rate options tend to be consistent with
beliefs about imperfect correlation of interest rates.

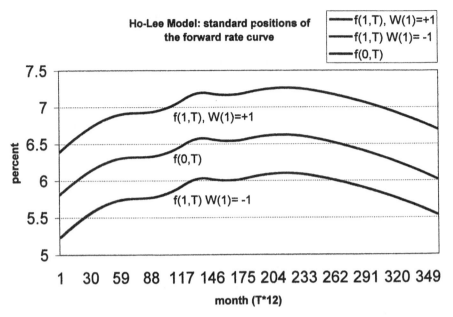

FIGURE 13.2
Possible positions for the forward-rate curve in 1 year assuming $\pm\sigma$ displacements. Notice that the displaced curves are nearly parallel to the initial forward curve (except for the effect of the term $0.5\sigma^2 t^2$). In the Ho–Lee model, all forward rates move essentially by the same amount.

obtain the following stochastic differential equation for the dynamics of the discount factors:

$$\frac{d\,P_t^T}{P_t^T} = (T - t)\,\sigma\,dW(t) + \left(f(0;\,t) + \sigma\,W(t) + \frac{\sigma^2 t^2}{2} \right) dt\,.$$

Therefore, in the Ho–Lee model, the prices of zero-coupon bonds follow a log-normal diffusion with a volatility proportional to the time-to-maturity. An explicit expression for the price of a zero-coupon bond maturing at date T, given the initial discount curve, is from (13.13),

$$P_t^T = e^{-\int_t^T f(t;\,s)\,ds}$$

$$= e^{-\int_t^T \left(f(0;\,s) + \sigma\,W(s) + \sigma^2 \left(st - \frac{1}{2}t^2 \right) \right) ds}$$

$$= P_0^T \cdot e^{-\int_t^T \sigma\,W(s)\,ds} \cdot e^{-\frac{\sigma^2 T t (T - t)}{2}}\,. \tag{13.16}$$

This formula can be used to price certain interest-rate derivatives, such as caps and floors, in closed-form. Details of option pricing calculations using the Ho–Lee model will be studied in Chapter 15.

13.3 Mean Reversion: The Modified Vasicek or Hull–White Model

Another arbitrage-free model for the term-structure that deserves attention is the case when $\nu = 1$ and

$$\dot{\sigma}(t; T) = \sigma\, e^{-\kappa\,(T-t)},$$

where κ is a positive constant. The volatility of forward rates decreases exponentially as the maturity increases. The main consequence of this assumption is that forward rates with short maturities (T small) are more volatile than forward rates with long maturities. The situation is depicted schematically in Figure 13.3.

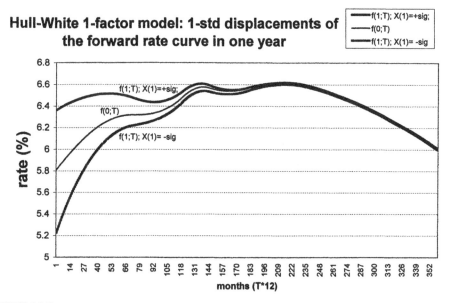

FIGURE 13.3
Positions for the forward rate curve in 1 year according to the Hull–White model. The parameters are as in Figures 13.1 and 13.2 with $\kappa = 0.25$. These curves are generated using Eq. (13.17) with $X(1) = \pm\sigma$, and an initial curve $f(0; T)$. Due to the mean-reversion coefficient κ, shocks are "damped" as T increases.

The corresponding HJM equation for the evolution of forward rates is

$$df(t; T) = \sigma\, e^{-\kappa\,(T-t)}\, dW(t) + \left(\sigma\, e^{-\kappa\,(T-t)} \int_t^T \sigma\, e^{-\kappa\,(s-t)}\, ds\right) dt$$

$$= \sigma\, e^{-\kappa\,(T-t)}\, dW(t) + \frac{\sigma^2}{\kappa} \left(e^{-\kappa\,(T-t)} - e^{-2\kappa\,(T-t)}\right) dt.$$

Integrating with respect to t, we obtain

$$f(t; T) = f(0; T) + \sigma \int_0^t e^{-\kappa\,(T-s)}\, dW(s) - \frac{\sigma^2}{2\kappa^2} \left(1 - e^{-\kappa\,(T-t)}\right)^2$$

$$+ \frac{\sigma^2}{2\kappa^2} \left(1 - e^{-\kappa T} \right)^2 . \qquad (13.17)$$

Introducing the state-variable

$$X(t) = \sigma \int_0^t e^{-\kappa (t-s)} \, dW(s) \,,$$

and using the fact that

$$\int_0^t e^{-\kappa (T-s)} \, dW(s) = e^{-\kappa(T-t)} X(t) \,,$$

we obtain the following formulas for $f(t; T)$ and r_t:

$$f(t; T) = f(0; T) + e^{-\kappa (T-t)} X(t) - \frac{\sigma^2}{2\kappa^2} \left(1 - e^{-\kappa (T-t)} \right)^2 + \frac{\sigma^2}{2\kappa^2} \left(1 - e^{-\kappa T} \right)^2 ,$$

and

$$r_t = f(0; t) + X(t) + \frac{\sigma^2}{2\kappa^2} \left(1 - e^{-\kappa t} \right)^2 . \qquad (13.18)$$

This model is known as the **modified Vasicek model** (Vasicek [17]; Hull and White (1989); Heath, Jarrow and Morton [7]). The above formulas show that the forward rates and the instantaneous short rate are linear functions of the same Gaussian process $X(t)$. In particular, all instantaneous forward rates are 100% correlated.

There are important differences between the Ho–Lee model and the present model. In the Ho–Lee model, the short-term rate tends to infinity as $t \to \infty$ with probability 1, due to the quadratic term in Eq. (13.15). In contrast, the short-term interest rate in the modified Vasicek model has a mean and a variance that are bounded independently of t. More precisely, the short rate process behaves asymptotically for $t \gg 1$ with

$$r_t \approx f(0; t) + \sigma X(t) + \frac{\sigma^2}{2\kappa^2} .$$

The right-hand side of this equation is a Gaussian random variable with mean

$$\mu_\infty(t) = f(0; t) + \frac{\sigma^2}{2\kappa^2}$$

and variance

$$\sigma_\infty^2(t) = \sigma^2 \, \mathbf{E} \left\{ \left(\int_0^t e^{-\kappa (t-s)} \, dW(s) \right)^2 \right\}$$

$$= \sigma^2 \int_0^t e^{-2\kappa (t-s)} \, ds$$

$$\approx \frac{\sigma^2}{2\kappa} \qquad (t \gg 1).$$

The short-rate $r(t)$ has a nontrivial **asymptotic probability distribution** for large t. We say that the short rate **reverts to its mean**. The parameter κ is usually called the **rate of mean reversion**.

Forward rates also exhibit mean reversion. In fact, for large T, we have

$$f(t; T) \approx f(0; T) + e^{-\kappa (T-t)} X(t), \quad T \gg 1,$$

Notice that the forward-rate curve is more volatile for small values of $T - t$. The short-term rate is asymptotic in long-term equilibrium about the mean μ_∞. Notice that the variance of the asymptotic distribution is inversely proportional to the rate of mean reversion. Thus, for large κ, the equilibrium distribution of the short rate is sharply concentrated about the forward-rate curve, with a small upward bias. For small rates of mean reversion, the short rate has wide fluctuations about the forward curve and a large positive bias.

As in the Ho–Lee model, zero-coupon bond prices are log-normal and can be computed explicitly. Thus, the model is quite tractable for computing option prices, as we shall see in Chapter 15.

13.4 Factor Analysis of the Term-Structure

So far, we have seen examples of **one-factor models** such as Ho–Lee and modified Vasicek, which are driven by a single Brownian motion. In these models, forward rates move essentially "in parallel." More precisely, the yield-curve performs a one-dimensional random motion. In reality, yield-curve deformations are more complicated.

Figure 13.4 exhibits a historical time-series for benchmark U.S. Treasury yields. It suggests that the yield curve can be "horizontal" for certain periods—meaning that the variation in yields is small across maturities—and at other periods, there can be strong differences between yields of different maturities. Figure 13.5 exhibits the same information in a different form; it displays the evolution of the U.S. Treasury curve as a function of maturity and the current date. Clearly, the correlation between rates is not perfect, and the curve can evolve into different shapes. Figure 13.6 gives three examples of the realized yield curves in the 1990s.

REMARK 13.4   If taken "very literally," a one-factor model can produce absurd results. In fact, the one-dimensional nature of shocks to the curve imply that the interest-rate risk of holding, for example, a 1-year option on a 30-year bond can be diversified with *any* interest-rate-sensitive security held in the appropriate hedge-ratio. On the other hand, common sense tells us that a good hedge would be to hold a fractional amount of 30-year bonds or bond futures against the option position.   ∎

One way to incorporate more complex deformations of the curve is to use **multifactor models** driven by several Brownian motions. In order to do this, we need to determine what is the correct value of ν—the "effective dimension" of the space of deformations. Second, we must determine the functional forms of the deformations of the term structure (for instance the functions $\sigma_k(t; T), k = 1, \ldots \nu$).

In this section we discuss the method of **Principal Component Analysis** (PCA) for estimating the variance–covariance matrix of shocks to the term structure (Judge et al. [11]). To associate the discussion with a particular dataset, we consider the historical series of bench-

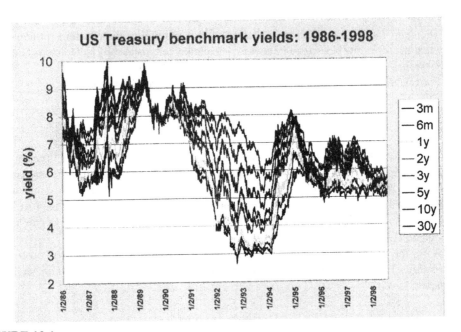

FIGURE 13.4
**Joint evolution of yields of benchmark Treasury instruments from January 1986 to July
1998. The yields correspond to 3-month, 6-month, and 1-year bills, 2-, 3-, and 5-year
notes, and the 10-year and 30-year bonds. These are the on-the-run, or more recently
issued, bonds corresponding to the quoted maturity. (For instance, the last 30-year bond
issued in this period has maturity date November 15, 2027. Its yield is the 30-year yield
quoted in the graph after it was issued).**

mark U.S. Treasury yields displayed in Figures 13.4 and 13.5. This dataset consists of daily
closing yields on eight benchmark U.S. government securities: 3-month, 6-month, and 1-year
T-bills, 2-, 3-, and 5-year notes, and 10- and 30-year bonds. The observations went from
January 1986 to July 1998.

Let us denote by $Y_1(\tau), Y_2(\tau), \dots T_8(\tau), \tau = 0, 1, \dots, N$ the different observations of the
yield curve in the dataset. For a given observation period $\Delta\tau$ (measured in days) we consider
the differences of successive yields

$$D_{i,n} = Y_i((n+1)\Delta\tau) - Y_i(n\Delta\tau), \quad i = 1, 2, \dots 8,$$

where n ranges over the number of consecutive periods of $\Delta\tau$ days in the dataset. Henceforth,
we will refer to these time series as the **differenced** data. Let $\widehat{C}_{ij}$ be the 8×8 empirical
covariance matrix

$$\widehat{C}_{ij} = \frac{1}{N} \sum_{n=1}^{N} \left(D_{i,n} - \overline{D_i}\right)\left(D_{j,n} - \overline{D_j}\right).$$

Here, N represents the number of periods and $\overline{D_i}$ is the sample mean of $D_{i,n}, n = 1, \dots, N$.

The principal component analysis consists in computing the eigenvalues λ_k and eigenvectors

$$v^{(k)} = \left(v_1^{(k)}, \dots v_8^{(k)}\right), \quad k = 1, 2, \dots 8$$

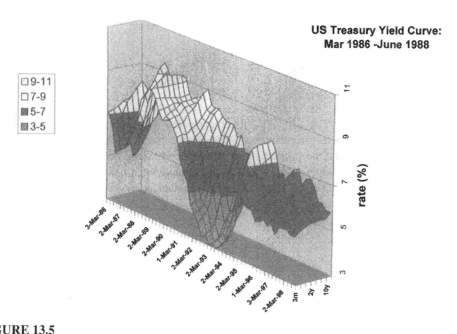

FIGURE 13.5
Evolution of the U.S. Treasury yield curve: 1986 to 1998.

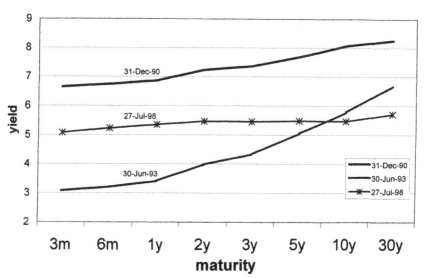

FIGURE 13.6
Three different yield curves in the 1990s.

of $\widehat{C}_{ij}$ and ranking the latter according to the magnitudes of the eigenvalues. From standard matrix theory, we have

$$\hat{C}_{ij} = \sum_{k=1}^{8} \lambda_k v_i^{(k)} v_j^{(k)} . \tag{13.19}$$

This is consistent with assuming that the yield differences satisfy

$$D_{i,n} = \overline{D_i} + \sum_{k=1}^{8} \sqrt{\lambda_k}\, v_i^{(k)}\, \xi_{k,n} \tag{13.20}$$

where the random variables $\xi_{1,n}, \ldots \xi_{8,n}$ are uncorrelated with mean zero and variance 1. In fact, the covariance of the left-hand side of (13.20) is given precisely by Eq. (13.19). Thus, the normalized eigenvectors of the correlation matrix can be viewed as "deformation eigenmodes" of the yield curve and the eigenvalues describe the "amplitude" of each mode.

The idea behind PCA consists in eliminating modes that have low eigenvalues and retaining the ones such that the corresponding eigenvalues $\lambda_1 > \lambda_2, \ldots, \lambda_\nu$ explain about 99% of the empirical covariance, in the sense that

$$\frac{\sum_{k=1}^{\nu} \lambda_k}{\sum_{k=1}^{8} \lambda_k} > 0.99\,.$$

Using only the reduced set of eigenvectors and the corresponding eigenvalues, we obtain a low-dimensional approximation to the statistics of yield differences, namely

$$D_{i,n} = \overline{D_i} + \sum_{k=1}^{\nu} \sqrt{\lambda_k}\, v_i^{(k)}\, \xi_{k,n}\,. \tag{13.21}$$

Using the dataset for the eight U.S. Treasury benchmark yields, we computed the covariance matrix of the yield differences over quarterly observation intervals ($\Delta\tau \approx 90$). The PCA shows that the first three eigenvectors explain 98% of the observed variance of quarterly shocks. Specifically, we obtained

Eigenvalue (λ_k)	Importance ($\lambda_k / \sum_j \lambda_j$)
2.5225	88.70%
0.2533	8.91%
0.0400	1.41%
$k \geq 3$	$< 0.8\%$

From the principal eigenvectors, we generated by linear interpolation three functions of the time to maturity $x = T - t$, which are plotted in Figure 13.7. It is clear from the magnitudes of the eigenvalues that the main contribution (88.7%) comes from "parallel displacements" of the yield curve, represented by the approximately horizontal curve. The two other deformation modes correspond to variations in the slope (8.91%) and "bending" motions (1.4%). This suggests that a reasonable effective dimension for the space of deformations of the Treasury yield curve is $\nu = 2$ or $\nu = 3$. This result was first derived in a pioneering paper by Litterman and Scheinkman [12].[7]

Having determined the components of the shocks to the yield curve according to the PCA with quarterly observation periods, the next step is to use this information to construct "local"

---

[7]We note, however, that the effective dimension may vary if we go to less developed markets or if we consider different datasets. For instance, European central banks have actively managed short-term interest rates in recent years in view of the impending European Monetary Union. This has resulted in forward-rate curve dynamics that are very different from the U.S. Treasury and U.S. LIBOR curves.

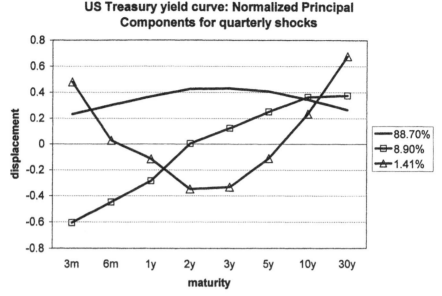

FIGURE 13.7
The three normalized eigenvectors with largest eigenvalues are plotted as functions of T, using linear interpolation between maturities. The numbers on the right represent the relative importance of the three eigenvalues, which explain 99% of the deformations.

volatility functions $\dot\sigma(t; T)$ for the HJM equations. A simple method is to "identify" the bond yield curve with the theoretical zero-coupon yield curve.[8]

For this we use the fact that the yield of a zero-coupon bond with maturity T has the stochastic differential

$$dY(t; T) = d\left(\frac{-1}{(T-t)}\log P_t^T\right)$$

$$= \frac{-1}{(T-t)}\sum_{k=1}^{\nu}\sigma_k(t; T)\,dZ_k(t) + \text{drift terms}. \qquad (13.22a)$$

Let $V_k(T), k = 1, 2, \ldots \nu$ be smooth interpolation functions associated with the principal components, as shown in Figure 13.7. The PCA suggests that the yield curve should satisfy approximately (under the identification between bond yields and zero yields) the difference equations

$$Y_{t+\Delta t}^T - Y_t^T = \sum_{k=1}^{\nu} V_k(T-t)\sqrt{\lambda_k}\,\xi_k(t) + \text{mean} \qquad (13.22b)$$

---

[8]This approximation is crude, but leads to a simple answer. In practice, we may have to work harder to go from market data to theoretical yield curves in this calibration procedure, perhaps adjusting the principal components slightly. Nevertheless, given that we expect zero-coupon yields to follow essentially the same statistics, we will proceed in this fashion.

where $\{\xi_k(t), k = 1, 2, \ldots \nu\}$ are uncorrelated standard normal random variables and where the factor $\Delta t = 0.25$ (quarterly observation periods). Matching the stochastic terms in (13.22a) and (13.22b), we obtain an estimator for the volatility structure of zero coupon bonds

$$\sigma_k(t; T) = (T - t) V_k(T - t) \sqrt{\frac{\lambda_k}{\Delta t}} \tag{13.23}$$

that can be used to construct the HJM dynamics.

REMARK 13.5   The use of $T - t$ instead of T in Eq. (13.22b) is due to the fact that the PCA was done using *relative* maturities and not fixed maturities (T) as in the HJM equations. The general form of the components obtained in this way is translation-invariant: they are functions of $T - t$.

The volatility and correlation of instantaneous forward rates is obtained by differentiating (13.23) with respect to the maturity variable T. The corresponding formula is

$$\dot{\sigma}_k(t; T) = \left((T - t)\, \dot{V}_k(T - t) + V_k(T - t)\right) \sqrt{\frac{\lambda_k}{\Delta T}} = \psi_k(T - t). \tag{13.24}$$

Summarizing, the PCA methodology applied to historical data determines estimators for the effective dimension ν and structure functions $\psi_1(T - t), \ldots \psi_\nu(T - t)$ such that a shock to the forward rate curve can be modeled as the SDE

$$df(t; T) = \sum_{k=0}^{\nu} \psi_k(T - t)\, dZ_k(t) + \mu(t; T)\, dt .$$

This equation can now be used to generate a risk-neutral measure by setting the drift equal to

$$\mu_{HJM}(t; T) = \sum_{k=1}^{\nu} \psi_k(T - t) \int_t^T \psi_k(s - t)\, ds ,$$

consistently with the HJM theorem.   ∎

REMARK 13.6   Notice that the latter equation can be rewritten in the form

$$df(t; T) = \sigma_f(T - t) \left[\sum_{j=1}^{\nu} \hat{\psi}_j(T - t)\, dZ_j \right] + \mu(t; T)\, dt \tag{13.25}$$

where

$$\hat{\psi}_k(x) = \frac{\psi_k(x)}{\left(\sum_{j=1}^{\nu} \psi_j^2(x)\right)^{1/2}}$$

$$\sigma_f^2(x) = \sum_{j=1}^{\nu} \psi_j^2(x) .$$

Some practitioners use a slightly different approach for modeling the volitility of instantaneous forward rates. In this approach, the functions $\hat{\phi}_j(x)$ are "fitted" to the empirical correlation matrix of the data and the volatility $\sigma_f(t; T)$ is determined from the prices of liquidly traded interest rate options (caps, floors, swaptions). Thus, in the latter approach, the modeler fits the correlations "historically" and "implies" the volatilities from market prices. This hybrid approach has the advantage that it can take into consideration the term-structure of implied volatilities. ∎

13.5 Example: Construction of a Two-Factor Model with Parametric Components

We implement a two-factor model in which the deformation modes for the yields and forward rates are represented using simple functional forms. For the sake of simplicity, we restrict ourselves to the case $\nu = 2$, but the method applies for any number of factors.

Based on the Principal Component Analysis and the results of the previous section, we assume that the interpolated deformation vectors $V_k(x)$ can be represented as

$$V_1(x) = A\, e^{-\kappa_1 x}$$

$$V_2(x) = B\left(1 - 2e^{-\kappa_2 x}\right),$$

where A, B, κ_1, and κ_2 are constants and $x = T - t$. Here, $V_1(x)$ represents a deformation in which all yields move in the same direction ("shift") and $V_2(x)$ represents a deformation in which the endpoints of the yield-curve move in opposite directions ("tilt"). It is implicitly assumed that κ_1 is small and $\kappa_1 < \kappa_2$.

From formula (13.23) and the latter expressions, we deduce that the volatility components of zero-coupon bond yields are given by

$$\sigma_1(t; T) = a\, x\, e^{-\kappa_1 x}$$

$$\sigma_2(t; T) = b\, x\left(1 - 2e^{-\kappa_2 x}\right) \tag{13.26}$$

where $a = A/\Delta t$ and $b = B/\Delta t$ are scaled factors, as in Eq. (13.23) and (13.24). Differentiating (13.26) with respect to the maturity parameter, we obtain expressions for the components of the forward-rate volatilities

$$\dot{\sigma}_1(t; T) = \psi_1(x) = a\left(1 - \kappa_1 x\right)e^{-\kappa_1 x}$$

$$\dot{\sigma}_2(t; T) = \psi_2(x) = b\left(1 - 2(1 - k_2 x))e^{-\kappa_2 x}\right). \tag{13.27}$$

The functions in (13.26) and (13.27) are *parametric* representations of the deformation of the forward-rate curve. By choosing the parameters a, b, κ_1, and κ_2 appropriately, we can fit the components to the shapes of deformations estimated by PCA. We can also choose the parameters in more subtle ways in order to reflect traders' beliefs about the option-implied volatility of interest rates.[9]

The equations in (13.27) imply that the covariance matrix of forward rates is given by

$$C(x, x') = \psi_1(x)\,\psi_1(x') + \psi_2(x)\psi_2(x')\,, \quad x = T - t\,, \ x' = T' - t\,. \qquad (13.28)$$

In particular, notice that the (absolute) volatility of the short-term rate is

$$\sigma_r = \sqrt{C(0; 0)} = \sqrt{\dot{\sigma}_1^2(t; t) + \dot{\sigma}_1^2(t; t)}$$

$$= \sqrt{a^2 + b^2}\,. \qquad (13.29)$$

From the equations in (13.26) and (13.27), we also conclude that the HJM drift is given by the expression

$$\mu_{HJM} = \dot{\sigma}_1(t; T)\sigma_1(t; T) + \dot{\sigma}_2(t; T)\sigma_2(t; T)$$

$$= a^2 x\,(1 - \kappa_1 x)\,e^{-2\kappa_1 x} + b^2 x\,(1 - 2(1 - k_2 x))\left(1 - 2\,e^{-\kappa_2 x}\right)\,.$$

The parameterization of the principal components as simple combinations of exponentials gives rise to a model that can be easily implemented numerically.

Numerical choice of parameters. The following choice of parameters gives rise to reasonable fit with data:

$$a = 0.008,\ \ b = 0.003,\ \ \kappa_1 = 0.0\ \ \kappa_2 = 0.35\ .$$

The volatility of the short-term interest rates is approximately 0.85%. To get a better idea of what this means, we note that at an interest-rate level of $r(0) = 5\%$ this corresponds to a log-normal volatility $\sigma = 0.85/5 = 17\%$.

We can also assess the relative importance of the two factors by analyzing the covariance of the 6-month and the 30-year forward rates.[10] The corresponding numerical values of the covariance are, from Eq. (13.28),

$$C(0.25, 0.25) = 6.81016E - 05$$

$$C(0.25, 30) = 5.79208E - 05$$

$$C(30, 30) = 7.30103E - 05\,.$$

This gives a correlation coefficient $p(0.25, 30) = 82\%$.

---

[9]Equation (13.29) below suggests that a term-structure of volatility for the short-term rate can be incorporated as well, by assuming that a and b are time-dependent. We shall not pursue this point here, but mention that it is important in practice.

[10]Since we have a two factor model, choosing the endpoints of the curve to do this "ex-post" PCA is reasonable, albeit somewhat arbitrary.

The eigenvalues are $\lambda_1 = 0.1285$, $\lambda_2 = 0.0126$, and the corresponding eigenvectors are (0.6920, 0.7219) ("parallel shift") and $(-0.7219, 0.6920)$ ("tilt mode"). The relative importances of the two modes are, respectively, 91.08% and 8.92%, consistent with the PCA of the previous section.

The rate of mean-reversion κ_2 for the second component was selected so that the "eigenvector" $V_2(x)$ crosses the axis approximately at $x = 2$ years (which is consistent with the historical PCA results in Figure 13.7 of the last section). Figure 13.8 displays the components for forward-rate curve deformations in this model.

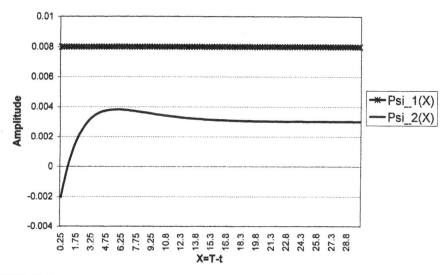

FIGURE 13.8
The two shocks to the forward-rate curve generated with the model, according to Eq. (13.27). One shock is perfectly parallel and contributes 91.08% of the variance, the other model corresponds to a "tilt" mode that contributes 8.92% of the variance.

Finally, we assumed that $f(0; T)$ is a smooth curve, constructed with data from the USD swap market on November 12, 1998.

To illustrate the properties of this model for pricing derivatives, we simulated 1000 random paths of forward-rate curves for the two-factor model. Using the same sequence of pseudo-random numbers, we also simulated 1000 paths with $a = 0.008544$ and $b = 0$, which corresponds to a Ho–Lee model with the same short-rate volatility. We then calculated the prices of a series of swaptions with various maturities and underlying swaps. The results are listed in Figure 13.1.

The simulations suggest that Ho–Lee produces higher swaptions premia. This can be justified by the fact that swaptions are options on a *series of cash-flows*. In the one-factor model, these flows are perfectly (and positively) correlated, whereas in the multifactor model interest-rate correlations are lower. The net effect is that the *volatility of the series of cash-flows* is higher under the one-factor model due to a lack of diversification. This implies a higher premium for swaptions under a one-factor model. The effect should become more accentuated for underlying swaps with longer tenors.

Table 13.1 Prices of swaptions obtained by Monte Carlo simulation for the two-factor model and a Ho–Lee (HL) model with the same volatility for the short-term rate. The maturities and "tails" (tenors of the underlying swaps at the exercise date) are in years. The tail of 0.25 corresponds to an option on a 3-month forward-rate agreement. Longer tails correspond to semiannual swaps. All prices are given in basis points (cents per $100 of notional). The HL model tends to give higher prices, especially for long-tail swaptions.

SWAPTION			MODEL	
maturity	tail	strike	2-factor	Ho-Lee
1	0.25	4.824	7.6	7.9
2	0.25	5.096	11.8	12.1
5	0.25	5.669	16.9	17.2
1	5	5.273	206.3	209.8
2	5	5.357	256	258.9
5	5	5.684	290.6	294.8
1	10	5.481	506.3	514.2
2	10	5.676	516.6	524.4
5	10	5.998	588.6	599.1

13.6 More General Volatility Specifications in the HJM Equation

So far, we have dealt essentially with Gaussian specifications of the evolution of forward rates. It is also possible to specify more general volatility structures, in which σ_f may be functions of the forward rates.

The most common choices for a volatility function for forward rates, aside from the ones discussed previously, are

$$\sigma_f(t;\, T) = \sigma_0(t;\, T)\, (f(t;\, T))^\alpha \, , \qquad (13.30)$$

where $\sigma_0(t;\, T)$ is a deterministic function of calendar time and the time-to-maturity and α is a positive exponent. The value $\alpha = 0$ corresponds to a Gaussian model and $\alpha = 1$ to a "log-normal" specification of volatility. Notice however that $\alpha = 1$ does not give rise exactly to a log-normal rate under the risk-neutral measure because the HJM drift,

$$\mu_f(t;\, T) = \sum_{k=1}^{\nu} f(t;\, T)\, \sigma_0(t;\, T)\, \hat{\phi}_k(T - t) \int_t^T f(t;\, s)\, \sigma_0(t;\, s)\, \hat{\phi}_k(s - t)\, ds \, ,$$

depends on the forward rate curve (in particular, it is not constant or linear). Functional forms of the type (13.30) are proposed and studied in Heath, Jarrow, and Morton [7] and Morton [16].

The functional form (13.30) with $\alpha = 1$ presents some interesting advantages for calibrating the model to (Black'76) implied volatility data for interest rate caps. On the other hand, it presents a serious pathology: forward rates can become infinite ("blow up") in finite time.[11]

This pathology can be demonstrated by the following calculation, for the case $\nu = 1$, $\alpha = 1$ with $\sigma_0(t; T) = \sigma_0 = $ constant. The solution of the corresponding HJM equation

$$\frac{df(t; T)}{f(t; T)} = \sigma_0 \, dZ + \left(\sigma_0^2 \int_t^T f(t; s) \, ds \right) dt$$

satisfies, formally,

$$f(t; T) = f(0; T) \, e^{\sigma_0 Z(t) - \frac{\sigma_0^2 t}{2} + \sigma_0^2 \int_0^t ds \int_s^T f(s; u) \, du}$$

$$= f(0; T) \cdot M(t) \cdot e^{\int_0^t ds \, \sigma_0^2 \int_s^T f(s; u) \, du} \ .$$

Assume, for simplicity, that $f(0; T) = f_0 = $ constant. Differentiating both sides of the last equation with respect to T, we obtain

$$\frac{\partial}{\partial T} f(t; T) = f(t; T) \sigma_0^2 \int_0^t f(s; T) \, ds$$

$$= \frac{\sigma_0^2}{2} \frac{\partial}{\partial t} \left(\int_0^t f(s; T) \, ds \right)^2 . \tag{13.31}$$

Hence, integration with respect to t, yields the differential equation

$$\frac{\partial}{\partial T} \int_0^t f(s; T) \, ds = \frac{\sigma_0^2}{2} \left(\int_0^t f(s; T) \, ds \right)^2 .$$

Setting

$$X(T) = \int_0^t f(s; T) \, ds \ ,$$

we conclude from Eq. (13.25) that

$$X(T) = \frac{2 X(t)}{2 - \sigma_0^2 X(t) (T - t)} \ .$$

Notice that $X(T)$ is a monotone function of T and that it blows up as T increases (along every Brownian path Z). Actually, suppose that there exists some date $t_0 > 0$ such that the HJM equation admits a finite solution for all $s \leq t_0$ and all forward dates T such that $s < T < \infty$. We shall see that this leads to a contradiction. In fact this would imply that $X(t_0) < +\infty$ with

---

[11]Of course, this does not happen when time is discretized in finite intervals Δt. This is why HJM has been implemented successfully in the financial industry for several years.

probability 1. Suppose, then, that we select a constant k such that the inequality $X(t_0) \geq k$ holds with nonzero probability $p(k)$. But then, by the above equation, we conclude that

$$X(T) > \frac{2k}{2 - \sigma_0^2 k (T - t_0)}$$

with probability at least $p(k)$. Therefore, if we choose

$$T_0 = \frac{2}{\sigma_0^2 k} + t_0 \,,$$

we conclude that for all $T \geq T_0$, we have $X(T) = +\infty$. This implies that some rate on the curve $\{f(t_0; T), \ T > t\}$ must be infinite—a contradiction! Hence, the solution of the above model blows up instantaneously: *for every $t > 0$ the probability that the forward rates blow up is finite.*[12]

Recently, Brace et al. [2] and Jamshidian [10] considered arbitrage-free models for the evolution of forward rates with *finite* tails (3 months, 6 months),

$$L(t; T) = \frac{1}{\Delta t} \left(\frac{P_t^T}{P_t^{T+\Delta t}} - 1 \right) \,, \qquad \Delta t = 0.25 \text{ or } 0.5 \,. \tag{13.32}$$

A striking result obtained by Brace et al. is that a "log-normal" specification the volatility of rates with finite tails, i.e.,

$$dL(t; T) = \sigma(t; T) L(t; T) \sum_{k=1}^{\nu} \hat{\psi}_k(t; T) dZ_k(t) + \quad \text{drift term} \,, \tag{13.33}$$

does *not* lead to blow-up.[13] Models for finite-tail rates (13.32), sometimes called **market models**, are becoming increasingly popular among practitioners. The main reason is that the volatility parameter in (13.33) is easy to calibrate to the prices interest-rate options. See Chapter 15, Section 15.4.

In practice, the choice of a volatility/correlation structure is dictated by the following considerations:

- The model should reflect the statistical structure of interest rates
- It should be computationally tractable
- It should be easy to calibrate to observed option prices

To gain a perspective on the question of computational tractability, the reader should note that in practice, $f(t; \bullet)$ is a high-dimensional vector. It can have dimension as large as 120 if

---

[12]Further computations along these lines, which we omit for the sake of brevity, show that the solutions of the HJM equation with the simple volatility model (13.24) blow up in finite time if $\alpha \geq 1$ and reach zero in finite time if $\alpha < 1$. This example shows that care must be taken when modeling the volatility process to avoid undesirable properties for the risk-neutral measure.

[13]This is consistent with the absence of blow-up in numerical discretization of HJM, which effectively corresponds to finite-tail rates.

we consider a 30-year forward rate curve with instantaneous forward rates discretized using 3-month intervals. This precludes, in most cases, the use of PDEs and lattice models that are restricted to dimensions 1 and 2. Monte Carlo simulation seems the only viable alternative for computing prices and sensitivities in such "full-blown" multifactor settings.

Nevertheless, multifactor term-structure models are often needed to value securities with time-optionality such as callable bonds, American-style bond options, or American swaptions. This requires lattice models or PDE methods that can only be implemented if the curve $f(t; T)$ can be represented as a function of a low-dimensional vector of state-variables.

The problem can be phrased as follows: assume that the number of factors is small (e.g., $v = 2$ or 3). Since the "dimension of the noise" is small, we would like to describe the forward rate curve as a function of a small number of parameters $\mathbf{X}(t) = (X_1(t), X_2(t), \dots, X_M(t))$, i.e.,

$$f(t; T) = F(\mathbf{X}(t), t, T),$$

where $F(\cdot, \cdot, \cdot)$ is a deterministic function, M is a number equal to or comparable to v, and $\mathbf{X}(t)$ is a diffusion process. In this case, the vector $\mathbf{X}$ acts as a "coordinate system" that permits the representation of the curve in terms of a low dimensional object. Since an $\mathbf{R}^M$-valued diffusion process can be represented numerically in terms of a lattice model, volatility formulations that lead to a reduced representation are ideal in terms of computational tractability.

The Ho–Lee model and the modified Vasicek models are examples of term-structure models that admit reduced representations (with $M = 1$). In the Ho–Lee case, the underlying state variable can be taken to be $X(t) = W(t)$, and in the modified Vasicek model we can take

$$X(t) = \int_0^t e^{-\kappa (t-s)} dW(s),$$

which is a diffusion process, since it satisfies the stochastic differential equation

$$dX = -\kappa X dt + dW.$$

The study of term-structure models that admit a reduced representation, also known as **Markov term-structure models**, will be carried out in the next chapter.

References and Further Reading

[1] Black and Scholes (1973)

[2] Brace, A., Gatarek, D., and Musiela, M. (1997), The Market Model of Interest Rate Dynamics, *Mathematical Finance, 7*, pp. 127–154.

[3] Chen, R.R. and Scott, L. (December 1993), Maximum Likelihood Estimation for a Multifactor Equilibrium Model of the Term Structure of Interest Rates, *Journal of Fixed Income*, pp. 53–72.

[4] Cox, J., Ingersoll, J., and Ross, S. (1985), A Theory of the Term Structure of Interest Rates, *Econometrica, 53*, pp. 385–407.

[5] Heath, D., Jarrow R., and Morton, A. (1990a), Bond Pricing and the Term Structure of Interest Rates: A Discrete Time Approximation, *Journal of Financial Quantitative Analysis,* **25**, pp. 419–440.

[6] Heath, D., Jarrow, R., and Morton, A. (1990b), Contingent Claim Valuation with a Random Evolution of Interest Rates, *Rev. Futures Markets,* **9**, pp. 54–76.

[7] Heath, D., Jarrow, R., and Morton, A. (1992), Bond Pricing and the Term Structure of Interest Rates: A New Methodology, *Econometrica,* **60**, pp. 77–105.

[8] Ho, T. and Lee, S. (1986), Term Structure Movements and Pricing Interest Rate Contingent Claims, *Journal of Finance,* **42**, pp. 1129–1142.

[9] Hull and White (1989)

[10] Jamshidian, F. (1997), LIBOR and Swap Market Models and Measures, *Finance Stochastic.*

[11] Judge, G.G., Griffiths, W.E., Hill, R.C., Lutkepohl, H., and Lee, T.C. (1988), *Introduction to the Theory and Practice of Econometrics,* 2nd ed., John Wiley, New York.

[12] Litterman, R. and Scheinkman, J. (June 1980), Common Factors Affecting Bond Returns, *Journal of Fixed Income,* pp. 55–61.

[13] Longstaff, F. and Schwartz, E. (1992a), Interest Rate Volatility and the Term Structure: A Two-Factor General Equilibrium Model, *Journal of Finance,* **47**, pp. 1259–1282.

[14] Longstaff, F. and Schwartz, E. (December 1992b), A Two-Factor Interest Rate Model and Contingent Claims Valuation, *Journal of Fixed Income,* **3**, pp. 16–23.

[15] Merton, R.C. (Spring 1973), Theory of Rational Option Pricing, *Bell Journal of Economics and Management Science,* **4**, pp. 141–183.

[16] Morton, A.J. (1989), Arbitrage and Martingales. Doctoral dissertation, Cornell University, Ithaca, NY.

[17] Vasicek, O.A. (1977), An Equilibrium Characterization of the Term Structure, *Journal of Financial Economics,* **5**, pp. 177–188.

Chapter 14

Exponential-Affine Models

In this chapter, we study term-structure models with the property that the forward rate curve is an **affine function** of Markov state-variables. Due to their computational tractability, these models are important for practical applications.

We henceforth assume that the forward-rate curve can be represented in the form

$$f(t;\ T) = \sum_{i=1}^{N} a_i(t;\ T)\, X_i(t) + b(t;\ T) \tag{14.1}$$

where $\mathbf{X}(t) = (X_1(t), X_2(t), \ldots X_N(t))$ is a vector of state variables satisfying a system of stochastic differential equations

$$dX_i = \sigma_{ik}(\mathbf{X},\ t)\, dZ_k + \mu_i(\mathbf{X},\ t)\, dt \quad 1 \le i \le N, \quad 1 \le k \le \nu. \tag{14.2}$$

Here, the Z_k's are independent Brownian motions.[1] Notice that Eq. (14.1) implies that the discount factors P_t^T are **exponential functions** of the state variables:

$$P_t^T = e^{-\sum_{i=1}^{N} \left(\int_t^T a_i(t;\ s)\, ds \right) X_i(t) - \int_t^T b(t;\ s)\, ds}$$

$$= e^{-\sum_{i=1}^{N} p_i(t;\ T)\, X_i(t) - q(t;\ T)}. \tag{14.3}$$

A major advantage of exponential-affine (EA) models is that they can be implemented numerically using finite-difference schemes on lattices with low dimensionality that describe the evolution of the state variables.[2] Another advantage of EA models is that there exist closed-from solutions for the prices of zero-coupon bonds and simple bond options. Notice that the affine dependence of the curve on the state variables allows us to express any cash-flow contingent on forward rates as an elementary function of the state variables—basically, a combination of exponential functions.[3] Exponential-affine models are thus well suited for pricing and hedg-

---

[1] We restrict this discussion to state variables which are Markov Ito processes, i.e., diffusions. State-variables that follow Markov jump processes or jump-diffusion processes are also used to model default risk. Their study is beyond the scope of this book.

[2] In practice, workstations can handle dimensions $N = 1$ and 2. Three-factor models can been implemented on "coarse grained" lattices but are computationally very expensive.

[3] In contrast, the use of a "general" state variable model requires solving partial differential equations to compute the values of the discount factors and forward rates in terms of the state variables. Hence, the valuation of interest rate derivatives, such as bond options, would require solving several PDEs instead of only one.

ing American bond options, American swaptions, callable bonds, and other derivatives that require dynamic programming algorithms best implemented on lattices (Figure 14.1).

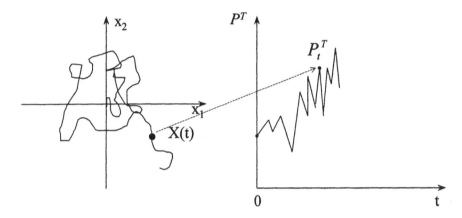

$$P_t^T = \exp\left[p(t,T)\cdot X(t) + q(t,T)\right]$$

FIGURE 14.1
Exponential-affine models work as follows: a vector of state-variables X_t follows a diffusion process (left). At any time t, the discount factors are explicit exponential-affine functions of the state-variables (right). Cash-flows of derivative securities that depend on the state of the term-structure in the future can be represented as explicit functions of the state variables. Finite-difference schemes or closed-form solutions are used to compute expectation values of functions of X.

Another advantage of exponential-affine models is that the volatilities and correlations of forward rates are easy to compute and thus to calibrate to a desired correlation/covariance structure. Applying Ito's Formula to Eq. (14.1), we obtain

$$df(t; T) = \sum_{k=1}^{\nu} \dot{\sigma}_k(t; T)\, dZ_k + m(t; T)\, dt,$$

where $\dot{\sigma}_k(t; T)$ and $m(t; T)$ are given, respectively, by

$$\dot{\sigma}_k(t; T) = \sum_{i=1}^{N} \sigma_{ik}\, a_i(t; T)$$

and

$$m(t; T) = \sum_{i=1}^{N} \left(\frac{\partial a_i(t; T)}{\partial t} X_i(t) + a_i(t; T)\, \mu_i \right) + \frac{\partial b(t; T)}{\partial t}. \tag{14.4}$$

In particular, the standard deviation of the forward rate $f(t; T)$ is given by

$$\sigma_f(t; T) = \sqrt{\sum_{k=1}^{\nu} \left(\sum_{i=1}^{N} \sigma_{ik}\, a_i(t; T) \right)^2},$$

and the normalized correlation components are

$$
\hat{\psi}_k(t; T) = \frac{\sum\limits_{i=1}^{N} \sigma_{ik} a_i(t; T)}{\sqrt{\sum\limits_{k=1}^{\nu} \left(\sum\limits_{i=1}^{N} \sigma_{ik} a_i(t; T) \right)^2}} \qquad 1 \le k \le \nu .
$$

In the following sections, we characterize the distributions for the state variables [i.e., the SDEs (14.2)] that give rise to exponential-affine models.

14.1 A Characterization of EA Models

The requirement that forward-rates are affine functions of the state variables imposes strong constraints on the process **X**. In other words, we cannot assign arbitrary probability distributions to the state variables **X**(t) and expect to generate an affine model that satisfies the HJM no-arbitrage conditions. Essentially, exponential-affine models must essentially have state variables that are either Gaussian or belong to the class of Cox–Ingersoll–Ross (CIR) square-root diffusions.[4]

Technically, the issue is that we require Eq. (14.1) or (14.3) to hold under a *risk-neutral measure*. We know that under such a measure the process that describes the evolution zero-coupon bond prices has a drift equal to the short-term interest rate—the endpoint of the curve $f(t; t)$. At the same time, we require that zero-coupon bond prices are exponential functions of the forward-rate curve, cf. Eq. (14.3). This gives rise to constraints of the model, i.e., on the distribution of the state variables.[5]

Observe that, since $r_t = f(t; t)$, the short rate satisfies

$$
r_t = \sum_{i=1}^{N} \bar{a}_i(t) X_i(t) + \bar{b}(t) , \tag{14.5}
$$

where, for simplicity, we introduced the notation[6]

$$
\bar{a}_i(t) = a_i(t; t) , \qquad \bar{b}(t) = b(t; t) .
$$

---

[4]Other models, such as the Li–Ritchken–Sankarasubhramanyam (LRS) model can also be viewed as belonging to the exponential-affine category. Such models are less widely used than Gaussian or square-root diffusion models presented here. See References and Further Reading at the end of this chapter for information about LRS models.

[5]These constraints are equivalent to assuming that the function $f(t; T)$ in (14.1) satisfies the Heath–Jarrow–Morton stochastic differential equation.

[6]It is often possible to "normalize" the functions $a_i(t; T)$ by imposing the condition $\bar{a}_i(t) = 1$. In fact, whenever $a_i(t; t) \neq 0$, this entails no loss of generality, since we can always redefine the state variables X_i using the transformation $X_i \longrightarrow a_i(t; t) X_i$. Nevertheless, there are situations in which $a_i(t; t) = 0$ for some indices i. This corresponds to models in which the short rate depends on a smaller number of state variables than the entire forward-rate curve. One important example of this is the Li–Ritchken–Sankarasubhramanyam model.

We now make use of the fact that each discount factor P_t^T satisfies, under any risk-neutral measure,

$$dP_t^T = P_t^T \left[\sum_{k=1}^{v} \sigma_k(t; T) dZ_k + r_t dt \right]. \tag{14.6}$$

Applying Ito's Formula to Eq. (14.3), we find that the (log-normal) drift of P_t^T is given by

$$\delta = \frac{1}{P_t^T} \left[\frac{\partial P_t^T}{\partial t} + \frac{1}{2} \sum_{i,j=1}^{N} A_{ij} \frac{\partial^2 P_t^T}{\partial X_i \partial X_j} + \sum_{i=1}^{N} \mu_i \frac{\partial P_t^T}{\partial X_i} \right],$$

where

$$A_{ij} = \sum_{k=1}^{v} \sigma_{ik} \sigma_{jk}$$

is the diffusion matrix associated with the process $\mathbf{X}$. Due to the exponential form of P_t^T, we can compute all partial derivatives in the equation for the drift explicitly. After some computation, we find that

$$\delta = -\left(\sum_{i=1}^{N} \dot{p}_i X_i + \dot{q} \right) + \frac{1}{2} \sum_{i,j=1}^{N} A_{ij} p_i p_j - \sum_{i=1}^{N} \mu_i p_i,$$

where dots represent derivatives with respect to "calendar time" t. Using Eq. (14.6), which states that $\delta = r_t$, and the formula for the short-term interest rate in (14.5), we conclude that

$$-\left(\sum_{i=1}^{N} \dot{p}_i X_i + \dot{q} \right) + \frac{1}{2} \sum_{i,j=1}^{N} A_{ij} p_i p_j$$

$$-\sum_{i=1}^{N} \mu_i p_i = \sum_{i=1}^{N} \bar{a}_i X_i + \bar{b},$$

or

$$\frac{1}{2} \sum_{i,j=1}^{N} A_{ij} p_i p_j - \sum_{i=1}^{N} \mu_i p_i = \sum_{i=1}^{N} (\dot{p}_i + \bar{a}_i) X_i + \dot{q} + \bar{b}. \tag{14.7}$$

We conclude therefore that

PROPOSITION 14.1

A necessary condition for the diffusion process $\mathbf{X}_t$ defined in (14.2) to give rise to an exponential-affine model with zero-coupon bonds defined by (14.3) is that the combination

$$\frac{1}{2} \sum_{i,j=1}^{N} A_{ij} p_i p_j - \sum_{i=1}^{N} \mu_i p_i \tag{14.8}$$

is an affine function of the variables X_i.

It is therefore natural to consider diffusions such that the coefficients A_{ij} and μ_i of the associated Fokker–Plank operator

$$\mathbf{L}_t = \frac{\partial}{\partial t} + \frac{1}{2} \sum_{ij} A_{ij} \frac{\partial^2}{\partial X_i \, \partial X_j} + \sum_i \mu_i \frac{\partial}{\partial X_i} \, ,$$

are affine functions of $\mathbf{X}$.[7]
We set, accordingly,

$$A_{ij} = A_{ij}^{(0)} + \sum_k A_{ij,k}^{(1)} X_k \tag{14.9a}$$

and

$$\mu_i = \mu_i^{(0)} + \sum_k \mu_{i,k}^{(1)} X_k \, . \tag{14.9b}$$

Equating the coefficients of the variables X_i and the constant terms in the resulting equation, we obtain a system of ordinary differential equations for the coefficients p_i and q, viz.,

$$\dot{p}_i + \sum_k \mu_{ki}^{(1)} p_k + \bar{a}_i = \frac{1}{2} \sum_{k,l} A_{kl,i}^{(1)} p_k \, p_l \quad 1 \le i \le N \, , \tag{14.10a}$$

$$\dot{q} + \sum_i \mu_i^{(0)} p_i + \bar{b} = \sum_{i,j} \frac{1}{2} A_{ij}^{(0)} p_i \, p_j \, . \tag{14.10b}$$

In view of the fact that $P_T^T = 1$ and Eq. (14.3), the functions p_i and q must also satisfy the boundary conditions

$$p_i(T; T) = 0 \, , i = 1, 2, \ldots N \, , \quad q(T; T) = 0 \, . \tag{14.11}$$

Equation (14.10a) and the first boundary condition in (14.11) constitute a system of ordinary differential equations for the vector $(p_1, \ldots p_N)$. The last equation is then used to obtain q by integrating with respect to t and using the corresponding boundary condition in (14.11).

REMARK 14.1   Equation (14.10a) represents a system of ordinary differential equations that is either linear in p or has a quadratic nonlinearity, according to whether $A_{kl,i}^{(1)}$ vanishes or not. In the latter case, the nonlinear (14.10a) belongs to the well-known class of **Ricatti differential equations**. In the case of time-independent coefficients $\mu^{(1)}$ and $A_{ij}^{(1)}$, Ricatti equations admit closed-form solutions. ∎

---

[7]As mentioned earlier, there exist models for which Eq. (14.8) holds but for which A_{ij} and μ_i are not *individually* affine in the state variables. We will not consider these more complicated models here.

14.2 Gaussian State-Variables: General Formulas

If we assume that $A_{ij} = A_{ij}^{(0)}$ in Eq. (14.9a) (with $A_{ij}^{(0)}$ possibly time-dependent but independent of $\mathbf{X}$), the process $\mathbf{X}$ satisfies a system of linear differential equations

$$dX_i = \sum_k \sigma_{ik}\,dZ_k + \mu_i^{(0)}\,dt + \sum_j \mu_{ij}^{(i)} X_j\,dt\,, \qquad (14.12)$$

where σ_{ik} is the *square root* of $A_{ij}^{(0)}$, i.e.,

$$A_{ij}^{(0)} = \sum_{k=1}^{v} \sigma_{ik}\,\sigma_{jk}\,.$$

We introduce the auxiliary matrix-valued function $\Psi(t;\,T)$, which solves the differential equation

$$\frac{d}{dT}\,\Psi(t;\,T) = \mu^{(1)}(T)\,\Psi(t;\,T)\,, \qquad \Psi(t;\,t) = \mathbf{I}\,.$$

Here $\mathbf{I}$ denotes the identity matrix ($\mathbf{I}_{ij} = 1$ if $i = j$ and $\mathbf{I}_{ij} = 0$ if $i \neq j$).[8] It is easy to verify that the solution of the SDE (14.12) satisfies

$$\mathbf{X}(t') = \Psi(t;\,t') \cdot X(t) + \cdot \int_t^{t'} \Psi(s;\,t') \cdot \mu^{(0)}(s)\,ds$$

$$+ \int_t^{t'} \Psi(s;\,t') \cdot \sigma(s) \cdot dZ(s)\,, \qquad (14.13)$$

for all $0 \leq t < t'$. In particular, this formula shows that $\mathbf{X}(t)$ has a Gaussian distribution.[9]

Consider the coefficients $p_i(t;\,T)$ arising in the zero-coupon bond formula, which satisfy (14.10a). Equations (14.10a) reduce to a system of linear ordinary differential equations

$$\frac{d\,p_i}{dt} + \sum_j p_j\,\mu_{ji}^{(1)} + \bar{a}_i = 0\,,$$

with boundary conditions $[\,p_i(t;\,T)\,]_{t=T} = 0$. The solution of this system can be computed as follows: let $\Phi(t;\,T)$ be the solution of the matrix-valued differential equation

$$\frac{d}{dt}\,\Phi(t;\,T) = -\,\Phi(t;\,T)\,\mu^{(1)}(t)$$

$$\Phi(T;\,T) = \mathbf{I}\,.$$

---

[8] The function Ψ is known as the **transfer function** of the linear system $\dot{x} = \mu^{(1)}\,x$.
[9] Solutions of linear stochastic differential equations such as (14.12) are called a **Gauss–Markov process**.

Using the method of variation of constants and condition (14.11) again, we find that $p(t; T) = (p_1(t; T), \ldots p_N(t; T))$ is given by

$$p(t; T) = \int_t^T \bar{a}(s) \cdot \Phi(t; s) \, ds \, , \tag{14.14}$$

where $\bar{a}(s) = (\bar{a}_1(s), \ldots \bar{a}_N(s))$.

It remains to compute the coefficient $q(t; T)$. This is done by integrating both sides of Eq. (14.10b) from t to T. For simplicity, we will assume for now that $\mu_i^{(0)} = 0$ and remove this assumption later. The result is

$$q(t; T) = \int_t^T \bar{b}(s) \, ds - \frac{1}{2} \int_t^T \sum_{i,j} A_{ij}^{(0)}(s) \, p_i(s; T) \, p_j(s; T) \, ds \, . \tag{14.15}$$

Notice, however, that we have not yet specified the function $\bar{b}$. We claim that this function is completely determined by the condition that the model prices correctly all zero-coupon bonds P_0^T, $T > 0$. In fact, equating the forward-rate curve at time $t = 0$ to the affine function (14.1) defining the forward-rate curve, we obtain

$$f(0; T) = b(0; T) = \frac{\partial q(0; T)}{\partial T} \, .$$

We shall use this relation and Eq. (14.15) to determine $\bar{b}$. Differentiating Eq. (14.15) with respect to T and setting $t = 0$, we have

$$\frac{\partial q(0; T)}{\partial T} = \bar{b}(T) - \frac{\partial}{\partial T} \frac{1}{2} \int_0^T \sum_{i,j} A_{ij}^{(0)}(s) \, p_i(s; T) \, p_j(s; T) \, ds$$

$$= \bar{b}(T) - \int_0^T \sum_{i,j} A_{ij}^{(0)}(s) \frac{\partial p_i(s; T)}{\partial T} \, p_j(s; T) \, ds \, ,$$

where we used the boundary conditions (14.11) and the symmetry of the matrix A_{ij}.

Hence, we conclude that the function $\bar{b}$ is given by

$$\bar{b}(T) = f(0; T) - \int_0^T \sum_{i,j} A_{ij}^{(0)}(s) \frac{\partial p_i(s; T)}{\partial T} \, p_j(s; T) \, ds \, .$$

The coefficient $q(t; T)$ is recovered by substituting this expression into formula (14.15). After some computation, we obtain the following expression for $q(t; T)$:

$$q(t; T) = \int_t^T f(0; s) \, ds - \frac{1}{2} \int_0^t \sum_{i,j} A_{ij}^{(0)} \left(p_i(s; T) \, p_j(s; T) - p_i(s; t) \, p_j(s; t) \right) ds$$

$$= \int_t^T f(0; s) \, ds + q_0(t; T) \, , \tag{14.16}$$

where we set

$$q_0(t;\ T)\ =\ -\frac{1}{2}\int_0^t \sum_{i,\,j} A_{ij}^{(0)}\left(p_i(s;\ T)\,p_j(s;\ T)\ -\ p_i(s;\ t)\,p_j(s;\ t)\right) ds\ . \qquad (14.17)$$

REMARK 14.2   This calculation shows explicitly that we can fit an arbitrary initial term-structure of interest rates $f(0;\ T)$ to an exponential-affine model with Gaussian factors. This is done by choosing the function $\bar{b}$ [or, equivalently, the coefficients $q(t;\ T)$, or $b(t;\ T)$] as specific functions of $\bar{a}_i$, σ_{ik}, $\mu^{(1)}$, and $f(0;\ T)$. The volatilities of the factors can be specified as arbitrary functions of time. Some practitioners call $\bar{b}(t)$ the "fudge factor"—it is the term that needs to be added to the linear combination of state variables in order to fit the current term-structure.   ∎

Reduction to the case $\mu_i^{(0)} = 0$. As we now show, the values of the parameters $\mu_i^{(0)}$ and $X_i(0)$ are irrelevant, after calibration to the forward-rate curve.

In fact, given the expressions obtained for X_i and p_i in (14.13) and (14.14), we conclude that the forward rates satisfy the SDE

$$df(t;\ T) = \sum_i \dot{p}_i(t;\ T)\,dX_i(t)$$

$$= \sum_{i,k} \dot{p}_i(t;\ T)\,\sigma_{ik}\,dZ_k(t)\ +\ \text{drift terms}\ .$$

In particular, the quantities $\mu_i^{(0)}$ and $X_i(0)$ do not appear in the variance–covariance matrix of forward rates. Therefore, they will not appear in the HJM drift for forward rates, which is determined completely by the covariance and the initial forward-rate curve. This allows us to set $X_i(0) = \mu_i^{(0)} = 0$ and to work with the "reduced" state variables $\mathbf{Y} = (Y_i(t), \dots, Y_N(t))$ defined by

$$\mathbf{Y}(t)\ =\ \int_0^t \Psi(s;\ t)\ \cdot\ \sigma(s)\cdot\ dZ(s)\ .$$

Notice that these state variables satisfy the reduced linear SDE

$$dY_i\ =\ \sum_k \sigma_{ik}\,dZ_k\ +\ \sum_k \mu_{ik}^{(i)}\,Y_k\,dt\quad,\quad Y_i(0)\ =\ 0.$$

Clearly, there is a unique EA model driven by these factors that fits the yield-curve [namely the one that has the HJM drift corresponding to the variance–covariance structure of $(Y_i(t), \dots, Y_N(t))$].

In particular, we have

$$f(t;\ T)\ =\ \sum_i a_i(t;\ T)\,Y_i(t)\ +\ b(t;\ T) \qquad (14.18)$$

and

$$P_t^T\ =\ \exp\left[-\,p_i(t;\ T)\,Y_i(t)\ -\ q(t;\ T)\right]\ , \qquad (14.19)$$

where $b(t; T)$ and $q(t; T)$ depend only on $\bar{a}_i$, σ_{ik}, $\mu_{ik}^{(1)}$, and the current forward rate curve $f(0; T)$.

We end this section by observing that the Gaussian model enjoys an interesting *factorization property*. Since we have

$$e^{-\int_t^T f(0; s)\, ds} = \frac{P_0^T}{P_0^t} ,$$

we conclude from Eqs. (14.16) and (14.19) that the value of the discount factor P_t^T is given by

$$P_t^T = \frac{P_0^T}{P_0^t} \cdot \exp\left[\sum_i - p_i(t; T)\, Y_i(t) - q_0(t; T) \right]. \tag{14.20}$$

In the last equation, the factor P_0^T / P_0^t depends only on the current term-structure of interest rates and is independent of the volatility assumptions. On the other hand, the second factor depends only on the *volatility structure* induced by the state variables. This factorization property extends to non-Gaussian models as well. It is very convenient for computing the values of interest-rate options, as we shall see in Chapter 15.

14.3 Gaussian Models: Explicit Formulas

Explicit expressions for the coefficients of Gaussian models can be derived if we assume that the matrix $\mu_{ij}^{(1)}$ is diagonal and that the parameters are constant in time.[10] Accordingly, we set

$$\mu_{ij}^{(1)} = - \kappa_i\, \delta_{ij} ,$$

and

$$A_{ij}^{(0)} = \sigma_i\, \sigma_j\, \rho_{ij} ,$$

where $\sigma_1, \ldots \sigma_N$ and ρ_{ij} are, respectively, the instantaneous volatilities and correlations of $Y_1, \ldots Y_N$.

Under these assumptions, it is easy to check that the resolvents $\Phi(t; T)$ and $\Psi(t; T)$ are diagonal matrices, with

$$\Phi_{ii}(t; T) = \Psi_{ii}(t; T) = e^{-\kappa_i(T-t)} , \quad i = 1, 2, .., N ,$$

and that the functions $p_i(t; T)$ and $a_i(t; T)$ are given by

$$p_i(t; T) = \frac{\bar{a}_i}{\kappa_i} \left(1 - e^{-\kappa_i(T-t)} \right) , \quad a_i(t; T) = \bar{a}_i\, e^{-\kappa_i(T-t)} . \tag{14.21}$$

Notice that the SDE for the (reduced) state variables is

$$dY_i = -\kappa_i\, Y_i\, dt + \sigma_i\, dW_i ,$$

---

[10]In practice, the latter assumption may not be appropriate if we wish to calibrate the model to a term-structure of volatilities. Nevertheless, we discuss the constant coefficients case, because it leads to simple mathematical expressions that give insight into the structure of the model.

where W_i are Brownian motions with $\mathbf{E}\left(dW_i\,dW_j\right) = \rho_{ij}\,dt$. Each state variable behaves like a *mean-reverting* or *mean-repelling* Gaussian process, according to the sign of κ_i. The corresponding multifactor model is a "superposition" of *exponentially shaped deformations* of the forward-rate curve driven by Gauss–Markov state variables.

It remains to compute the function $q(t;\ T)$ needed to calibrate the model to the initial term structure. Using Eq. (14.15), we conclude after a straightforward (but tedious) calculation that

$$q(t;\ T) = \int_t^T f(0;\ s)\,ds$$

$$-\sum_{ij}\frac{\bar{a}_i\,\bar{a}_j\,\sigma_i\,\sigma_j\,\rho_{ij}}{2\,\kappa_i\,\kappa_j}\left[\frac{(1 - e^{-\kappa_i\,(T-t)})\,(1 - e^{-\kappa_i\,t})}{\kappa_i}\right.$$

$$+\frac{(1 - e^{-\kappa_j\,(T-t)})\,(1 - e^{-\kappa_j\,t})}{\kappa_j}$$

$$\left.-\frac{(1 - e^{-(\kappa_i+\kappa_j)\,(T-t)})(1 - e^{-(\kappa_i+\kappa_j)t})}{\kappa_i + \kappa_j}\right]. \tag{14.22}$$

The coefficient $b(t;\ T)$ is obtained by differentiating with respect to T. Accordingly,

$$b(t;\ T) = f(0;\ T)$$

$$+\sum_{ij}\frac{\bar{a}_i\,\bar{a}_j\,\sigma_i\,\sigma_j\,\rho_{ij}}{2\,\kappa_i\,\kappa_j}\left[(e^{-\kappa_i\,(T-t)})(1 - e^{-\kappa_i\,t})\right.$$

$$+ (e^{-\kappa_j\,(T-t)})(1 - e^{-\kappa_j\,t})$$

$$\left.- e^{-(\kappa_i+\kappa_j)\,(T-t)}(1 - e^{-(\kappa_i+\kappa_j)t})\right]. \tag{14.23}$$

The "fudge factor" for the short-rate process, $\bar{b}$, is obtained by setting $T = t$ in this last expression. It is given by

$$\bar{b}(t) = f(0;\ t)$$

$$+\sum_{ij}\frac{\bar{a}_i\,\bar{a}_j\,\sigma_i\,\sigma_j\,\rho_{ij}}{2\,\kappa_i\,\kappa_j}(1 - e^{-\kappa_i\,t})(1 - e^{-\kappa_j\,t}). \tag{14.24}$$

We conclude that the short rate process for the Gaussian model with constant coefficients and diagonal $\mu^{(1)}$ has the form

$$r_t = f(0;\ t) + \sum_i \bar{a}_i\,\sigma_i\int_0^t e^{-\kappa_i(t-s)}\,d\,W_i$$

$$+ \sum_{i,j} \frac{\bar{a}_i \, \bar{a}_j \, \sigma_i \, \sigma_j \, \rho_{ij}}{2 \, \kappa_i \, \kappa_j} \, (1 - e^{-\kappa_i t})(1 - e^{-\kappa_j t}) \,. \tag{14.25}$$

These formulas generalize the ones obtained when we discussed the modified Vasicek model [cf. Chapter 12, Eq. (12.19)]. Observe, in particular that Eq. (14.25) implies that

$$\mathbf{E}\{r_t\} \; = \; f(0; t) + \sum_{i,j} \frac{\bar{a}_i \, \bar{a}_j \, \sigma_i \, \sigma_j \, \rho_{ij}}{2 \, \kappa_i \, \kappa_j} \, (1 - e^{-\kappa_i t})(1 - e^{-\kappa_j t}) \,.$$

Thus, the expected value of the short rate under the risk-neutral measure is always greater than the corresponding forward rate.[11]

REMARK 14.3 principal components. The "mean-reversion parameters" κ_i determine the shape of the components (or eigendirections) of the covariance matrix of instantaneous forward rates. In fact, from (14.21),

$$\mathbf{Cov}\left(df(t; T), \, df(t; T')\right) \; = \; \sum_{i,j=1}^{N} \sigma_i \, \sigma_j \, \rho_{ij} \, \bar{a}_i \, \bar{a}_j \, e^{-\kappa_i (T-t)} \, e^{-\kappa_j (T'-t)} \, dt \,. \tag{14.26}$$

A principal component analysis sheds light on the role played by the correlation matrix driving the Brownian motions. Denote by θ_k^2 and $\mathbf{v}^{(k)} = \left(v_1^{(k)}, \dots v_\nu^{(k)}\right)$ the eigenvalues and eigenvectors of the correlation matrix ρ_{ij}. Hence, we have

$$\rho_{ij} \; = \; \sum_{k=1}^{\nu} \theta_k^2 \, v_i^{(k)} \, v_j^{(k)} \,.$$

Substituting this into (14.26), and taking the square root of the covariance matrix, we find that the components driving the instantaneous forward rates are

$$\psi_k(x) \; = \; \theta_k \sum_{i=1}^{N} \bar{a}_i \, \sigma_i \, v_i^{(k)} \, e^{-\kappa_i x} \,, \quad x = T - t \,, \quad k = 1, 2 \dots \nu, \tag{14.27}$$

i.e., that the dynamics are given by

$$df(t; T) \; = \; \sum_{k=1}^{\nu} \psi_k(T - t) \, dW_k(t) + \text{HJM drift terms}$$

where W_k are independent Brownian motions. Thus, the use of correlations between the Brownian motions allows us to generate components that are linear combinations of exponentials. This class of functions can be used to model quite general covariance structures. The two-factor model described in the previous chapter is a special case of a covariance structure realizable within an exponential-affine Gaussian model. ∎

---

[11]This inequality is consistent with the fact that futures-implied rates (expected short-rates) are higher than forward rates if the term rate and the funding rates are positively correlated (cf. Chapter 12).

14.4 Square-Root Processes and the Non-Central Chi-Squared Distribution

The other important class of processes for the state-variables that gives rise to exponential-affine models are the so-called **square-root diffusions**, which correspond to

$$A_{ij} = A_{ij}^{(0)} + \sum_{k=1}^{N} A_{ij,k}^{(1)} X_k$$

with $A_{ij,k}^{(1)} > 0$ for some k. Square-root processes were introduced in the field of finance by Cox et al. [4] (see also Longstaff and Schwartz [10], Scott [1987], and Duffie and Kan [5]).

In this section, we consider a one-dimensional diffusion process X_t such that the diffusion coefficient $a = A_{11}$ is an affine function of X, i.e.,

$$a = a^{(0)} + a^{(1)} X.$$

Accordingly, X satisfies the SDE

$$dX = \sqrt{a^{(0)} + a^{(1)} X} \, dZ + \left(\mu^{(0)} + \mu^{(1)} X \right) dt. \tag{14.28}$$

For simplicity, we assume that the coefficients are constant. It is convenient to set

$$a^{(1)} = \sigma^2 > 0$$

and

$$Y = \frac{a^{(0)}}{a^{(1)}} + X.$$

From Eq. (14.28), we see that the stochastic process Y satisfies formally the SDE

$$dY = \sigma \sqrt{Y} \, dZ + \kappa (\theta - Y) \, dt \tag{14.29}$$

where

$$\kappa = -\mu^{(1)} \quad \text{and} \quad \theta = -\frac{\mu^{(0)}}{\mu^{(1)}} + \frac{a^{(0)}}{a^{(1)}}. \tag{14.30}$$

Equation (14.29) can be viewed as the "canonical form" of a one-dimensional square-root process. It defines a stochastic process $Y(t)$ in the obvious way provided that $Y(t)$ remains positive.[12] The issue of what happens if and when the process hits zero is more delicate and must be considered separately.

The first question of interest is to determine conditions on the coefficients σ, κ, and θ which ensure that $Y(t)$ is well defined for all $t > 0$. Unlike the Gaussian case, the solution of Eq. (14.29) cannot be expressed in a simple form using Ito integrals. Nevertheless, its behavior can be analyzed using PDE techniques.

---

[12]For details, see the existence and uniqueness theorems for solutions of SDEs in Karatzas and Shreve [8].

PROPOSITION 14.2
Suppose that $Y(0) > 0, \kappa \geq 0, \theta \geq 0$.
(i) If

$$\kappa\,\theta \;\geq\; \frac{1}{2}\,\sigma^2\,,$$

the SDE (14.29) admits a solution $Y(t)$ that is strictly positive for all $t > 0$.
(ii) If

$$0 \;<\; \kappa\,\theta \;<\; \frac{1}{2}\,\sigma^2\,,$$

the SDE (14.29) admits a unique solution, which is nonnegative but occasionally hits $Y = 0$.[13]
(iii) If $\kappa\,\theta = 0$, the process $Y(t)$ vanishes at a finite time and remains equal to zero thereafter.
(iv) If $\kappa\,\theta > 0$, then, as $t \to \infty$, $Y(t)$ has a long-term equilibrium Gamma distribution with density function

$$p(Y) \;=\; \frac{\left(\frac{2\kappa}{\sigma^2}\right)^{\frac{2\kappa\theta}{\sigma^2}}}{\Gamma\left(\frac{2\kappa\theta}{\sigma^2}\right)} \cdot Y^{\frac{2\kappa\theta}{\sigma^2} - 1}\, \exp\left(-\frac{2\kappa}{\sigma^2}\, Y \right)\,, \tag{14.31}$$

where $\Gamma(p) = \int_0^\infty x^{p-1}\, e^{-x}\, dx$ is the Gamma function.[14]

PROOF   See Appendices A and B of this chapter (see also Figure 14.2).   ∎

REMARK 14.4   The asymptotic density for $t \to 0$ in (14.31) is controlled by the ratio $\frac{\kappa\theta}{\sigma^2}$. If $\kappa\theta > 0.5\sigma^2$, the distribution vanishes at zero. In the "marginal case" $\kappa\theta = \sigma^2/2$, the Gamma density is an exponential and thus has a finite value at $Y = 0$. If $0 < \kappa\theta < \sigma^2/2$, the density has a singularity at $Y = 0$—this is the regime in which the diffusion visits zero but is reflected back into the into $\{Y : Y > 0\}$ immediately. Finally, it is easy to verify that $p(Y)$ converges to a Dirac mass at $Y = 0$ as $\kappa\,\theta \to 0$. This is consistent with the fact that the process is absorbed at zero if $\kappa\,\theta = 0$.   ∎

We analyze the distribution of $Y(t)$ for finite times t. Set $Y(0) = Y_0$. The transition probability density can be represented as a function of three variables:

$$p\,(Y_0,\, t;\; Y) \;=\; \mathbf{P}\{Y(t) = Y | Y(0) = Y_0\}\,.$$

This function is easily characterized by its moment-generating function (Laplace transform)

$$\hat{p}(Y_0,\, t;\; \lambda) \;=\; \int_0^\infty e^{-\lambda Y}\, p\,(Y_0,\, t;\; Y)\, dY \;=\; \mathbf{E}\left\{e^{-\lambda Y(t)} \,|\, Y(0) = Y_0\right\}\,. \tag{14.32}$$

---

[13]More precisely, if $Y(t) = 0$ for some value of t, it is immediately reflected back into the half-line $\{Y : Y > 0\}$ [7].

[14]This is a Gamma distribution with shape parameter $a = \frac{2\kappa\theta}{\sigma^2} - 1$ and scale parameter $b = \frac{\sigma^2}{2\kappa}$. See Abramowitz and Stegun [1].

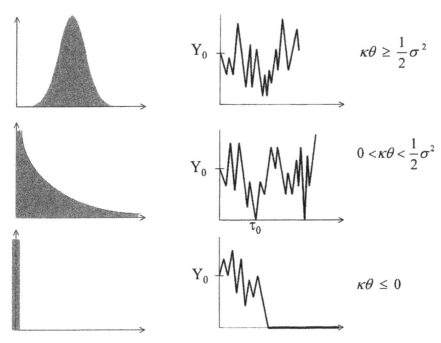

FIGURE 14.2
The three alternatives for square-root diffusions. If $\kappa\theta > \frac{1}{2}\sigma^2$, the path never hits the
level $Y = 0$ (top). If $0 < \kappa\theta \leq \frac{1}{2}\sigma^2$, the path hits the level zero and immediately returns
to the half-line $Y > 0$ (middle). If $\kappa\theta \leq 0$, the process hits $Y = 0$ and is absorbed at zero
(bottom). The explanation for this behavior has to do with the relation between the drift
at $Y = 0$, which is $\kappa\theta$, and the volatility. A "strong" drift (top) pushes the process away
from zero, a weak drift lets it hit zero but the process is not absorbed at zero. A zero
drift necessarily implies that the process stays at zero after hitting for the first time. The
approximate shapes of the asymptotic densities for $t \to +\infty$ are shown on the left.

PROPOSITION 14.3
Assume that $Y_0 > 0$ and that $\kappa\,\theta > 0$. The moment-generating function of $Y(t)$ is given by

$$\hat{p}(Y_0, t; \lambda) = \frac{1}{\left[1 + \frac{\lambda\,\sigma^2}{2\kappa}\left(1 - e^{-\kappa t}\right)\right]^{\frac{2\kappa\theta}{\sigma^2}}} \cdot \exp\left[\frac{-\lambda\,Y_0\,e^{-\kappa t}}{1 + \frac{\lambda\,\sigma^2}{2\kappa}(1 - e^{-\kappa t})}\right]. \quad (14.33)$$

PROOF   See Appendix B.   ∎

   Probability distribution functions with moment-generating functions of the form (14.33)
belong to the class of *non-central Chi-square distributions*.[15] More precisely, we have.

---

[15] The terminology comes from the fact that if ν is an integer, $x_1, x_2, \ldots x_\nu$ are standard normal random variables

(with mean zero and variance 1), and a and b are real numbers, then the random variable $\sum_{i=1}^{\nu}(x_i + b)^2$ has

a moment-generating function $\dfrac{1}{(1+2\lambda)^{\frac{\nu}{2}}} \cdot e^{-\frac{\lambda\,\nu\,b^2}{(1+2\lambda)}}$. The parameters ν and $c = \nu\,b^2$ are called the **number**
of degrees of freedom and the **non-centrality parameter**, respectively. The case $b = 0$ gives the standard

COROLLARY 14.1
For $\kappa > 0, \theta > 0$, the random variable

$$\frac{4\kappa}{\sigma^2 \left(1 - e^{-\kappa t}\right)} \cdot Y(t) \tag{14.34}$$

has a non-central Chi-squared distribution with

$$\nu = \frac{4\kappa\theta}{\sigma^2} \tag{14.35}$$

degrees of freedom and non-centrality parameter

$$c = \frac{4\kappa Y_0 e^{-\kappa t}}{\sigma^2 \left(1 - e^{-\kappa t}\right)} \, . \tag{14.36}$$

There is an important difference with respect to the Gaussian case: the non-centrality parameter depends on the initial value Y_0. This implies that the choice of initial value for the state variable affects the shape of the distribution of interest rates. In particular, we have

COROLLARY 14.2
For $\kappa > 0, \ \theta > 0$, we have

$$\mathbf{E}\{Y(t)\,|\,Y(0) = Y_0\} = e^{-\kappa t} Y_0 + \theta \left(1 - e^{\kappa t}\right) \tag{14.37}$$

and

$$\mathbf{Var}\{Y(t)\,|\,Y(0) = Y_0\} = \frac{\sigma^2}{2\kappa} \left(1 - e^{-\kappa t}\right)$$

$$\left[\frac{\sigma^2}{2\kappa}\left(1 - e^{-\kappa t}\right) + \frac{4\kappa\theta}{\sigma^2} Y_0 e^{-\kappa t}\right] . \tag{14.38}$$

Moreover,

$$p(Y_0, t; Y) \propto Y^{\frac{2\kappa\theta}{\sigma^2} - 1}, \quad Y \ll 1 . \tag{14.39}$$

Figures 14.3 and 14.4 display different snapshots of the distribution of $Y(t)$ using values of the parameters σ, κ, and θ estimated by Chen and Scott from historical values for U.S. dollar rates. The picture were generated using formulas (14.34), (14.35), and (14.36) and a program that generates the non-central Chi-squared distribution. The parameters estimated by Chen and Scott [3] for one-factor term-structure models are $\sigma = 3\%$, $\kappa = 25\%$, and $\theta = 6.8\%$. Notice that the estimated number of degrees of freedom, $\nu \approx 75.56$, corresponds to the case when zero is not attained.

Chi-square distribution (sum of squares of ν independent normals). The concept of a fractional number of degrees of freedom ν comes from analytic continuation of the dimension ν to arbitrary positive real numbers. Inversion of the Laplace transform gives a formula for the density of the non-central χ^2 as a power series (cf. Abramowitz and Stegun [1]): $nc \, \chi^2(x) = \sum_{i=0}^{\infty} \frac{e^{-c/2} \, (c/2)^i}{i!} \frac{x^{(\nu+2i/2)-1} \, e^{-x/2}}{2^{(\nu+2i)/2} \, \Gamma\left(\frac{\nu+2i}{2}\right)}$.

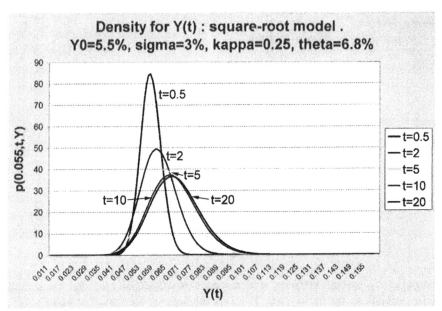

FIGURE 14.3
Probability density function of $Y(t)$ **at different times** $t = 0.5, 2, 5, 10, 20$. **The parameters are** $Y_0 = 5.5\%$, $\sigma = 3\%$, $\kappa = 0.25, \theta = 6.8\%$.

14.5 One-Factor Square-Root Model: Discount Factors and Forward Rates

Having characterized the distribution of the state-variable Y, we compute the coefficients $p(t; T)$ and $q(t; T)$ arising in the exponential formula for discount factors

$$P_t^T = \exp[-p(t; T) Y(t) - q(t; T)] .$$

From the general considerations in Section 14.1 [Eqs. (14.10a) and (14.10b)], the functions $p(t; T)$ and $q(t; T)$ satisfy the ordinary differential equations:

$$\dot{p} - \kappa p + \bar{a} = \frac{\sigma^2}{2} p^2 \quad p(T; T) = 0, \tag{14.40}$$

$$\dot{q} + \kappa\theta p + \bar{b} = 0 \quad , \quad q(T; T) = 0. \tag{14.41}$$

The transformations

$$Y \leftrightarrow \bar{a} Y, \quad \sigma \leftrightarrow \bar{a}^{1/2} \sigma, \quad \kappa \leftrightarrow \bar{a} \kappa \tag{14.42}$$

have the effect of reducing the computation to the case $\bar{a} = 1$. We assume in the rest of this section that $\bar{a} = a(t; t) = 1$.

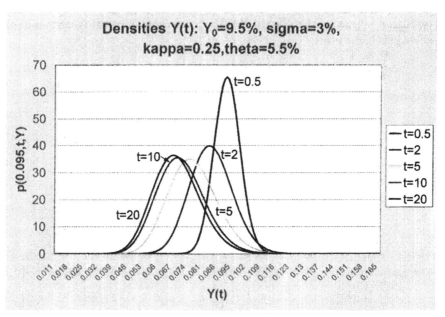

FIGURE 14.4
Same as in Figure 14.3 with $Y_0 = 9.5$.

Ricatti equations such as (14.40) can be "linearized" via the transformation

$$p = \frac{x}{y},$$

where x and y satisfy a linear system of ordinary differential equations of the form

$$\dot{x} = A_1 x + B_1 y$$

$$\dot{y} = A_2 x + B_2 y. \tag{14.43}$$

In fact, it follows from the two equations that

$$\left(\frac{x}{y}\right)^{\cdot} = \frac{\dot{x}}{y} - \frac{x \dot{y}}{y^2}$$

$$= \frac{A_1 x + B_1 y}{y} - \frac{x (A_2 x + B_2 y)}{y^2}$$

$$= B_1 + (A_1 - B_2)\left(\frac{x}{y}\right) - A_2 \left(\frac{x}{y}\right)^2.$$

Hence, setting

$$A_1 = -B_2 = \frac{\kappa}{2} \ ; \quad B_1 = -1 \ ; \quad A_2 = -\frac{\sigma^2}{2},$$

the function $p = x/y$ satisfies Eq. (14.40). The solution of this equation can thus be obtained by solving the linear system

$$\dot{x} = \frac{\kappa}{2} x - \frac{\sigma^2}{2} y \tag{14.44a}$$

$$\dot{y} = -x - \frac{\kappa}{2} y \tag{14.44b}$$

with the boundary conditions

$$x(T; T) = 0 \quad \text{and} \quad \dot{x}(T; T) + y(T; T) = 0.$$

(The latter boundary condition arises from the fact that we assume that $\bar{a} = -[\dot{p}]_{t=T} = 1$.)
   A straightforward computation of the solution of the system (14.44a) and (14.44b) yields

$$p(t; T) = \frac{2\left(1 - e^{-\nu(T-t)}\right)}{\alpha + \kappa + (\alpha - \kappa)e^{-\alpha(T-t)}}, \tag{14.45}$$

where

$$\alpha = \sqrt{\kappa^2 + 2\sigma^2}. \tag{14.46}$$

   The function $q(t; T)$ is obtained by integrating both sides of Eq. (14.43), which gives the result

$$q(t; T) = \int_t^T \bar{b}(s)\, ds + \kappa\theta \int_t^T p(s; T)\, ds$$

$$= \int_t^T \bar{b}(s)\, ds + \kappa\theta \left(\frac{2\alpha}{\sigma^2}\right) \log\left[\frac{\alpha + \kappa + (\alpha - \kappa)e^{-\alpha(T-t)}}{2\alpha\, e^{-\frac{\sigma^2(T-t)}{\kappa + \alpha}}}\right].$$

As in the analysis of Gaussian models, the fudge factor $\bar{b}$ is determined by matching the value of the zero-coupon bonds P_0^T to the market prices. From Eq. (14.42), we obtain

$$q(0; T) + p(0; T) Y(0) = \int_0^T f(0; s)\, ds. $$

Setting $t = 0$ in Eq. (14.47) and solving for $\int_0^T \bar{b}$ gives

$$\int_0^T \bar{b}(s)\, ds = \int_0^T f(0; s)\, ds - p(0; T) Y(0)$$

$$-\frac{2\alpha\kappa\theta}{\sigma^2} \log\left[\frac{\alpha + \kappa + (\alpha - \kappa)e^{-\alpha T}}{2\alpha\, e^{-\frac{\sigma^2 T}{\kappa + \alpha}}}\right].$$

Using this identity, we conclude that

$$q(t; T) = \int_t^T f(0; s), ds - (p(0; T) - p(0; t)) Y(0) - \log A(t; T) \qquad (14.47)$$

where

$$A(t; T) = \left[\frac{\alpha + \kappa + (\alpha - \kappa) e^{-\alpha T}}{(\alpha + \kappa + (\alpha - \kappa) e^{-\alpha t}) \cdot (\alpha + \kappa + (\alpha - \kappa) e^{-\nu (T-t)})} \right]^{\frac{2\alpha\kappa\theta}{\sigma^2}}.$$

The value at time t of a zero-coupon bond paying \$1 at time T is therefore

$$P_t^T = \left(\frac{P_0^T}{P_0^t} \right) \cdot A(t; T) \cdot \exp[- p(t; T) Y(t) + (p(0; T) - p(0; t)) Y(0)] . \qquad (14.48)$$

REMARK 14.5 The stochastic discount factor P_t^T can be expressed as the product of the forward price for delivery at time t of a zero-coupon bond maturing at time T and a model-dependent quantity that depends on the volatility of the forward-rate curve. In contrast with the case of Gaussian state variables, the initial value of the state variable $Y(0)$ and the long-term mean θ appears in the expression of the discount factor. ∎

Let us derive formulas for the instantaneous forward rates and the short-term rate. We have

$$f(t; T) = a(t; T) Y(t) + b(t; T)$$

where $a(t; T) = \frac{\partial p(t; T)}{\partial T}$ and $b(t; T) = \frac{\partial q(t; T)}{\partial T}$. Differentiating Eq. (14.45) with respect to T, we obtain

$$a(t; T) = \frac{4 \alpha^2 e^{-\alpha (T-t)}}{\left[\alpha + \kappa + (\alpha - \kappa) e^{-\alpha T} \right]^2} , \qquad (14.49)$$

and, from (14.47),

$$b(t; T) = f(0; T) - \frac{4 \alpha^2 e^{-\alpha T} Y(0)}{\left[\alpha + \kappa + (\alpha - \kappa) e^{-\alpha T} \right]^2} + \left(\frac{2\alpha^2 \kappa (\alpha - \kappa) \theta}{\sigma^2} \right)$$

$$\times \left[\frac{e^{-\alpha T}}{\alpha + \kappa + (\alpha - \kappa) e^{-\alpha T}} - \frac{e^{-\alpha(T-t)}}{\alpha + \kappa + (\alpha - \kappa) e^{-\alpha(T-t)}} \right] . \qquad (14.50)$$

Setting $T = t$ in this last formula, we conclude that the short rate process is given by

$$r_t = Y(t) + f(0 t) - \frac{4 \alpha^2 e^{-\alpha t} Y(0)}{\left[\alpha + \kappa + (\alpha - \kappa) e^{-\alpha t} \right]^2}$$

$$- \frac{2\alpha \kappa \theta (1 - e^{-\nu t})}{\alpha + \kappa + (\alpha - \kappa) e^{-\alpha t}} . \qquad (14.51)$$

Notice that this function satisfies

$$a(t;\ T) \approx \frac{4\alpha^2}{(\kappa + \alpha)^2}\, e^{-\alpha\,(T-t)} \quad \text{for } T - t \gg 1\,.$$

This is in qualitative agreement with the Gaussian models, in which indexgaussian models@Gaussian modelsα is replaced by κ. Thus, the standard deviation of forward rates decays exponentially with rate α as the maturity increases. The functions in (14.49) correspond to the exponentials $e^{-\kappa_i(T-t)}$ in the context of Gaussian models. They determine the shape of the deformations of the forward-rate curve.

References and Further Reading

[1] Abramowitz, M. and Stegun, I. (1964), *Handbook of Mathematical Functions, with Formulas, Graphs and Mathematical Tables*, National Bureau of Standards, Washington D.C.

[2] Chen, R.R., and Scott, L. (December 1993), Maximum Likelihood Estimation for a Multi-factor Equilibrium Model of the Term Structure of Interest Rates, *Journal of Fixed Income*, pp. 14–31

[3] Chen, R.R. and Scott, L. (Winter 1995), Interest Rate Options in Multi-factor Cox-Ingersoll-Ross Models of the Term Structure, *Journal of Fixed Income*, pp. 53–72.

[4] Cox, J., Ingersoll, J., and Ross, S. (1985), A Theory of the Term Structure of Interest Rates, *Econometrica,* **53**(2), pp. 385–407.

[5] Duffie, D. and Kan, R. (1995), Multi-Factor Interest Rate Models, *Philosophical Transactions of the Royal Society, Series A*, **317**, pp. 577–586.

[6] Feller, W. (1971), *An Introduction to Probability Theory and Its Applications, vol 2*, John Wiley & Sons, New York.

[7] Ikeda, N. and Watanabe, S. (1989), *Stochastic Differential Equations and Diffusion Processes,* North Holland-Kodansha, Amsterdam.

[8] Karatzas, I. and Shreve, S. (1988), *Brownian Motion and Stochastic Calculus*, Springer, Berlin.

[9] Li, A., Ritchen, P., and Sankarasubramanian. (1995), Lattice Models for Pricing American Interest Rate Claims, *Journal of Finance,* **50**, pp. 719–737.

[10] Longstaff, F. and Schwartz, E. (1992), Interest Rate Volatility and the Term Structure: A Two-Factor General Equilibrium Model, *Journal of Finance,* **47**, pp. 1259–1282.

Appendix A: Behavior of Square-Root Processes for Large Times

We sketch a proof of the statements in Proposition 14.2 regarding the probability that the square-root diffusion hits zero. Consider the square-root process

$$dY(t) = \sigma \sqrt{Y(t)} \, dZ(t) + \kappa(\theta - Y(t)) \, dt \, .$$

We will assume that the initial value $Y(0) = Y$ is a positive number. Let M and ϵ be such that $0 < \epsilon < Y < M$. To study the question of whether the process hits zero or not, we introduce a stopping time τ, representing the first time that the process $\{Y(t), \, t > 0\}$ exists the strip $\{\xi \, : \, \epsilon < \xi < M\}$. Consider the probability that the process exits the strip for the first time by crossing $Y = \epsilon$:

$$\mathbf{P}\{Y(\tau) = \epsilon \mid Y(0) = Y\} \, . \tag{A.1}$$

We claim that this probability, viewed as a function of the starting point Y, is equal to the solution $u(Y)$ of the two-point boundary-value problem

$$\begin{cases} \frac{1}{2}\sigma^2 \, Y \, \frac{\partial^2 u}{\partial Y^2} + \kappa \, (\theta - Y) \, \frac{\partial u}{\partial Y} = 0 \\[2mm] u(M) = 0, \quad u(\epsilon) = 1 \, . \end{cases} \tag{A.2}$$

To verify this, we apply Ito's Formula to the solution of (A.2), namely, to observe that if $t < \tau$, then

$$du(Y(t)) = \sigma \sqrt{(Y(t))} \frac{\partial u}{\partial Y}(Y(t)) \, dZ + \left[\frac{1}{2}\sigma^2 Y \frac{\partial^2 u}{\partial Y^2} + \kappa \, (\theta - Y) \frac{\partial u}{\partial Y} \right]_{Y=Y(t)} dt \, .$$

On account of (A.2), we conclude that if $t < \tau$, we have

$$u(Y(t)) = u(Y) + \int_0^t \sigma \sqrt{(Y(s))} \frac{\partial u}{\partial Y}(Y(s)) \, dZ(s) \, .$$

By continuity of the diffusion path, this equation holds for $t = \tau$ as well. Taking expected values on both sides of the last equation and using the Optional Stopping theorem for continuous-time martingales (see Karatzas and Shreve [8]), we have

$$\mathbf{E}\{u(Y(\tau))|Y(0) = Y\} = u(Y) \, .$$

On the other hand, the boundary conditions satisfied by u at the endpoints of the interval $[\epsilon, \, M]$ imply that

$$\mathbf{E}\{u(Y(\tau)) \, |Y(0) = Y\} = \mathbf{P}\{Y(\tau) = \epsilon \mid Y(0) = Y\} \, ,$$

which establishes our claim.

Now, consider the solution of the boundary-value problem (A.2), which is given by

$$u(Y) = \frac{\displaystyle\int_Y^M x^{-\frac{2\kappa\theta}{\sigma^2}} e^{\frac{2\kappa}{\sigma^2}} dx}{\displaystyle\int_\epsilon^M x^{-\frac{2\kappa\theta}{\sigma^2}} e^{\frac{2\kappa}{\sigma^2}} dx} . \tag{A.3}$$

Let us study the behavior of the solution as $\epsilon \to 0$, keeping Y and M fixed. Notice first that if

$$\frac{2\kappa\theta}{\sigma^2} \geq 1 \tag{A.4}$$

the denominator tends to infinity. It follows that under (A.4) we must have

$$\lim_{\epsilon \to 0} \mathbf{P}\{Y(\tau) = \epsilon \mid Y(0) = Y\} = 0 .$$

Since the latter equation holds for any value of M, we conclude that

$$\mathbf{P}\{Y(t) > 0 \text{ for all } t \mid Y(0) = Y\} = 1 ,$$

i.e., that the process does not hit zero (Figure A.5).
   Assume next that

$$0 < \frac{2\kappa\theta}{\sigma^2} < 1 .$$

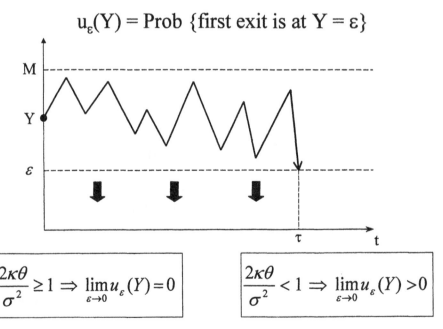

$$u_\epsilon(Y) = \text{Prob } \{\text{first exit is at } Y = \epsilon\}$$

$$\frac{2\kappa\theta}{\sigma^2} \geq 1 \Rightarrow \lim_{\epsilon \to 0} u_\epsilon(Y) = 0 \qquad \frac{2\kappa\theta}{\sigma^2} < 1 \Rightarrow \lim_{\epsilon \to 0} u_\epsilon(Y) > 0$$

FIGURE A.5
Diagram corresponding to the calculation of the probability of hitting zero. We compute the probability of exiting the strip first through $Y = \epsilon$ and then let ϵ tend to zero.

In this case, we have

$$\mathbf{P}\{Y(t) = 0 \text{ for some } t \text{ before } Y(t) = M|\ Y(0) = Y\}$$

$$= \lim_{\epsilon \to 0} \mathbf{P}\{Y(\tau) = \epsilon\ |\ Y(0) = Y\}$$

$$= \frac{\displaystyle\int_{Y}^{M} x^{\frac{-2\kappa\theta}{\sigma^2}}\ e^{\frac{2\kappa}{\sigma^2}}\ dx}{\displaystyle\int_{0}^{M} x^{\frac{-2\kappa\theta}{\sigma^2}}\ e^{\frac{2\kappa}{\sigma^2}}\ dx},$$

which is a positive quantity less than unity because the integral in the denominator converges. By letting M tend to infinity in both sides of the last equation, the integrals will diverge at infinity, and we can conclude that

$$\mathbf{P}\{Y(t) = 0 \text{ for some } t|\ Y(0) = Y\}\ =\ 1\ .$$

The analysis of the behavior of Y_t after the fist hitting time of zero is more delicate. The reader is referred to Feller [6] or Ikeda and Watanabe [7] for a rigorous study of this question.

Appendix B: Characterization of the Probability Density Function of Square-Root Processes

As indicated before the statement of Proposition 14.3, it is convenient to analyze the moment-generating function of the diffusion. Suppose first that $\kappa\theta > \frac{\sigma^2}{2}$, so that the process never hits $Y = 0$. In this case, we can apply the standard theory (cf. Chapter 12, section on Fokker–Planck theory) or Ito calculus to the function

$$\pi(Y,\ t,\ T;\ \lambda)\ =\ \mathbf{E}_t\left\{e^{-\lambda Y(T)}\ |\ Y(t)\ =\ Y\right\}$$

to conclude that it satisfies the partial differential equation

$$\pi_t\ +\ \frac{\sigma^2 Y}{2}\ \pi_{YY}\ +\ \kappa\ (\theta\ -\ Y)\ \pi_Y\ =\ 0$$

$$\pi(Y,\ T,\ T;\ \lambda)\ =\ e^{-\lambda Y}\ . \tag{B.1}$$

A solution of this final-value problem can be sought in the form

$$\pi(Y,\ t,\ T;\ \lambda)\ =\ \exp\left(-r(t,\ T;\ \lambda)\ Y\ -\ s(t,\ T;\ \lambda)\right)\ .$$

Substitution of this function into the PDE (B.1) shows that the functions $r(t, T; \lambda)$ and $s(t, T; \lambda)$ satisfy the ordinary differential equations

$$\dot{r} - \kappa r = \tfrac{1}{2}\sigma^2 r^2$$

$$\dot{s} + \kappa \theta r = 0 \,,$$

with boundary conditions

$$r(T, T; \lambda) = \lambda \,, \quad s(T, T; \lambda) = 0 \,.$$

These Ricatti ordinary differential equations can be solved in closed form (the transformation $y = \frac{1}{r}$ is used to linearize the first equation). We find, after some calculation, that

$$r(t, T; \lambda) = \frac{\lambda e^{-\kappa (T-t)}}{1 + \frac{\lambda \sigma^2}{2\kappa}(1 - e^{-\kappa (T-t)})}$$

and

$$s(t, T; \lambda) = \frac{2\kappa \theta}{\sigma^2} \cdot \log\left[1 + \frac{\lambda \sigma^2}{2\kappa}\left(1 - e^{-\kappa t}\right)\right] \,.$$

Setting $t = 0$ and $T = t$, this implies that the moment-generating function of $Y(t)$ is given by

$$\pi(Y, 0; t) = \frac{1}{\left[1 + \frac{\lambda \sigma^2}{2\kappa}\left(1 - e^{-\kappa t}\right)\right]^{\frac{2\kappa \theta}{\sigma^2}}}$$

$$\cdot \exp\left[\frac{-\lambda Y_0 e^{-\kappa t}}{1 + \frac{\lambda \sigma^2}{2\kappa}(1 - e^{-\kappa t})}\right] \,, \tag{B.2}$$

which corresponds to Eq. (14.33).

Notice that this moment-generating function, associated with the non-central χ^2 distribution, is well-defined for all parameter values such that $0 < \frac{\kappa \theta}{\sigma^2} < 0.5$. It can also be shown that Eq. (B.2) characterizes the statistics of square-root processes even for the case where $\kappa \theta < \sigma^2/2$ (i.e., when the process hits zero).

The asymptotic distribution for large t associated with the process $Y(t)$ is obtained by taking the limit $t \to \infty$ in (B.2). The result is

$$\pi(Y, 0, \infty; \lambda) = \frac{1}{\left[1 + \frac{\lambda \sigma^2}{2\kappa}\right]^{\frac{2\kappa \theta}{\sigma^2}}} \cdot \exp\left[\frac{-\lambda Y_0}{1 + \frac{\lambda \sigma^2}{2\kappa}}\right] \,.$$

It can be easily checked this is the moment-generating function of the Gamma distribution of Eq. (14.31).

Appendix C: The Square-Root Diffusion with $\nu = 1$

According to the claim made in Proposition 14.2, square-root diffusions with $\kappa\theta < \frac{\sigma^2}{2}$ hit zero occasionally (actually, infinitely many times), but immediately reenter the half-line $\{Y : Y > 0\}$. In Appendix 14.5 of this chapter we did not discuss the behavior of such processes after the first time that zero is hit. It turns out that the case when

$$\kappa\theta = \frac{\sigma^2}{4} \quad \text{i.e.,} \quad \nu = 1$$

is simple to understand. It reveals the structure of the paths of square-root diffusions that hit zero but are not "absorbed."

Consider a Gauss–Markov process satisfying the stochastic differential equation

$$dX_t = a\,dZ_t - b\,X_t\,dt \qquad (C.1)$$

where a and b are positive constants.[16] We have thoroughly analyzed this class of diffusions in Sections 14.2 and 14.3. Let us define a new process Y_t by

$$Y_t = X_t^2. \qquad (C.2)$$

Applying Ito's Formula to X_t^2 and using (C.1), we find that

$$dY_t = 2\,X_t\,dX_t + (dX_t)^2$$

$$= 2\,X_t\,a\,dZ_t - 2b\,X_t^2\,dt + a^2\,dt$$

$$= 2a\,|X_t|\,\text{sign}(X_t)\,dZ_t - 2b\,X_t^2\,dt + a^2\,dt$$

$$= \sigma\,\sqrt{Y_t}\,dW_t + \kappa\,(\theta - Y_t)\,dt. \qquad (C.3)$$

To obtain the last equation, we introduced the notation

$$\sigma = 2a, \quad \kappa = 2b, \quad \theta = \frac{a^2}{2b}$$

and defined the process

$$W_t = \int_0^t \text{sign}(X_s)\,dZ_s.$$

This process is a martingale that satisfies

$$\mathbf{E}\left\{W_t^2\right\} = \mathbf{E}\left\{\int_0^t (\text{sign}(X_s))^2\,ds\right\} = t.$$

---

[16]This process is usually called an **Ornstein–Uhlenbeck** process.

Therefore, W_t is a Brownian motion and, on account of (C.3), Y_t is a square-root diffusion. It is a special case of a square-root diffusion because the number of degrees of freedom is

$$\frac{4\kappa\theta}{\sigma^2} = \frac{4\,a^2}{(2a)^2} = 1\,.$$

This is perhaps the simplest example of a square-root diffusion that accesses the point $Y = 0$ but is not absorbed—falling under category (ii) of Proposition 14.2. The structure of the process Y_t is transparent: it is the square of a Gaussian diffusion. This gives an independent justification of why the $\nu = 1$ square-root process has a χ^2 distribution with one degree of freedom.

Dynamically, the evolution of the process Y_t is obtained by a mapping the dynamics of X_t via (C.2). This mapping idea can be generalized to arbitrary values of the ratio $\frac{\kappa\theta}{\sigma^2}$ by introducing other processes X_t, known as modified Bessel processes.[17]

---

[17]For more information on the correspondence between square-root diffusions and Bessel processes, see Feller [6].

Chapter 15

Interest-Rate Options

This chapter describes techniques for pricing caps, bond options, and swaps. We discuss the general principles first and derive pricing formulas in the context of exponential-affine models. Closed-form solutions are presented in some simple cases.

Throughout the chapter, we make systematic use of the concept of **forward measures**. This technique simplifies considerably the calculations of prices of European-style derivative securities, especially when the value of the underlying asset is correlated with funding costs. In a sense, the use of forward measures allows us to "separate" (computationally speaking) the effect of discounting cash-flows with the computation of the option's premium due to volatility. In Section 15.1 we give a general introduction of this idea along with some examples.

The second part of this chapter concerns the pricing of caps. We present explicit pricing formulas and discuss the relation between "cap volatility"—the Black–Scholes implied volatility of caps—and the volatility parameters of Gaussian models. This relation is crucial for calibrating the models to cap prices.

The third section deals with bond options and swaptions. In particular, we describe a result of F. Jamshidian for one-factor models whereby the pricing of bond options is reduced to pricing a series of options on zero-coupon bonds.

The last section gives a brief introduction to the Brace–Gatarek–Musiela model. This approach is particularly well suited for pricing interest-rate options and is easy to calibrate to the prices of caps.

15.1 Forward Measures

Definition and Examples

Consider a security that pays a series of dividends $D_1, D_2, \ldots D_N, \ldots$ on the dates $T_1 < T_2 < \ldots T_N < T$ and denote by V_t its price process.[1] Let π denote a risk-neutral measure. The quantity

$$\widetilde{V}_t = V_t - \sum_{i=1}^{N} \mathbf{E}_t^{\pi} \left\{ e^{-\int_t^{T_i} r_s \, ds} D_i \right\}$$

---

[1] The term "dividends" includes any coupon payments for a bond, for example, as well as more general cash-flows. Here, T is a given date.

represents the value of this security minus the present value of the future cash-flows up to time T. By the standard cash-and-carry replication argument,[2] the forward price of the security for delivery at date T is given by

$$(f_T V)_t = \frac{\tilde{V}_t}{P_t^T}.$$ (15.1)

We want to study the evolution of $(f_T V)_t$ under a general model with ν factors. Notice that the numerator of (15.1) can be interpreted as the price of a synthetic security that pays no dividends and is worth V_T dollars at date T (the original security "stripped" of the N dividends). Hence, it follows that

$$d\tilde{V}_t = \tilde{V}_t \left(\sigma^{\tilde{V}} \cdot dZ_t + r_t \, dt \right),$$ (15.2)

where r_t is the short-term interest rate, $\sigma^{\tilde{V}} = \left(\sigma^{\tilde{V}_1}, \sigma^{\tilde{V}_2}, \ldots \sigma^{\tilde{V}_\nu} \right)$ is the vector of volatility "components" for the price of the security and $Z_t = (Z_t^1, \ldots Z_t^\nu)$ represents a ν-dimensional Brownian motion.

Recall that under the risk-neutral measure we have

$$dP_t^T = P_t^T \left(\sigma(t; T) \cdot dZ_t + r_t \, dt \right),$$ (15.3)

since P_t^T is the price of a zero that pays no dividends.[3] Let us compute the Ito differential of the forward price, $\tilde{V}_t / P_t^T$. From Ito's Lemma, we have

$$d\left(\frac{\tilde{V}_t}{P_t^T} \right) = \frac{d\tilde{V}_t}{P_t} - \frac{\tilde{V}_t}{(P_t^T)^2} dP_t^T - \frac{d\tilde{V}_t \, dP_t^T}{(P_t^T)^2} + \frac{\tilde{V}_t \, (dP_t^T)^2}{(P_t^T)^3}$$

$$= \frac{\tilde{V}_t}{P_t} \left(\sigma^{\tilde{V}} \cdot dZ_t + r_t \, dt \right) - \frac{\tilde{V}_t}{P_t} \left(\sigma(t; T) \cdot dZ_t + r_t \, dt \right)$$

$$- \frac{\tilde{V}_t}{P_t} \sigma^{\tilde{V}} \cdot \sigma(t; T) \, dt + \frac{\tilde{V}_t}{P_t} \| \sigma(t; T) \|^2 \, dt$$

$$= \frac{\tilde{V}_t}{P_t} \left(\sigma^{\tilde{V}} - \sigma(t; T) \right) \cdot dZ_t$$

$$- \frac{\tilde{V}_t}{P_t} \left(\sigma^{\tilde{V}} \cdot \sigma(t; T) - \| \sigma(t; T) \|^2 \right) dt.$$ (15.4)

Hence, the forward price satisfies the equation

$$d(f_T V)_t = (f_T V)_t \left[\left(\sigma^{\tilde{V}} - \sigma(t; T) \right) \cdot dZ_t - \left(\sigma^{\tilde{V}} \cdot \sigma(t; T) - \| \sigma(t; T) \|^2 \right) dt \right]$$

---

[2] By borrowing $\tilde{V}_t$ dollars and issuing "notes" corresponding to the intermediate cash-flows, a trader can purchase the security initially. The cash-flows are then used to pay the notes. Delivery of the security at the forward price $\tilde{V}_t / P_0^T$ allows the trader to pay back the loan with accrued interest.

[3] We use the same notation as in Chapters 13 and 14. We denote the norm of a ν-dimensional vector x by $\|x\| = \sqrt{x_1^2 + \ldots x_\nu^2}$ and the inner product of two vectors by $x.y = \sum_{k=1}^\nu x_k y_k$.

$$= (f_T V)_t \left(\sigma^{\widetilde{V}} - \sigma(t; T) \right) \cdot (dZ_t - \sigma(t; T) dt) . \tag{15.5}$$

Consequently, if we define a new process W_t by

$$W_t = Z_t - \int_0^t \sigma(t; T) ds \tag{15.6}$$

—or, in differential form,

$$dW_t = dZ_t - \sigma(t; T) dt \tag{15.6a}$$

—we obtain the stochastic differential equation for the forward price

$$d(f_T V)_t = (f_T V)_t \left(\sigma^{\widetilde{V}} - \sigma(t; T) \right) \cdot dW_t . \tag{15.7}$$

Recall now that, by Girsanov's theorem, the process W_t is a Brownian motion under a new probability measure π^{*T} equivalent to π.[4] Therefore, the forward price is a martingale under π^{*T}.

We have obtained the following important result:

PROPOSITION 15.1
Assume that π is a risk-neutral pricing measure and let T be a fixed date. Then, there exists a probability measure π^{*T}, that is equivalent to π, such that the forward price for delivery at time T of any traded security is a martingale *under π^{*T}.*

The probability π^{*T} is called the **forward measure for delivery at date T** (associated with π). The main application of forward measures is to simplify the computation of European-style option prices. Consider, for example, a call option with strike K and expiration date T on a security with price V. We know that the fair value of this option at time $t = 0$ is equal to the expectation of the discounted cash-flows under π:

$$C_0 = \mathbf{E}^\pi \left\{ e^{-\int_0^T r_s \, ds} \max(V_T - K, 0) \right\} . \tag{15.8}$$

Notice that this expression involves the random variable V_T as well as the stochastic discounting term $\exp(-\int_0^T r_s \, ds)$. In principle, the computation of the expected value in the last equation requires the knowledge of the *joint distribution* of these two random variables. Forward measures allow us to take a "shortcut:" all we need to know is the distribution of $(f_T V)_T$ under the forward measure and the term interest rate from time 0 to time T.

---

[4]This probability is defined on events occurring up to time T. It is characterized by its Radon–Nykodym derivative

$$\frac{d\pi^{*T}}{d\pi} = \exp\left(\int_0^T \sigma(t; T) \cdot dZ_t - \frac{1}{2} \int_0^T \|\sigma(t; T)\|^2 \, dt \right) .$$

This follows from Girsanov's theorem (see Appendix to Chapter 9). We will not make explicit use of this formula in this chapter.

In fact, observe that by Proposition 15.1 $(f_T C)_t$ is a martingale *under* π^{*T}. In particular, we have

$$(f_T C)_0 = \mathbf{E}^{\pi^{*T}} \{ (f_T C)_T \}$$

$$= \mathbf{E}^{\pi^{*T}} \{ \max(V_T - K, 0) \} \, ,$$

or, in dollars at time $t = 0$,

$$C_0 \;=\; P_0^T \, \mathbf{E}^{\pi^{*T}} \{ \max(V_T - K, 0) \} \, . \tag{15.9}$$

Thus, if we know the statistics of V_T under π^{*T} and the discount factor P_0^T, we can compute the value of the option without having to consider explicitly the term $\exp(-\int_0^T r_s \, ds)$.

The advantage of working with forward prices instead of spot prices is that we can "separate" the effect of volatility from the effects of discounting cash-flows. The funding costs (associated with discounting by the short-term rate) are implicitly taken into account in the forward prices. The procedure is quite general, and can be applied to different situations.

15.2   Commodity Options with Stochastic Interest Rate

Consider a commodity in an economy with stochastic interest rates. We are interested in pricing an option on this asset with maturity T, assuming a one-factor Ho–Lee interest rate model. We assume that, under π, the joint dynamics of the commodity price and of a zero-coupon bond with maturity T are given by stochastic differential equations

$$\begin{cases} \dfrac{dS_t}{S_t} = \sigma^S \, dZ + (r_t - d_t) \\[2ex] \dfrac{dP_t^T}{P_t^T} = \sigma^r \, (T - t) \, dZ' + r_t \, dt \, . \end{cases} \tag{15.10}$$

Here d_t is the dividend process, σ^S and σ^r are constant, and Z and Z' are Brownian motions such that $\mathbf{E}^\pi \{ dZ \, dZ' \} = \rho \, dt$. We can write

$$Z'(t) \;=\; \rho Z(t) + (1 - \rho^2)^{1/2} Z''(t) \, ,$$

where $Z''(t)$ is a Brownian motion that is independent of $Z(t)$. Notice that the equation for the evolution of P_t^T can be rewritten in the form

$$\frac{dP_t^T}{P_t^T} \;=\; \sigma^r \, (T - t) \, \rho \, dZ + \sigma^r \, (T - t) \, (1 - \rho^2)^{1/2} \, dZ'' + r_t \, dt \, . \tag{15.11}$$

From Eq. (15.7), the volatility of $(f_T S)_t$ is equal to the difference between the volatility vectors for dS/S and dP^T/P^T. Hence, the stochastic differential equation for the forward price of the commodity under π^{*T} can be written as

$$\frac{d(f_T S)_t}{(f_T S)_t} = \left(\sigma^S - \rho\sigma^r (T-t)\right) dW_1 - (1-p^2)^{1/2}\sigma^r(T-t)dW_2 \qquad (15.12)$$

where W_1 and W_2 are independent one-dimensional Brownian motions.

This implies that the instantaneous volatility of $f_T S_t$, $\sigma_f(t)$, satisfies

$$\left(\sigma_f(t)\right)^2 = \left(\sigma^S - \rho\sigma^r(T-t)\right)^2 + (1-p^2)(\sigma^r)^2(T-t)$$

$$= \left(\sigma^S\right)^2 - 2\rho\sigma^S\sigma^r(T-t) + (\sigma^r)^2(T-t)^2 .$$

The *term volatility* of $(f_T S)_T$ (from 0 to T), $\overline{\sigma}_T$, is obtained by averaging the instantaneous variance $(\sigma_f(t))^2$ on the interval from 0 to T. The result is

$$\overline{\sigma}_T = \sqrt{\frac{1}{T} \int\limits_0^T (\sigma_f(t))^2 dt}$$

$$= \sqrt{(\sigma^S)^2 - \rho\sigma^r\sigma^S T + \frac{(\sigma^r)^2 T^2}{2}} . \qquad (15.13)$$

Using the forward measure, we deduce that the value of the call option is given explicitly by

$$C_0 = P_0^T \mathbf{E}^{\pi^{*T}} \{\max((f_T S)_T - K, 0)\} .$$

From Eq. (15.11), $(f_T S)_t$ is a log-normal martingale with term-volatility $\overline{\sigma}_T$. We can therefore apply the Black–Scholes formula to obtain

$$C_0 = P_0^T \left((f_T S)_0 N(d_1) - K N(d_2) \right)$$

with

$$d_1 = \frac{1}{\overline{\sigma}_T \sqrt{T}} \log\left(\frac{(f_T S)_0}{K}\right) + \frac{1}{2}\overline{\sigma}_T \sqrt{T} , \quad d_2 = d_1 - \overline{\sigma}_T \sqrt{T}$$

and $\overline{\sigma}_T$ as in (15.13). This gives a concise expression for the option value in terms of the strike, forward price, discount factor, and term volatility. The way in which this formula was derived is much more efficient than a computation of the expectation of discount cash-flows using π.

15.3 Options on Zero-Coupon Bonds

Consider a zero-coupon bond with face value $1 maturing at date $T + \Delta T$, with $\Delta T > 0$. Let us compute the value of a call on this bond with expiration date T and strike K. According to Eq. (15.7), the forward price satisfies

$$\frac{d(f_T \, P^{T+\Delta T})_t}{(f_T \, P^{T+\Delta T})_t} = (\sigma(t; \, T + \Delta T) - \sigma(t; \, T)) \cdot dW_t \tag{15.14}$$

where W is a Brownian motion under π^{*T}. The solution of this SDE is, formally,

$$(f_T \, P^{T+\Delta T})_t = \left(\frac{P_0^{T+\Delta T}}{P_0^T}\right) M_t \tag{15.15}$$

where $\frac{P_0^{T+\Delta T}}{P_0^T}$ is the forward price at time 0 and M_t is the exponential martingale

$$\exp\left\{\int_0^t (\sigma(s; \, T + \Delta T) - \sigma(s; \, T)) \cdot dW_s \right.$$

$$\left. - \frac{1}{2} \int_0^t \|\sigma(s; \, T + \Delta T) - \sigma(s; \, T)\|^2 \, ds \right\}. \tag{15.15a}$$

If we assume, for example, that $\sigma(t; \, T)$ is a deterministic vector (which corresponds to the case of Gaussian rates), then M_t is log-normal. To compute the value of the call, which is given by the expectation value

$$C_0 = P_0^T \, \mathbf{E}^{\pi^{*T}} \left\{ \max((f_T \, P^{T+\Delta T})_T - K, 0) \right\},$$

we can use the Black–Scholes formula

$$C_0 = P_0^{T+\Delta T} \, N(d_1) - K \, P_0^T \, N(d_2)$$

where

$$d_1 = \frac{1}{s(T; \, T + \Delta T)\sqrt{T}} \log\left(\frac{P_0^{T+\Delta T}}{K \, P_0^T}\right) + \frac{1}{2} s(T; \, T + \Delta T)\sqrt{T}$$

$$d_2 = d_1 - s(T; \, T + \Delta T)\sqrt{T}$$

and

$$s(T; \, T + \Delta T) = \left(\frac{1}{T} \int_0^T \|\sigma(s; \, T + \Delta T) - \sigma(s; \, T)\|^2 \, ds \right)^{1/2}$$

is the term volatility of M_t. We can evaluate explicitly this formula in the context of the one-factor Vasicek model, where

$$\sigma(t; \, T) = \sigma \frac{1 - e^{\kappa(T-t)}}{\kappa}.$$

Here σ is the volatility of the short rate and κ is the rate of mean-reversion (Figure 15.1). Using this formula, we obtain the implied volatility of a T-year option on a zero-coupon bond with maturity $T + \Delta T$:

$$s(T; \, T + \Delta T) \; = \; \sigma \left(\frac{1 - e^{-\kappa \, \Delta T}}{\kappa} \right) \sqrt{\frac{1 - e^{-2\kappa T}}{2 \kappa T}} \, .$$

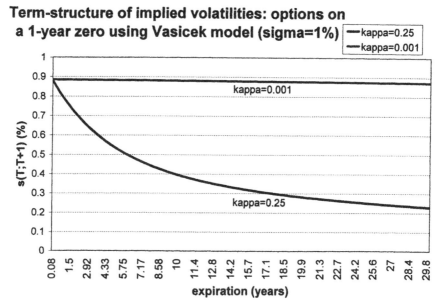

FIGURE 15.1
Implied volatilities for options on 1-year zero-coupon bond ($\Delta T = 1$) for different expiration dates T. We use a modified Vasicek model with $\sigma = 1\%$ and compare different values of κ.

15.4 Money-Market Deposits with Yield Protection

Banks often offer clients certificates of deposit indexed to LIBOR (for example LIBOR - 100 bps) and also guarantee a minimum rate of return (for example 3%) regardless of interest-rate fluctuations. This investment vehicle has an embedded option, since the holder is guaranteed a minimum return even if market rates drop below the "floor" rate. For simplicity, we assume that the CD has a fixed maturity of T years and that the short rate follows a modified Vasicek model.

We denote the guaranteed minimum rate by $\underline{r}$, and assume that interest accrues in the account at the rate $d_t = r_t - \delta$, where δ is positive number. The evolution of the value of an account initially worth \$1 is given by

$$\frac{dB_t}{B_t} \; = \; (r_t - \delta) \, dt \quad B_0 = 1 \, .$$

By cash-and-carry replication, the "forward price" of the account at time $t = 0$ is $(f_T B)_0 = e^{-\delta T}/P_0^T$.[5]

Since the instantaneous volatility of B_t is zero, we deduce from (15.7) that the SDE for $(f_T B)_t$ under π^{*T} is

$$\frac{d(f_T B)_t}{(f_T B)_t} = -\sigma(t; T)\,dW_t \,.$$

Therefore,

$$(f_T B)_t = \frac{e^{-\delta T}}{P_0^T}\,\exp\left(-\int_0^t \sigma(s; T)\cdot dW_s - \frac{1}{2}\int_0^t \|\sigma(s; T)\|^2\,ds\right).$$

We now use the fact that the payoff for an investor that deposits \$1 in his account is

$$\max\left(B_T, e^{r T}\right) = e^{r T} + \max\left(B_T - e^{r T}, 0\right).$$

Let A denote the present value of this cash flow. Using the forward measure π^{*T}, we can write

$$A = P_0^T\left(e^{r T} + \mathbf{E}^{\pi^{*T}}\max\left((f_T B)_T - e^{r T}, 0\right)\right).$$

We obtain a more explicit answer by introducing the term rate $R(0, T) = -\ln(P_0^T)/T$, and applying the Black–Scholes formula. In fact, the latter equation can be written in the form

$$A = P_0^T\, e^{r T} + e^{-\delta T} N(d_1) - P_0^T\, N(d_2)$$

$$= e^{-(R(0,T)-r)\,T} + e^{-\delta T} N(d_1) - e^{-R(0,T)\,T}\, N(d_2)$$

with

$$d_1 = \frac{R(0, T) - \delta - r}{s\sqrt{T}} + \frac{s\sqrt{T}}{2}, \quad d_2 = d_1 - s\sqrt{T},$$

and

$$s = s(0; T) = \sigma\,\frac{1 - e^{-\kappa T}}{\kappa}.$$

Finally, to obtain the "fair value" of δ for a given floor rate r (Figure 15.2), we set $A = 1$ and solve for δ in terms of r in the equation

$$e^{-(R(0,T)-r)\,T} + e^{-\delta T} N(d_1) - e^{-R(0,T)\,T}\, N(d_2) = 1. \qquad (15.16)$$

See Figure 15.2 for numerical values of r and δ.

Forward Rates and Forward Measures

The following propositions characterize the dynamics of forward rates under forward measures.

---

[5] In fact, $e^{-\delta T}$ dollars invested at date 0 at the short-term market rate r_t will replicate the final value, B_T.

Insured CD: 2-year maturity

2-year rate = 6%   Sigma = 1%   Kappa = 10%

r	delta
1.00%	0.78%
1.25%	1.00%
1.50%	1.14%
1.75%	1.30%
2.00%	1.44%
2.25%	1.58%
2.50%	1.70%
2.75%	1.80%
3.00%	1.90%
3.25%	2.04%
3.50%	2.14%
3.75%	2.23%
4.00%	2.33%
4.25%	2.43%
4.50%	2.59%
4.75%	2.73%
5.00%	2.91%

FIGURE 15.2
Fair value of δ for a 2-year insured certificate of deposit, according to Eq. (15.16). The CD guarantees a "floor" yield $\underline{r}$ and accrues at the continuous rate $r_t - \delta$. The interest rate follows a modified Vasicek process with $\sigma = 0.01$ and $\kappa = 0.1$.

PROPOSITION 15.2

Let π represent a risk-neutral pricing measure. Under the forward probability measure $\pi^{*,T}$, the instantaneous forward rate $f(t, T)$ satisfies the stochastic differential equation

$$df(t, T) = \dot{\sigma}(t, T) \cdot dW_t , \qquad (15.17)$$

where $\dot{\sigma}(t, T) = (\dot{\sigma}_1(t, T), \ldots \dot{\sigma}_v(t, T))$ are the components of the standard deviation of $f(t; T)$. In particular, $f(t; T)$ is a martingale under π^{*T}.

PROOF   Recall that $f(t, T) = -\frac{\partial \ln(P_t^T)}{\partial T}$. It follows from (15.3) that[6]

$$\dot{\sigma}(t, T) = -\frac{\partial (\sigma(t; T))}{\partial T} ,$$

is the standard deviation of $df(t; T)$. Equivalently, we have

$$\sigma(t; T) = -\int_t^T \dot{\sigma}(t, s) \, ds .$$

---

[6]See the proof of the HJM theorem in Chapter 13 if necessary.

Consider now the HJM equation (i.e., the dynamics of forward rates in the risk-neutral measure π)

$$df(t, T) = \dot{\sigma}(t, T) \cdot dZ_t + \left(\dot{\sigma}(t; T) \cdot \int_t^T \dot{\sigma}(t, s)\, ds \right) dt$$

$$= \dot{\sigma}(t, T) \cdot dZ_t - \dot{\sigma}(t; T) \cdot \sigma(t; T)\, dt \;.$$

Making the substitution (15.6a) in this last equation, we obtain

$$df(t, T) = \dot{\sigma}(t, T) \cdot (dW_t + \sigma(t; T)\, dt) - \dot{\sigma}(t; T) \cdot \sigma(t; T)\, dt = \dot{\sigma}(t, T) \cdot dW_t \;,$$

where W is a Brownian motion with respect to π^{*T}. This is what we wanted to show. ∎

For the sake of completeness, we present in the next proposition the dynamics for *any* instantaneous forward rate $f(t, T')$ under π^{*T}.

PROPOSITION 15.3
Under π^{*T}, the forward rate $f(t, T')$ satisfies the stochastic differential equation

$$df(t, T') = \dot{\sigma}(t, T') \cdot dW_t + \left(\dot{\sigma}(t; T') \cdot \int_T^{T'} \dot{\sigma}(t, s)\, ds \right) dt$$

$$= \dot{\sigma}(t, T') \cdot dZ_t + \dot{\sigma}(t; T') \cdot \big(\sigma(t; T) - \sigma(t; T') \big)\, dt \;. \quad (15.18)$$

The proof is similar to that of Proposition 15.2.

REMARK 15.1   In general, $f(t, T')$ is not a martingale under π^{*T} unless $T = T'$. It is important to remember that forward measures are *maturity-specific*: the martingale property for forward rates under π^{*T} holds only for forwards that deliver cash-flows at date T. ∎

The next proposition concerns forward rates with *finite tails*, i.e., **forward term rates**.

PROPOSITION 15.4
The forward rate process
$$F(t; T, T + \Delta T) \;, \quad t < T,$$
associated with a FRA for the time period $(T, T + \Delta T)$, is a martingale under $\pi^{*T+\Delta T}$.

PROOF   The forward rate can be written in the form

$$F(t; T, T + \Delta T) = \frac{1}{\Delta T} \left(\frac{P_t^T}{P_t^{T+\Delta T}} - 1 \right). \quad (15.19)$$

Notice that the fraction in the right-hand side represents the "forward price for delivery at date $T + \Delta T$ of the bond with maturity T".[7] Applying Eq. (15.7) to this fraction, we find that

$$dF(t;\, T, T + \Delta T) \;=\; \frac{1}{\Delta T}\, \frac{P_t^T}{P_t^{T+\Delta T}}\, (\sigma(t;\, T) - \sigma(t;\, T + \Delta T))\, dW_t\,, \qquad (15.20)$$

which shows that $F(t;\, T, T + \Delta T)$ is a martingale under $\pi^{*\,T}$. We notice also that this last equation can be rewritten, using Eq. (15.20), in the form

$$\frac{dF(t;\, T, T + \Delta T)}{F(t;\, T, T + \Delta T)} =$$

$$\left(\frac{1 + \Delta T\, F(t;\, T, T + \Delta T)}{\Delta T\, F(t;\, T, T + \Delta T)}\right)\, (\sigma(t;\, T) - \sigma(t;\, T + \Delta T))\, dW_t\,. \qquad (15.21)$$

REMARK 15.2   Equation (15.21) is the basis of the "market models" of Brace, Gatarek, and Musiela [1] and Jamshidian [5]. In these models, the forward rates $F(t;\, T, T + \Delta T)$ are the variables of interest (cf. Section 15.4).   ∎

15.5   Pricing Caps

General Considerations

A cap is equivalent to a series of calls on the floating rate ("caplets") with different maturities and a common strike. Usually, cash-flows are exchanged **in arrears:** the floating rate is set a date T for the period $(T,\, T + \Delta T)$ and the corresponding cash-flows are exchanged at date $T + \Delta T$.[8]

We consider in this section "idealized" caps with the following terms: the reset dates are $T_0 = 0, T_1, \ldots T_{N-1}$ and the cash-flow dates are $T_1, T_2, \ldots T_N$. We set $\Delta T_n = T_{n+1} - T_n$ for $n = 0, \ldots N - 1$. We assume that the floating rate is the term rate $R(T_n, T_{n+1})$ and denote the strike by $\overline{R}$. The notional amount is set at \$1.

The payoff corresponding to date T_{n+1}, $n \geq 1$, is

$$c_n \;=\; \Delta T_n\, \max\left(R(T_n,\, T_{n+1}) - \overline{R}\right)\,, \qquad (15.22)$$

(cf. Eq. (12.12) Chapter 12). As a consequence, we have

$$Cap \;=\; \sum_{n=0}^{N-1} P^{T_{n+1}}\, \Delta T_n\, \mathbf{E}^{\pi^{*\,T_{n+1}}}\left\{\max(F(T_n;\, T_n,\, T_{n+1}) - \overline{R},\, 0)\right\}\,, \qquad (15.23)$$

---

[7] This may appear a bit awkward at first glance, since $T + \Delta T > T$, but it is nevertheless correct. The key observation is that the fraction on the right-hand side of (15.19) has denominator $P_t^{T+\Delta T}$ and that the numerator is the price of a security at time t.

[8] This mimics, to some extent, the fact that the cap insures the holder against a rise in interest rates. The contract has the effect of capping the floating interest rate payments of an issuer of floating rate debt.

where $\pi^{*\,T_{n+1}}$ is the forward measure for delivery at date T_{n+1}. By Proposition 15.4, $F(t;\,T_n,\,T_{n+1})$ is a martingale under this probability measure. Therefore, we can write

$$Cap = \sum_{n=0}^{N-1} P^{T_{n+1}} \Delta T_n \, \mathbf{E}^{\pi^{*\,T_{n+1}}} \left\{ \max(F(0;\,T_n,\,T_{n+1})\,\xi_{T_n}^{(n)} - \overline{R},\,0) \right\}, \qquad (15.24)$$

where $\{\xi_t^{(n)}\; t \leq T_{n+1}\}$ is a martingale under $\pi^{*\,T_{n+1}}$ satisfying $\xi_0^{(n)} = 1$.[9]

An alternative formula can be derived if we view caplets as *puts on zero-coupon bonds*. Accordingly, we observe that the present value at date T_n of the payoff (15.22) can be written as

$$c_n = P_{T_n}^{T_{n+1}} \Delta T_n \, \max\left(R(T_n,\,T_{n+1}) - \overline{R},0\right)$$

$$= P_{T_n}^{T_{n+1}} \Delta T_n \, \max\left(\frac{1}{\Delta T_n}\left(\frac{1}{P_{T_n}^{T_{n+1}}} - 1\right) - \overline{R},0\right)$$

$$= \max\left(1 - (1 + \Delta T_n \overline{R})\,P_{T_n}^{T_{n+1}},0\right).$$

Therefore, the value of the cap can be written in the form

$$Cap = \sum_{n=0}^{N-1} P^{T_n} \mathbf{E}^{\pi^{*\,T_n}} \left\{ \max\left(1 - (1 + \Delta T_n \overline{R})\,(f_{T_n}P^{T_{n+1}})_{T_n},\,0\right) \right\}$$

$$= \sum_{n=0}^{N-1} P^{T_n} \mathbf{E}^{\pi^{*\,T_n}} \left\{ \max\left(1 - \left(\frac{1 + \Delta T_n \overline{R}}{1 + \Delta T_n F_n}\right) M_{T_n}^{(n)},\,0\right) \right\}. \qquad (15.25)$$

Here $F_n = F(0;\,T_n,\,T_{n+1})$ is the forward rate for the period $(T_n,\,T_{n+1})$. $M_t^{(n)}$ is a martingale under $\pi^{*\,T_n}$ such that

$$(f_{T_n}P^{T_{n+1}})_t = \frac{P_0^{T_{n+1}}}{P_0^{T_n}} M_t^{(n)} = \frac{1}{1 + \Delta T_n F_n} M_t^{(n)}. \qquad (15.26)$$

A glance at the cap pricing formulas (15.24) and (15.25) shows that the random variables used to express the payoffs are different in each case. In Eq. (15.24), the underlying indices are the *forward rates*. In Eq. (15.25), the *discount factors* are used. In the following sections, analytic solutions for the prices of caps are obtained by assuming specific distributions for these random variables.

REMARK 15.3   At-the-money-forward caps. Typically, "benchmark" caps quoted in the broker market have strikes equal to the swap rate with the same tenor. These are the **at-the-**

---

[9]The reader should notice that the first term in the sums (15.23) and (15.24) are equal to the intrinsic values of the caplets. This is because the payments are made in arrears: therefore the value of the payoff $\max(R(T_0;\,T_1) - \overline{R},\,0)$ is known at time zero. For at-the-money swaps, this first term is zero most of the time, as we will see later.

money-forward caps because caps and floors have equal value at this strike level. We can verify that this is a direct consequence of the cap/floor/swap parity (cf. Chapter 13).[10] The relation between the swap rate and the forward rates $F_n = F(0; T_n, T_{n+1})$ determines by how much each caplet is in or out of the money. Let us denote by S_n the swap rate for a swap with tenor T_n. Then, we have

$$0 = \sum_{m=0}^{n-1} \mathbf{E}^\pi \Delta T_m \left\{ (S_n - R(T_m, T_{m+1})) e^{-\int_0^{T_{m+1}} r_s \, ds} \right\}$$

$$= \sum_{m=0}^{n-1} \Delta T_m \, \mathbf{E}^\pi \left\{ (F_m - R(T_m, T_{m+1})) e^{-\int_0^{T_{m+1}} r_s \, ds} \right\}$$

$$+ \sum_{m=0}^{n-1} \Delta T_m \, (S_n - F_m) \, \mathbf{E}^\pi \left\{ e^{-\int_0^{T_{m+1}} r_s \, ds} \right\}$$

$$= \sum_{m=0}^{n-1} \Delta T_m \, (S_n - F_m) \, P_0^{T_{m+1}}.$$

Hence,

$$S_n = \frac{\sum\limits_{m=0}^{n-1} \Delta T_m \, P_0^{T_{m+1}} F_m}{\sum\limits_{m=0}^{n-1} \Delta T_m \, P_0^{T_{m+1}}}.$$

This formula shows that the swap rate is a weighted average of forward rates. Hence, if the forward rate curve is upward-sloping (the "normal" situation in U.S. LIBOR, for example), swap rates are lower than forward rates. This implies that short-maturity caplets are out-of-the money and long-maturity caplets are in-the-money.[11] Figure 15.3 gives an example drawn from market data in which the forward-rate curve is upward-sloping. The pricing of caps is very sensitive to the relative positions of the two curves. ∎

---

[10] In practice, the cap/floor/swap parity relation does not hold exactly, for the following reason: "plain vanilla" caps and floors reset *quarterly* and are indexed to the 3-month LIBOR rate. Vanilla swaps, on the other hand, reset *semiannually,* but are also indexed to 3-month LIBOR because the latter rate is much more widely quoted than 6-month LIBOR and is the "market benchmark." (For instance, Eurodollar futures settle on 3-month LIBOR.) The rate for which the parity holds is the theoretical swap rate for a *quarterly* resetting swap. We will neglect this subtle difference.

[11] Of course, this remark also applies to the cash-flows of a swap: the floating-payer has cheap financing at the beginning of the swap and expensive financing later. Similarly, the "floorlets" of a floor are in-the-money initially and out-of-the-money for longer maturities.

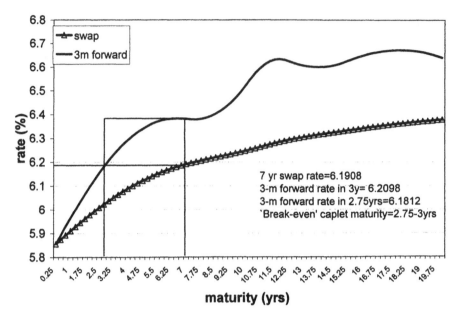

FIGURE 15.3
Curves corresponding to swap rates and 3-month forward rates in late November 1997
for the U.S. LIBOR market. We analyze the hypothetical case of a 7-year cap with strike
$\overline{R} = 6.19$. Caplets with maturities of 2.75 years or less are out-of-the money, and caplets
with maturities of 3 years or more are in-the-money.

Cap Pricing with Gaussian Models

We derive closed-form solutions for cap prices assuming a Gaussian term-structure model.
From Example 15.2 in Section 15.1 and Eq. (15.25), we obtain

$$Cap = \sum_{n=1}^{N-1} P_0^{T_n} BSP\left(\left(\frac{1 + \Delta T_n \overline{R}}{1 + \Delta T_n F_n}\right), \ 1, \ T_n, \ s_n\right), \tag{15.27}$$

with

$$s_n = \left(\frac{1}{T_n} \int^{T_n} \|\sigma(t; T_{n+1}) - \sigma(t; T_n)\|^2 \, dt\right)^{1/2}, \quad n = 1, 2, \ldots N-1, \tag{15.28}$$

where $BSP(S, K, T, \sigma)$ is the Black–Scholes value of a put with forward price S, strike K,
expiration date T, and volatility σ.

The simplicity of this formula and its derivation is remarkable. Notice, for example, that
we did not have to deal with the "fudge factor" $\overline{b}(t)$ associated with the short rate or with the
statistics of the short rate. The use of forward measures is clearly more "natural" than using
the risk-neutral measure, which involves explicit discounting.

To illustrate the use of formula (15.27) under different conditions we priced a series of
at-the-money forward caps with Gaussian one-factor models (Figure 15.4). In this case, we
have

$$s_n = \sigma \left(\frac{1 - e^{-\kappa \, \Delta T_n}}{\kappa}\right) \sqrt{\frac{1 - e^{-2\kappa \, T_n}}{2\kappa T_n}}.$$

Figures 15.4 and 15.5 show the effect of pricing caps with different rates of mean-reversion.

CAP PRICES	(in basis pts.)	sigma=0.003	Flat fwd rates=5%
maturity (yrs.)	**Ho-Lee**	**k=0.05**	**k=0.5**
1	6.03	5.91	4.99
2	19.05	18.42	14
3	35.4	33.74	23.5
4	53.97	50.74	32.84
5	74.09	68.74	41.83
6	95.31	87.29	50.43
7	117.31	106.1	58.62
8	139.82	124.92	66.41
9	162.65	143.6	73.83
10	185.63	162.01	80.89

FIGURE 15.4
Pricing caps with Gaussian term-structure models. We assume a flat 3-month forward curve and that caps are at-the-money forward. The models used were (i) Ho–Lee with $\sigma = 0.003$; (ii) modified Vasicek with same value of σ and $\kappa = 0.05$; (iii) with $\kappa = 0.5$. Cap prices decrease as we increase the mean-reversion rate κ. This is because the volatilities s_n decrease as κ increases; cf. Figure 15.1.

REMARK 15.4 **Forward rate correlations are irrelevant.** Notice that the volatilities in (15.28) depend only on the *magnitude* of the vector $\sigma(t; T_{n+1}) - \sigma(t; T_n)$ and not on the values of its individual components. Correlations between successive forward rates do not affect the value of a cap. This observation applies to any term-structure model. ∎

Cap Pricing with Square-Root Models[12]

The pricing of caps with square-root models also uses the fact that zero-coupon bonds are exponential functions of the state variables. However, the characterization of the probability distribution of the state variables Y_{T_n} under the forward measures π^{*T_n} is more delicate than in the Gaussian case.

For simplicity, we consider the case of a one-factor model with a single state variable Y_t satisfying the stochastic differential equation

$$dY_t = \sigma \sqrt{Y_t} \, dZ_t + \kappa \, (\theta - Y_t) \, dt . \tag{15.29}$$

---

[12]This section is more technical than the rest of the chapter and can be skipped in a first reading.

CAP PRICES in bps		sigma=0.003	initial fwd=5% dF=1bp	
maturity (yrs)	Ho-Lee	k=0.05	k=0.5	
1	5.86	5.74	4.82	
2	18.33	17.7	13.3	
3	34.99	33.33	23.15	
4	53.56	50.36	32.63	
5	73.79	68.5	41.96	
6	95.22	87.29	51.11	
7	117.51	106.45	60.1	
8	140.41	125.73	68.96	
9	163.85	144.96	77.71	
10	187.14	164.02	86.38	

FIGURE 15.5
Same as in Figure 15.4 but assuming a forward rate curve of the form $F_n = 0.05 + n\,0.0001$.

From the results of Chapter 14 [cf. Eq. (14.45)], we know that the discount factor $P_t^{T_n}$ can be written in the form

$$P_t^{T_{n+1}} = \frac{P_0^{T_{n+1}}}{P_0^t} \exp\left(-p(t;\, T_{n+1})\, Y_t - q_0(t;\, T_{n+1})\right) \tag{15.30}$$

where $p(t;\, T)$ and $q_0(t;\, T)$ are functions derived explicitly in Chapter 14. This implies, in particular, that

$$(f_{T_n} P^{T_{n+1}})_t = \frac{P_t^{T_{n+1}}}{P_t^{T_n}}$$

$$= \frac{P_0^{T_{n+1}}}{P_0^{T_n}} \exp\left(-p_n(t)\, Y_{T_n} - q_n(t)\right)$$

$$= \frac{1}{1 + \Delta T_n F_n} \exp\left(-p_n(t)\, Y_{T_n} - q_n(t)\right), \tag{15.31}$$

where F_n is the forward rate and

$$
\begin{cases}
p_n(t) = p(t; T_{n+1}) - p(t; T_n) \\
q_n(t) = q_0(t; T_{n+1}) - q_0(t; T_n) .
\end{cases}
$$

For the purpose of pricing successive caplets, we need to determine the distributions of Y_{T_n} under the forward measure π^{*T_n} for all n. To this end, we establish the following auxiliary result:

PROPOSITION 15.5

Assume that, under a risk-neutral pricing measure π, the process Y_t satisfies Eq. (15.29) and that the discount factors satisfy Eq. (15.30). Then, Y_t satisfies

$$
dY_t = \sigma \sqrt{Y_t}\, dW_t + \kappa\, (\theta - Y_t)\, dt - \sigma^2 Y_t\, p(t, T_n)\, dt , \tag{15.32}
$$

where W_t is a Brownian motion under π^{*T_n}.

PROOF   From Eq. (15.30), we have

$$
\frac{dP_t^T}{P_t^T} = - p(t; T)\sigma \sqrt{Y_t}\, dZ_t + r_t\, dt ,
$$

where r_t is the short rate. In particular, the volatility of the zero-coupon bond is

$$
\sigma(t; T) = - p(t; T)\,\sigma \sqrt{Y_t} . \tag{15.33}
$$

We now make use of the basic change of variables in Eq. (15.6), which states that if Z_t is a Brownian motion under π then

$$
W_t = Z_t - \int_0^t \sigma(t; T_n)\, dt
$$

is a Brownian motion under π^{*T_n}. Substituting dZ_t by $dW_t - \sigma(t; T_n)\, dt$ in Eq. (15.31) we obtain

$$
dY_t = \sigma \sqrt{Y_t}\, dW_t + \kappa\, (\theta - Y_t)\, dt + \sigma \sqrt{Y_t}\, \sigma(t; T_n)\, dt .
$$

The desired result is obtained by substituting the value of $\sigma(t; T_n)$ using Eq. (15.33).

We conclude therefore that the process Y_t is a square-root diffusion with *time-dependent coefficients* under the forward measure π^{*T}. The distribution of Y_t can be characterized as in the case of constant coefficients. We omit the proof of the following result.[13]   ∎

---

[13] See, for instance, Chen [2]. Following the method of Chapter 14, one can compute the moment-generating function of the distribution of Y_{T_n} under π^{*T_n} by solving a Ricatti equation. This calculation is left to the interested reader.

PROPOSITION 15.6

Given $T > 0$, define the parameters

$$\alpha = \sqrt{\kappa^2 + 2\sigma^2},$$

$$\xi_T = e^{\alpha T} \left(\frac{\alpha + \kappa + (\alpha - \kappa) e^{-\alpha T}}{2\alpha} \right)^2$$

$$\eta_T = \frac{\kappa}{\alpha} + \frac{\sigma^2 T}{\alpha^2} + \left(\frac{\alpha + \kappa}{2\alpha} \right)^2 e^{\alpha T} - \left(\frac{\alpha - \kappa}{2\alpha} \right)^2 e^{-\alpha T}.$$

Under the forward measure π^{*T_n}, the random variable

$$\left(\frac{4\kappa}{\sigma^2 \eta_T} \right) Y_{T_n}$$

has a non-central χ^2 distribution with

$$\nu = \frac{4\kappa\theta}{\sigma^2}$$

degrees of freedom and non-centrality parameter

$$c = \frac{4\kappa Y_0 \xi_{T_n}}{\sigma^2 \eta_{T_n}}.$$

From these considerations, we can price the nth caplet in Eq. (15.25) as follows:

$$Caplet = P_0^{T_n} \left(\int_{Y_n^*}^{\infty} g_n(y)\, dy - \left(\frac{1 + \Delta T_n \overline{R}}{1 + \Delta F_n} \right) \int_{Y_n^*}^{\infty} e^{-p_n y - q_n} g_n(y)\, dy \right) \qquad (15.34)$$

where $g_n(y)$ is the non-central χ^2 probability density of Y_{T_n} characterized in the latter proposition and

$$Y_n^* = \frac{1}{p_n} \ln \left(\frac{1 + \Delta T_n \overline{R}}{1 + \Delta F_n} \right) - \frac{q_n}{p_n}.$$

The prices of caps are obtained by applying formula (15.34) to each caplet in Eq. (15.25). For more information about pricing caps with square-root processes, we refer the reader to Chen [2].

REMARK 15.5   An alternative approach for computing caplet prices in the square-root framework is to evaluate the density $g_n(x)$ numerically using the forward Fokker–Planck partial differential equation and then to compute the integrals in (15.34) by quadrature (e.g., Simpson's rule).   ∎

Cap Pricing and Implied Volatilities

Cap prices are quoted in the market in terms of implied **cap volatilities**. Let $Cap(N)$ denote the market value of a cap with maturity T_N that is at-the-money-forward. By definition, the cap volatility $\overline{\sigma}_N$ is the number that makes the following equation true:

$$Cap\ (N)\ =\ \sum_{n=0}^{N-1} P^{T_{n+1}}\, \Delta_n\, BSC(F_n,\ S_N,\ T_{n+1}, \overline{\sigma}_N)\,, \tag{15.35}$$

where S_N is the swap rate, $BSC(F, S, T, \sigma)$ is the Black–Scholes formula for a call with forward price F, strike S, maturity T, and volatility σ. Cap volatility is the volatility *implied* by cap prices.[14]

Another useful set of quantities for cap pricing and for model calibration are the **caplet volatilities**. To give a precise definition of these quantities, consider a sequence of at-the-money-forward caps with maturities $T_1, T_2, \ldots T_N$. We assume that the prices of these caps are known.[15] For each $n = 1, 2, \ldots N$, we write

$$Cap(n)\ =\ \sum_{m=0}^{n-1} Caplet(n,\ m)\,, \tag{15.36}$$

where $Caplet(n,\ m)$ represents the "model value" of the caplet corresponding to the $(m+1)^{st}$ cash-flow date. Notice that the value of the first caplet is unequivocally known at time zero, since the cash-flow to be received at time T_1 is known with certainty. The other caplets must be priced with the model, consistently with (15.36).[16] We define the caplet volatilities $\sigma_n, n = 0, 1, \ldots N - 1$ recursively, as follows:

$$\sigma_0\ =\ 0\ =\ \text{implied volatility}\,(Cap(1))\,,$$

and

$$\sigma_1\ =\ \text{implied volatility}\,(Cap(2)\ -\ Caplet(2, 0))\,.$$

The next step is to define σ_2. For this purpose, let us assume that σ_1 is the term volatility of the 3-month spot rate for the period $(T_1; T_2)$. Accordingly, we define the value of the first nontrivial caplet of the 3-period swap by

$$Caplet(3,\ 1)\ =\ P_0^{T_2}\, \Delta T_1\, BSC\,(F_1,\ S_3, \sigma_1)\,.$$

Now, we define σ_2 as

$$\sigma_2\ =\ \text{implied volatility}\,(Cap(3)\ -\ Caplet(3,\ 0)\ -\ Caplet(3,\ 1))\,.$$

---

[14] We emphasize the number of periods of the cap by using the notation $Cap(N)$. Notice that the strike of the cap is taken to be the swap rate. The use of a single volatility parameter $\overline{\sigma}_N$, produces a cap price. Conversely, every cap price can be expressed in terms of an implied cap volatility. Equation (15.35) is the most parsimonious way of expressing the value of a cap in Black–Scholes terms. Notice also that the use of the Black–Scholes formula makes the convention that forward rates are log-normal.

[15] For example, they can be determined by interpolating the implied volatilities of liquidly traded at-the-money-forward caps.

[16] In practice, only caps and floors are traded. Caplet prices derived by the "bootstrapping" method are presented here.

This procedure is repeated each time a new cap is introduced: for the cap with n periods, we price the first $n - 1$ caplets using the term volatilities $\sigma_0 = 0, \sigma_1, \ldots \sigma_{n-2}$. The caplet volatility σ_n is then defined as

$$\sigma_{n-1} = \text{ implied volatility} \left(Cap(n) - \sum_{m=0}^{n-2} Caplet(n, m) \right)$$

$$= \text{ implied volatility} \left(Cap(n) - \sum_{m=0}^{n-2} P_0^{T_{m+1}} \Delta T_m \, BSC(F_m, S_n, T_{m+1}, \sigma_m) \right).$$

This defines the sequence of caplet volatilities. For any given maturity, we have

$$Cap(n) = \sum_{m=0}^{n-1} P_0^{T_{m+1}} \Delta T_m \, BSC(F_m, S_n, T_{m+1}, \sigma_m).$$

When using term-structure models to price caps and other interest-rate options, we must "translate" dollar prices into cap volatilities.

In the case of Gaussian models, the relation between the caplet volatilities σ_n and the volatilities s_n used in formula (15.27) is given by the identity

$$P^{T_{n+1}} \Delta_n BSC(F_n \, S_N, T_{n+1}, \sigma_{n+1}) = P_0^{T_n} BSP \left(\frac{1 + \Delta T_n S_N}{1 + \Delta T_n F_n}, 1, s_n \right). \qquad (15.37)$$

An approximate relation between the "Gaussian volatilities" s_n and the caplet volatilities σ_n can be obtained using the approximations

$$BSC(S, S, T, \sigma) = BSP(S, S, T, \sigma) \approx \frac{S \sigma \sqrt{T}}{\sqrt{2\pi}} \qquad (15.38)$$

and

$$F_n \approx S_n.$$

Accordingly, we obtain from (15.37)

$$s_n \approx \sigma_n \frac{\Delta T_n \, F_n}{1 + \Delta T_n \, F_n} \sqrt{1 + \frac{\Delta T_n}{T_n}}. \qquad (15.39)$$

This formula can be used to make a first guess of the volatility parameters s_n in a Gaussian model based on cap prices (Figure 15.6). In the U.S. dollar swap market, the error in caplet prices arising from this approximation is often less than 1 bp (Figure 15.7). However, the differences in prices between the approximation (15.39) and the true price can be quite significant for long-term caps, so formula (15.39) should be used to calibrate Gaussian models to cap prices.

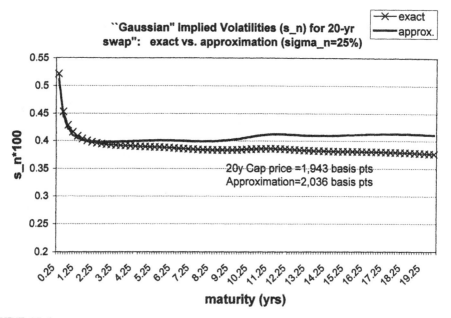

FIGURE 15.6

Approximate and exact values of s_n for a 20-year at-the-money cap with $S_{80} = 6.159\%$. We assumed a flat caplet volatility curve, $\sigma_n = 0.25$ for all n.

15.6 Bond Options and Swaptions

General Pricing Relations

This section discusses the pricing of bond options and swaptions. To be specific, we consider a put option with strike K and maturity date T_0 on an instrument that matures at a date $T_N > T_0$. We denote the cash-flow dates by $T_1 < T_2 < \cdots < T_N$. The cash-flow received by the bondholder at each date is assumed to be

$$\begin{cases} \Delta T_n \, a_n \, , & n \neq N \\ \\ \Delta T_N \, a_N \, + \, 1, & n = N \, , \end{cases}$$

where $\Delta T_n = T_n - T_{n-1}$.

If the underlying security is a fixed-coupon bond, we have

$$\Delta T_n \, a_n \; = \; \frac{C}{\omega}, \;\; i.e., \;\; a_n \; = \; \frac{C}{\Delta T_n \, \omega} \, ,$$

where C is the coupon rate expressed in decimals and ω is the frequency. Payer swaptions can be treated with the same formalism, by setting $K = 1$ and $a_n = \overline{R} =$ strike rate.[17] In this

---

[17] See Chapter 12, Eq. (12.23) for the equivalence between payer swaptions and bond puts. Heretofore, we referred to the underlying security as the "bond."

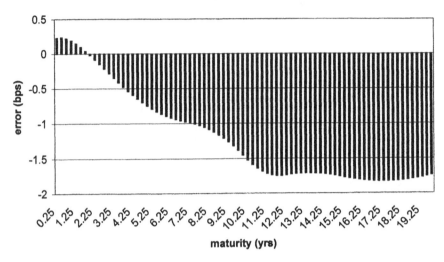

FIGURE 15.7
Errors in pricing caplets (in basis points) using the approximate formula (15.39) for the volatilities s_n. Data as in Figure 15.6.

case the option is equivalent to a payer swaption. We will derive formally pricing formulas for such options, using the forward measure π^{*T_0} corresponding to the option's expiration date.

Notice, first of all, that the payoff of the put option at time T_0 is

$$\max(K - B, \ 0) = \max\left(K - \sum_{n=1}^{N} \Delta T_n \, a_n \, P_{T_0}^{T_n} - P_{T_0}^{T_N}, \ 0 \right)$$

$$= \max\left(K - \sum_{n=1}^{N} \Delta T_n \, a_n \, (f_{T_0} P^{T_n})_{T_0} - (f_{T_0} P^{T_N})_{T_0}, \ 0 \right).$$

We know from the previous section that we can write

$$(f_{T_0} P^{T_n})_t = (f_{T_0} P^{T_n})_0 \, m_t^{(n)},$$

where $m_t^{(n)}$ is a martingale under π^{*T_0} normalized such that $m_0^{(n)} = 1$. Notice that, unlike the case of caps, which involve a series of forward measures and one martingale for each measure, here we use a single forward measure (π^{*T_0}) and a series of martingales with respect to this measure. A bond option can be viewed as an option on a *basket of zero-coupon bonds*. These martingales $m_t^{(n)}, n = 1, 2, \ldots N$ model the fluctuations of the forward prices of the zero-coupon bonds in the "basket."

The bond option price is given formally by

$$Bond \, Put = \tag{15.40}$$

$$P_0^{T_0} \, \mathbf{E}^{\pi^* \, T_0} \left\{ \max \left(K - \sum_{n=1}^{N} \Delta T_n \, a_n \, (f_{T_0} \, P^{T_n})_0 \, m_{T_0}^{(n)} - (f_{T_0} \, P^{T_N})_0 \, m_{T_0}^{(N)} , \, 0 \right) \right\} .$$

This expression can be evaluated once we specify the *joint distribution* of the random variables $\left(m_{T_0}^{(1)}, \ldots m_{T_0}^{(N)} \right)$. To better grasp the difficulty involved in pricing bond options, we consider the *forward bond price* process

$$(f_{T_0} \, B)_t = \sum_{n=1}^{N} \Delta T_n \, a_n \, (f_{T_0} \, P^{T_n})_0 \, m_t^{(n)} + (f_{T_0} \, P^{T_N})_0 \, m_t^{(N)}$$

$$\equiv (f_{T_0} \, B)_0 \, \overline{m}_t^N$$

where

$$\overline{m}_t^N = \frac{\sum\limits_{n=1}^{N} \Delta T_n \, a_n \, (f_{T_0} \, P^{T_n})_0 \, m_t^{(n)} + (f_{T_0} \, P^{T_N})_0 \, m_t^{(N)}}{\sum\limits_{n=1}^{N} \Delta T_n \, a_n \, (f_{T_0} \, P^{T_n})_0 + (f_{T_0} \, P^{T_N})_0} . \tag{15.41}$$

The process $\overline{m}_t^N$ is also a martingale normalized such that $\overline{m}_0^N = 1$.

The price of the bond put option can therefore be written in the form

$$Bond \, Put = P_0^{T_0} \, \mathbf{E}^{\pi^* \, T_0} \left\{ \max \left(K - (f_{T_0} \, B)_0 \, \overline{m}_{T_0}^N, \, 0 \right) \right\} . \tag{15.42}$$

If, for instance, we assume a Gaussian term-structure model for interest rates, the forward zero-coupon bonds are log-normal, which implies that the forward bond price is distributed like a *weighted sum of (dependent) lognormals*. Similarly, the assumption of a square-root dynamics for the underlying state variables leads to "log-χ^2" forward zero-coupon bond prices under $\pi^* \, T_0$. The evaluation of these expectation values must be done numerically, with the notable exception of one-factor models, as we now show.

Jamshidian's Theorem

Jamshidian [4] discovered a major simplification in the evaluation of bond option prices in the framework of one-factor exponential-affine models. His result, which we discuss below, shows that the bond option price is equal to the price of a series options on zero-coupon bonds. It leads therefore to closed-form solutions for Gaussian models. For square-root models, the option price can be expressed as a sum of integrals over the non-central χ^2 distribution, analogous to the ones used for pricing caps.

We begin by observing that in the case of one-factor models, the forward price of a zero-coupon can be expressed in the form

$$(f_{T_0} \, P^{T_n})_t = (f_{T_0} \, P^{T_n})_0 \exp \left(-p(t; T_n) \, Y_t - q_0(t; T_n) \right) .$$

In particular we have

$$m_t^{(n)} = \exp \left(-p(t; T_n) \, Y_t - q_0(t; T_n) \right) .$$

The payoff of the bond option becomes

$$
\max \left(K - \sum_{n=1}^{N} \Delta T_n a_n (f_{T_0} P^{T_n})_0 e^{-p_n Y_{T_0} - q_n} - (f_{T_0} P^{T_N})_0 m_{T_0}^{(N)} e^{-p_N Y_{T_0} - q_N} , 0 \right) ,
$$

where $p_n = p(T_0; T_n)$ and $q_n = q_0(T_0; T_n)$. We note that p_n is *positive* for all n (cf. Chapter 14 if necessary).

Jamshidian's key observation is that the bond forward price is equal to the strike price if and only if we have $Y_{T_0} = Y^*$, where Y^* is the solution of the equation

$$
K = \sum_{n=1}^{N} \Delta T_n a_n (f_{T_0} P^{T_n})_0 e^{-p_n Y^* - q_n} + (f_{T_0} P^{T_N})_0 e^{-p_N Y^* - q_N} . \tag{15.43}
$$

Moreover, the right-hand side of this equation is a decreasing function of Y^*. This has two important consequences: first, Y^* is *unique*. Second, the option is in-the-money if and only if $Y_{T_0} > Y^*$. Let us define the following auxiliary quantities:

$$
\begin{cases}
K_n = \Delta T_n a_n (f_{T_0} P^{T_n})_0 e^{-p_n Y^* - q_n} & \text{for } n \neq N \\[2mm]
K_N = (1 + \Delta T_N a_N) (f_{T_0} P^{T_N})_0 e^{-p_N Y^* - q_N} .
\end{cases} \tag{15.44}
$$

It follows that the option is in-the-money if and only if the following inequalities hold simultaneously:

$$
\begin{cases}
K_n > \Delta T_n a_n (f_{T_0} P^{T_n})_0 e^{-p_n Y_{T_0} - q_n} & \text{for } n \neq N \\[2mm]
K_N > (1 + \Delta T_N a_N) (f_{T_0} P^{T_N})_0 e^{-p_N Y_{T_0} - q_N} .
\end{cases}
$$

The conclusion is therefore that

$$
\max(K - B, 0) = \sum_{n=1}^{N-1} \max \left(K_n - \Delta T_n a_n (f_{T_0} P^{T_n})_{T_0}, 0 \right)
$$

$$
+ \max \left(K_N - (1 + \Delta T_N a_N) (f_{T_0} P^{T_N})_{T_0}, 0 \right) .
$$

Hence,

$$
Bond\,put = \sum_{n=1}^{N-1} P_0^{T_0} \mathbf{E}^{\pi^* T_0} \left\{ \max \left(K_n - \Delta T_n a_n (f_{T_0} P^{T_n})_{T_0} 0 \right) \right\}
$$

$$
+ P_0^{T_0} \mathbf{E}^{\pi^* T_0} \left\{ \max \left(K_N - (1 + \Delta T_N a_N)(f_{T_0} P^{T_N})_{T_0}, 0 \right) \right\} . \tag{15.45}
$$

We have shown

PROPOSITION 15.7

(**Jamshidian's theorem**) *The value of a bond option under a one-factor exponential-affine model is equal to the value of a portfolio of options on zero-coupon bonds with strikes defined by Eq. (15.44).*

In particular, the value of a bond option in a one-factor Gaussian model is

$$Bond\,Put = \sum_{n=1}^{N-1} BSP\left(\Delta T_n\, a_n\, P_0^{T_n}, \, , \, K_n\, P_0^{T_0}, \, T_n, \, \tilde{s}_n\right)$$

$$+ BSP\left((1 + \Delta T_N\, a_N)\, P_0^{T_N}, \, K_N\, P_0^{T_0}, \, T_N, \, \tilde{s}_N\right), \qquad (15.46)$$

where

$$\tilde{s}_n = \left(\frac{1}{T_0}\int_0^{T_0} (\sigma(t; T_n) - \sigma(t; T_0))^2\, dt\right)^{1/2}. \qquad (15.47)$$

This formula is very simple to compute: one first solves Eq. (15.43) to determine the critical value of the state-variable, Y^*. The strikes K_n for the zero-coupon bond options are computed using formula (15.44). Finally, the price of the bond option is obtained using formula (15.46).

We obtain a similar result for the case of square-root models, with the Black–Scholes formula replaced by integrals over the non-central χ^2 distribution.

REMARK 15.6   Jamshidian's theorem applies *only to one-factor models*. In the case of multifactor models, the state variable Y_t is a vector. The range of values of Y_t for which the option is in-the-money is no longer a "half-line" but rather a "region" in N-dimensional space. In general, there is no longer an exact relation between the bond option payoff and the payoffs on options of zero-coupon bonds. As suggested by the numerical example at the end of Chapter 14, the imperfect correlations between rates imply that, all other things equal, bond option prices will be lower with multifactor models than with one-factor models.[18] This is because the correlations between forward rates affect the prices of bond options and swaptions.

Example 15.1

To illustrate the use of Jamshidian's theorem, we priced a few 5-year puts on a 30-year bond. We assumed that the bond coupon rate is 5% and that the 6-month forward rate curve is flat at 5%. We used a Ho–Lee model with $\sigma = 0.003$, which corresponds roughly to a flat caplet volatility of 25%. The results are exhibited in Figures 15.8–15.12.   □

Volatility Analysis

The market convention is to quote option prices in Black–Scholes terms, i.e., in terms of the implied volatility that is obtained if the random variable defined in Eq. (15.41), $\overline{m}_{T_0}^N$, is

---

[18]By "all other things equal" we mean, for instance, that the one-factor model has a zero-coupon bond volatility $\sigma(t; T)$ that is equal to the magnitude of the vector $(\sigma_1(t; T), \dots \sigma_\nu(t; T))$ of the multifactor model.

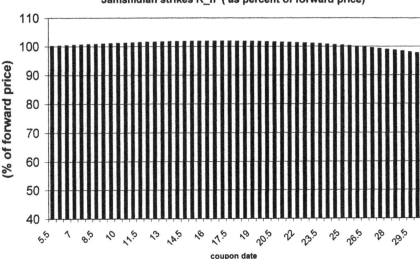

FIGURE 15.8
The series of strikes K_n, $n = 1, \ldots 50$ for a 5-year at-the-money option on 30-year bond.
(The "tail" is 25 years.) We assume that the forward rate curve is flat at 5% and that the
bond has a 5% coupon. The strikes are represented as percentages of the forward prices of
the cash-flows corresponding to each maturity. Notice that all options in formula (15.46)
are essentially at-the-money.

assumed to be log-normal.[19] It is therefore necessary to focus on the relation between the
BS implied volatility and the instantaneous volatilities of exponential-affine models. For this
purpose, we go back to the formula from the forward bond price

$$\frac{(f_{T_0} B)_{T_0}}{(f_{T_0} B)_0} = \frac{\sum_{n=1}^{N-1} a_n \, \Delta T_n \, (f_{T_0} P^{T_n})_0 \, m_{T_0}^{(n)} + (1 + a_N \Delta T_N)(f_{T_0} P^{T_N})_0 \, m_{T_0}^{(N)}}{\sum_{n=1}^{N-1} a_n \, \Delta T_n \, (f_{T_0} P^{T_n})_0 + (1 + a_N \Delta T_N)(f_{T_0} P^{T_N})_0}$$

$$= \sum_{n=1}^{N} w_n \, m_{T_0}^{(n)} \,, \tag{15.48}$$

---

[19]Notice, however, that this assumption does not necessarily represent the market's belief about the bond's
forward price distribution.

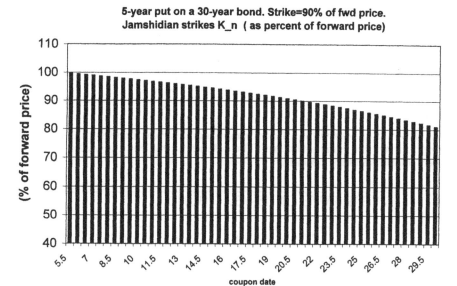

FIGURE 15.9
Same as in Figure 15.8, for a put with strike equal to 90% of the forward bond price. Notice that the puts in the Jamshidian formula become progressively more out-of-the-money at longer maturities.

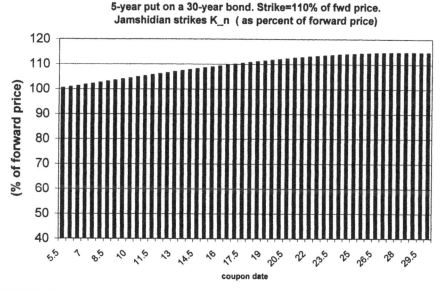

FIGURE 15.10
Same as in Figure 15.9, but with a strike equal to 110% of the forward price. The individual options become increasingly in-the-money as the maturity increases.

5-YEAR PUTS ON 30-YEAR BOND		
strike (% of par)	price (basis points)	implied volatility
90	50	10.5
95	141	11.7
100	305	13.5
105	554	15.6
110	860	17.7

FIGURE 15.11
Prices of 5-year options with different strikes on a 30-year bond with 5% coupon. The assumptions are the same as before. The column on the right gives the Black–Scholes implied volatilities of the options.

5-YEAR PUTS ON 15-YEAR BOND		
strike (% of par)	price (basis points)	implied volatility
90	3.5	7.5
95	37.16	8.2
100	168.45	10.7
105	435.54	13.89
110	789.13	16.79

FIGURE 15.12
Prices of 5-year options with different strikes on a 15-year bond with 5% coupon. The assumptions are the same as before. The column on the right gives the Black–Scholes implied volatilities of the options.

where the weights w_n are defined by

$$
w_n = \begin{cases}
\dfrac{a_n \, \Delta T_n (f_{T_0} P^{T_n})_0 m_{T_0}^{(n)}}{\displaystyle\sum_{n=1}^{N-1} a_n \, \Delta T_n (f_{T_0} P^{T_n})_0 + (1+a_N \, \Delta T_N)(f_{T_0} P^{T_N})_0} & \text{if } n \neq N \\[4ex]
\dfrac{(1+a_N \, \Delta T_N)(f_{T_0} P^{T_N})_0}{\displaystyle\sum_{n=1}^{N-1} a_n \, \Delta T_n (f_{T_0} P^{T_n})_0 + (1+a_N \, \Delta T_N)(f_{T_0} P^{T_N})_0} & \text{if } n = N .
\end{cases}
\tag{15.49}
$$

The simplest way to obtain an approximation for the implied volatility of the bond option is to assume that

$$
\sum_{n=1}^{N} w_n \, m_{T_0}^{(n)} \approx \exp\left(\sigma_B \, N \, \sqrt{T_0} - \frac{1}{2} \sigma_B^2 \, T_0^2 \right) ,
\tag{15.50}
$$

where N is a standard normal. Of course, this formula is not exact, because a weighted sum of log-normal random variables is not log-normally distributed in general. Notice, however, that the approximation made here is such that the first moments of both sides of the equation match. To obtain an approximation for the bond option volatility, we equate the second moments of both sides of (15.50). The result is

$$
\sigma_B^2 \approx \frac{1}{T_0} \ln \mathbf{E}^{\pi^{*T_0}} \left\{ \left(\sum_{n=1}^{N} w_n \, m_{T_0}^{(n)} \right)^2 \right\}
$$

$$
= \frac{1}{T_0} \ln \left(\sum_n \sum_{n'} w_n \, w_{n'} \, \mathbf{E}^{\pi^{*T_0}} \left(m_{T_0}^{(n)} \, m_{T_0}^{(n')} \right) \right)
$$

$$
= \frac{1}{T_0} \ln \left(\sum_n \sum_{n'} w_n \, w_{n'} \, e^{\tilde{s}_n \tilde{s}_{n'} \rho_{n,n'} T_0} \right) \tag{15.51}
$$

where

$$
\tilde{s}_n^2 = \frac{1}{T_0} \int_0^{T_0} \| \sigma(t; T_n) - \sigma(t; T_0) \|^2 \, dt
$$

and

$$
\rho_{n,n'} = \frac{1}{T_0 \, \tilde{s}_n \, \tilde{s}_{n'}} \int_0^{T_0} (\sigma(t; T_n) - \sigma(t; T_0)) \cdot (\sigma(t; T_{n'}) - \sigma(t; T_0)) \, dt \,.
$$

Formula (15.51) gives an approximation for the BS implied volatility of the bond option in terms of the volatility parameters of the Gaussian model.[20] To further simplify this expression, we use the approximations $e^x \approx 1 + x$ and $\ln(x) \approx x - 1$, which are valid for $x \ll 1$.

$$
\sigma_B \approx \sqrt{\sum_n \sum_{n'} w_n \, w_{n'} \, \tilde{s}_n \, \tilde{s}_{n'} \, \rho_{n,n'}} \,. \tag{15.52}
$$

Notice that Eq. (15.52) is consistent with the fact that the holder of a bond option is "long" the correlations between the forward prices of zero-coupon bonds: if the correlation increases, the bond volatility increases and so does the price of the option. It is important to note, however, that the above approximation cannot be valid uniformly for all strikes (see Figure 15.12).

REMARK 15.7 One important consideration regarding the use of term-structure models concerns the **volatility skew.** As exemplified in the above tables, Gaussian interest rate models produce a "volatility skew" when pricing bond options, whereby out-of-the-money bond puts have a lower implied volatility than out-of-the-money bond calls. ∎

At issue here is the matter of controlling the probability distribution for asset prices induced by term-structure models. Out-of-the-money options in fixed-income markets often trade at different implied volatilities than at-the-money or in-the-money options. An intuitive way of thinking about the skew is to note that bond prices vary in opposite direction to bond yields. The latter are positively correlated with interest rates in general. Therefore, a higher volatility for out-of-the-money *bond calls* is consistent with a higher volatility for OTM *interest*

---

[20]Another possible way of approximating the bond option volatility, especially for options that are not at-the-money, would be to generate a polynomial approximation of the option payoff max$(K - B, 0)$ and to use the higher-order moments of the random variables $\left(m_{T_0}^{(1)} \ldots m_{T_0}^{(N)} \right)$.

rate puts. Since yields vary in the same direction as interest rates, one can assume that the volatility skew presented in Figures 15.11 and 15.12 would be less pronounced if we used square-root processes or log-normal processes for interest rates, given that the corresponding distributions have shorter tails for $r \ll 1$. Such approaches have been adopted by practitioners to attempt to better model the observed implied volatility skew in interest-rate options. The reader can find an interesting study of bond option pricing in the recent work of Chen [2] and Sun and Jagannathan [7]. For a thorough account of some of the elementary models used by practitioners for pricing fixed income-derivatives, we recommend, for instance, Wong [8].

15.7   Epilogue: The Brace–Gatarek–Musiela model

We end this chapter with a brief description of a new approach for modeling the dynamics of interest rates that is particularly well-suited for pricing interest-rate options, the **Brace-Gatarek-Musiela model** (BGM).

In the previous sections we studied the pricing of caps and bond options in the framework of exponential-affine models. Despite their simplicity, the calibration of exponential-affine models to the prices of options traded in the market can be delicate. For example, matching a series of cap volatilities quoted in the market to the internal parameters of the model requires finding a series of "Gaussian implied volatilities" by formula (15.37). The point is that in Gaussian models zero-coupon bonds are log-normally distributed; forward rates are not.

The BGM approach is based on modeling directly the evolution of the forward rates $F_n(t) = F(t; T_n, T_{n+1})$ (Figure 15.13). This approach leads to elegant equations for the evolution of the term-structure that facilitate the calibration to option prices.

As usual, we assume that π is a risk-neutral measure and that $\pi^{* \, T_n}$ are the forward measures associated with different delivery dates. Recall from Proposition 15.4 that F_n satisfies the differential equation

$$\frac{dF_n}{F_n} = \left(\frac{1 + \Delta T_n \, F_n}{\Delta T_n \, F_n} \right) (\sigma(t; \, T_n) - \sigma(t; \, T_{n+1})) \cdot dW \qquad (15.53)$$

where W_t is a Brownian motion under the forward measure $\pi^{* \, T_{n+1}}$.

We wish to prescribe the volatility of F_n directly, as opposed to using the volatility vector of zero-coupon bonds $\sigma(t; \, T)$.[21] For this purpose, we set, in accordance with Eq. (15.53),

$$\widehat{\sigma}^{(n)}(t) = \frac{1 + \Delta T_n \, F_n}{\Delta T_n \, F_n} (\sigma(t; \, T_n) - \sigma(t; \, T_{n+1})) \, , \quad n = 1, 2, \ldots N - 1 , \qquad (15.54)$$

so that, under $\pi^{* \, T_{n+1}}$ we have, formally,

$$\frac{dF_n(t)}{F_n(t)} = \widehat{\sigma}^{(n)}(t) \cdot dW_t \, .$$

The following fundamental question arises:

---

[21]This is a multifactor analysis. We assume here that $\sigma(t; \, T) = (\sigma_1(t; \, T), \ldots \sigma_\nu(t, \, T))$ and that W is a ν-dimensional Brownian motion.

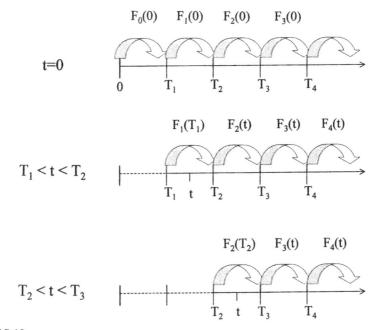

FIGURE 15.13
Schematic representation of the forward rates $F_n(t) = F(t; T_n, T_{n+1})$. The maturities T_n are fixed. If $T_n < t < T_{n+1}$ all rates F_m with $m \leq n$ are known.

Define an arbitrage-free term-structure model that is consistent with a given sequence of volatility vectors for forward rates:

$$\widehat{\sigma}^{(1)}(t) = \left(\widehat{\sigma}_1^{(1)}(t), \ldots, \widehat{\sigma}_\nu^{(1)}(t) \right)$$

$$\cdots\cdots\cdots\cdots\cdots$$

$$\widehat{\sigma}^{(N-1)}(t) = \left(\widehat{\sigma}_1^{(N-1)}(t), \ldots, \widehat{\sigma}_\nu^{(N-1)}(t) \right) . \tag{15.55}$$

To solve this problem, we observe that Eq. (15.54) implies formally that

$$\sigma(t; T_{n+1}) = \sigma(t; T_n) - \left(\frac{\Delta T_n F_n}{1 + \Delta T_n F_n} \right) \widehat{\sigma}^{(n)}(t), \quad n = 1, 2, \ldots N - 1 .$$

Based on this relation, we define recursively a zero-coupon bond volatility (*as in* HJM) which is consistent with (15.55). Namely, we set

$$\begin{cases} \sigma(t; T) = \sigma_0 \text{ (given)}, \text{ for } T \leq T_1 \\ \\ \sigma(t; T) = \sigma(t; T_n) - \left(\frac{\Delta T_n F_n(t)}{1+\Delta T_n F_n(t)} \right) \widehat{\sigma}^{(n)}(t), \text{ for } T_n \leq T < T_{n+1} . \end{cases} \tag{15.56}$$

This defines an adapted process $\sigma(t; T)$ in the variable t. By the Heath–Jarrow–Morton theorem, there exists an arbitrage-free probability measure π such that

(i)   The model is consistent with an initial instantaneous forward-rate curve $f(0; T)$.

(ii)   The model is consistent with the specification of forward-rate volatilities in (15.55).

To wit, this measure π is obtained by solving the HJM stochastic differential equations associated with the volatility structure and the initial curve.

Let us now characterize the statistics of the rates

$$(F_1(t),\ F_2(t),\ \ldots\ F_{N-1}(t))$$

under the measure π. For this purpose, we shall derive the evolution of these rates under a particular forward measure, namely $\pi^{* T_2}$.[22]

PROPOSITION 15.8

Assume that the zero-coupon-bond volatility is given by (15.56) and let π denote the risk-neutral measure on paths $f(t; T)$ that is consistent with this volatility specification and with an initial forward-rate curve. Then, for all $n = 1, 2, \ldots N - 1$, we have

$$\frac{dF_n(t)}{F_n(t)} = \widehat{\sigma}^{(n)}(t) \cdot dW(t)$$

$$+ \left(\sum_{m=2}^{n} \left(\frac{\Delta T_m\, F_m(t)}{1 + \Delta T_m\, F_m(t)},\ \right) \widehat{\sigma}^{(m)}(t) \cdot \widehat{\sigma}^{(n)}(t) \right) dt \qquad (15.57)$$

where $W(t)$ is a Brownian motion under $\pi^{* T_2}$.

PROOF   We write

$$F_n = \frac{1}{\Delta T_n} \left(\frac{P_t^{T_n}}{P_t^{T_{n+1}}} - 1 \right)$$

$$= \frac{1}{\Delta T_n} \left(\left(\frac{P_t^{T_n}}{P_t^{T_2}} \right) / \left(\frac{P_t^{T_{n+1}}}{P_t^{T_2}} \right) - 1 \right)$$

$$= \frac{1}{\Delta T_n} \left(\frac{(f_{T_2} P^{T_n})_t}{(f_{T_2} P^{T_{n+1}})_t} - 1 \right).$$

Let us apply Ito's Formula to this equation. From Eq. (15.7) in Section 15.1, we have

$$\frac{d(f_{T_2} P^{T_k})_t}{(f_{T_2} P^{T_k})_t} = (\sigma(t; T_k) - \sigma(t; T_2))\, dW, \quad k = n, n+1,$$

---

[22]The reason for choosing $\pi^{* T_2}$ as opposed to another forward measure is arbitrary, except for the fact that F_0 is known precisely at time $t = 0$. The choice of $\pi^{* T_2}$ makes the first stochastic rate, F_1, a martingale.

where W is a Brownian motion under π^{*T_2}. Using these equations and Ito's Formula, we find after a straightforward calculation that

$$dF_n = \frac{1}{\Delta T_n} \frac{P_t^{T_n}}{P_t^{T_{n+1}}} \left(\sigma(t; T_n) - \sigma(t; T_{n+1})\right) dW$$

$$+ \frac{1}{\Delta T_n} \frac{P_t^{T_n}}{P_t^{T_{n+1}}} \left(\sigma(t; T_n) - \sigma(t; T_{n+1})\right) \cdot \left(\sigma(t; T_2) - \sigma(t; T_{n+1}\right) dt \,.$$

We now use Eq. (15.54) to express this result in terms of the volatilities $\widehat{\sigma}^{(m)}$. Accordingly,

$$dF_n = F_n \widehat{\sigma}^{(n)}(t) \cdot dW(t) + F_n \sum_{m=2}^{n} \widehat{\sigma}^{(n)}(t) \cdot \left(\sigma(t; T_m) - \sigma(t; T_{m+1}\right) dt$$

$$= F_n \widehat{\sigma}^{(n)}(t) \cdot dW(t) + F_n \sum_{m=2}^{n} \widehat{\sigma}^{(n)}(t) \cdot \widehat{\sigma}^{(m)}(t) \left(\frac{\Delta T_m F_m}{1 + \Delta T_m F_m}\right) dt \,.$$

This is the result that we wanted to obtain.

Equation (15.57) gives the dynamics of the vector of forward rates $(F_1(t), F_2(t), \ldots F_{N-1}(t))$ under the probability measure π^{*T_2}. It can be compared with the Heath–Jarrow–Morton equation derived in Chapter 13. Notice that the drift

$$\mu_n = \sum_{m=2}^{n} \left(\frac{\Delta T_m F_m(t)}{1 + \Delta T_m F_m(t)}, \right) \widehat{\sigma}^{(m)}(t) \cdot \widehat{\sigma}^{(n)}(t)$$

resembles a "discretization" of the HJM drift for instantaneous forward rates, where the integral $\int_t^T \dot{\sigma}(t; s) \, ds$ is approximated by a discrete sum. Notice that the drift remains bounded for all positive values of F_m. It can be shown (cf. Ikeda and Watanabe [3], for example) that the solutions of the stochastic differential equation (15.57) remain finite for all times.

The BGM model resolves two important issues. The first one is theoretical: as shown in Chapter 13, the specification of *instantaneous* forward rate volatilities

$$\sigma^f(t; T) = f(t; T) \sigma(t)$$

gives rise to equations that "blow-up" in finite time. The BGM model does not have this pathology because its drift is bounded.

The second advantage has to do with the *calibration of the model to cap prices*. Since we can prescribe the instantaneous volatilities of the forward rates F_n, we can easily match a series of caplet volatilities. In fact, the relation between caplet volatilities and the instantaneous forward rate volatilities is given by

$$\sigma_n = \sqrt{\frac{1}{T_{n+1}} \int_{t=0}^{T_{n+1}} \|\widehat{\sigma}^{(n)}(t)\|^2 \, dt} \,, \quad n = 1, 2, \ldots N \,.$$

The caplet volatilities are therefore obtained as integrals over the magnitudes of the vectors $\widehat{\sigma}^{(n)}(t)$.

The BGM model can be used to price any contingent claim that depends on a vector of forward rates $(F_1, \ldots F_{N-1})$ and has cash-flow dates $T_1, T_2, \ldots T_N$. To be specific, let

$$\Phi(F_n(T_n), \ F_{n+1}(T_n), \ \ldots, \ F_{N-1}(T_n)) \tag{15.58}$$

represent a cash-flow determined at date T_n and paid at date T_{n+1}. The discounted cash-flow along a path until date T_2 is

$$X = \left(\Pi_{j=0}^{n} \frac{1}{1 + \Delta T_j \ F_j(T_j)}\right) \Phi(F_n(T_n), \ F_{n+1}(T_n), \ \ldots, \ F_{N-1}(T_n)) \ .$$

Therefore, the fair value of the contingent claim with value (15.58) is

$$P_0^{T_2} \ \mathbf{E}^{\pi^* T_2} \{X\}$$

where the rates $\{F_n(t)\}$ evolve according to Eq. (15.57). ∎

References and Further Reading

[1] Brace, A., Gatarek, D., and Musiela, M. (1997), The Market Model of Interest Rate Dynamics, *Mathematical Finance,* **7**, pp. 127–154.

[2] Chen, R.R. (1996), *Understanding and Managing Interest Rate Risks,* World Scientific, Singapore.

[3] Ikeda, N. and Watanabe, S. (1981), *Stochastic Differential Equations and Diffusion Processes,* North-Holland, Amsterdam.

[4] Jamshidian, F. (1989), An Exact Bond Option Pricing Formula, *Journal of Finance,* **44**, pp. 205–209.

[5] Jamshidian, F. (1997), LIBOR and Swap Market Models and Measures, *Finance and Stochastics,* pp. 293–330.

[6] Musiela, M. and Rutkowski, M. (1997), *Application of Martingale Methods in Financial Modelling,* Applications of Mathematics, vol. 36, Springer-Verlag, Berlin.

[7] Sun, G. and Jagannathan, R. (1998), An Evaluation of Multi-Factor CIR Models Using LIBOR, Swaps Rates, and Cap and Swaptions Prices, Working paper.

[8] Wong, M.A. (1991), *Trading and Investing in Bond Options,* John Wiley & Sons, New York.

Index

313

Printed and bound by CPI Group (UK) Ltd, Croydon, CR0 4YY

24/10/2024

01778287-0007